AF683819

CRAYFISH IN EUROPE AS ALIEN SPECIES

HOW TO MAKE THE BEST OF A BAD SITUATION?

CRUSTACEAN ISSUES

11

General editor:
FREDERICK R. SCHRAM
Zoological Museum, University of Amsterdam

A.A. BALKEMA / ROTTERDAM / BROOKFIELD / 1999

CRAYFISH IN EUROPE AS ALIEN SPECIES

How to make the best of a bad situation?

Edited by

FRANCESCA GHERARDI

Department of Animal Biology and Genetics 'Leo Pardi', University of Florence

DAVID M. HOLDICH

School of Biological Sciences, University of Nottingham

A.A. BALKEMA / ROTTERDAM / BROOKFIELD/ 1999

Authorization to photocopy items for internal or personal use, or the internal or personal use of specific clients, is granted by A.A. Balkema, Rotterdam, provided that the base fee of US$1.50 per copy, plus US$0.10 per page is paid directly to Copyright Clearance Center, 222 Rosewood Drive, Danvers, MA 01923, USA. For those organizations that have been granted a photocopy license by CCC, a separate system of payment has been arranged. The fee code for users of the Transactional Reporting Service is: 90 5410 469 4/99 US$1.50 + US$0.10.

Published by
A. A. Balkema, P.O. Box 1675, 3000 BR Rotterdam, Netherlands
Fax: +31.10.4135947; E-mail: balkema@balkema.nl; Internet site: http://www.balkema.nl

A. A. Balkema Publishers, Old Post Road, Brookfield, VT 05036-9704, USA
Fax: 802.276.3837; E-mail: info@ashgate.com

ISSN 0168-6356
ISBN 90 5410 469 4

©1999 A. A. Balkema, Rotterdam
Printed in the Netherlands

Contents

Case studies of alien crayfish in Europe

Conclusions

Preface

Freshwater crayfish are known to many only as a luxury food, but to others they also represent a keystone species in the freshwater environment, as well as an interesting experimental animal. Europe has far fewer crayfish species than most regions possessing them, and consequently any threat to their survival has been viewed with great concern. The elimination of many European crayfish populations by human activities since the Industrial Revolution and disease, particularly crayfish plague, has led to measures to conserve and increase the remaining stocks. One such method was to introduce alien crayfish to Europe which were better able to cope with the situation (Holdich 1988 see Introduction). To some extent this has worked and crayfish production in some water bodies is now back to levels experienced at the turn of the century, although in general production is below pre-plague levels. However, there has been a price to pay and all of the alien crayfish species introduced, and which are now established in the wild, are causing environmental problems, not least to the remaining native crayfish populations.

This gave rise to the idea that an international workshop should be convened so that interested parties could discuss the problem. The only previous international workshops relating mainly to European crayfish were those held in 1987 in Trondheim, Norway on their culture (Skurdal et al. 1988, see introductory chapter) and in 1991 in Kuopio, Finland on their management and stocking (see Finnish Fisheries Research 14, 1992). Funding was obtained by Francesca Gherardi to run the present workshop and, in particular, to provide funds to allow some East European colleagues to attend. The resulting 1997 workshop entitled 'The introduction of alien species of crayfish in Europe – how to make the best out of a bad situation?' was held at the University of Florence, Italy from September 24-27, 1997, and was attended by over 70 delegates from 15 countries. David Holdich, Nottingham, and Marco Vannini, Florence, were supervisors and Francesca Gherardi, Florence, was coordinator of the workshop. Silvia Barbaresi and Fabio Villanelli, both of Florence, formed the organising committee and were ably assisted by Patrizia Acquistapace, Chiara Benvenuto, Patrizia Busato, Ilaria Cipriani, Maddalena Giuggioli, Andrea Raddi, Silvia Russo and Gabriele Salvi (Florence).

The Workshop was held under the auspices of: the Centre for the Study of Tropical Faunistic and Ecology of the C.N.R. and the Italian Ministry of Environment. We would like to thank the following for their financial and other support: the European

Union, the International Association of Astacology, the Ministry of the University and of Scientific and Technological Research, the Provincia of Pistoia, the Regione Toscana, the Tuscan Regional Agency for the Environment (ARPAT) and the University of Florence, together with the Area of the Research of Florence of the C.N.R. for editing and printing Abstract Volumes and playbills, 'Ditta Robiglio' for the Welcome Party, 'Argenteria Brandimarte' and 'U. Gherardi S.p.A.' for the gifts, the Museum of 'La Specola', the Department of Animal Biology and Genetics and the Rector for use of their facilities, and the choir of the University of Florence directed by Maestro Valerio Del Piccolo for performing the farewell concert. Funds for editing the articles were provided by the Italian Ministry for the Agricultural Policy.

Francesca Gherardi, Florence
David Holdich, Nottingham
September 1998

Editors' summary

There have been some positive commercial, socio-economic, recreational, and biological benefits resulting from the introduction of alien crayfish into Europe, but there have also been many negative aspects. Alien crayfish are spreading at an alarming rate in many European countries. Their presence not only results in the decimation of native crayfish populations through disease and competition, but also adversely affects the freshwater environment. Crayfish plague is still rampant, and yet its presence is not easily identified in living crayfish.

Within 100 years the freshwater environment of many European countries may be dominated by alien crayfish unless measures are taken to counteract their spread and that of crayfish plague. When the plague fungus can be easily identified it should be made a reportable disease. At present, however, most European countries still have populations of native crayfish, some of which are sufficiently large to be harvested. On the one hand the European Habitats Directive affords some protection to native crayfish, but on the other hand removing trade barriers encourages the movement of alien crayfish. Protection should be given to all native European crayfish by the setting up of 'native crayfish reserves', not just for *Austropotamobius pallipes* within Special Areas of Conservation (SACs), but for whole countries. Encouragement should be given to reintroduction programmes, but genetic studies should be carried out so as to maintain as much genetic diversity as possible amongst populations. Methods into eradicating nuisance populations of alien crayfish should be researched, preferably by means friendly to the environment. Both the public, and government authorities should be informed via informative leaflets about the dangers of allowing the introduction of fast-growing invasive crayfish, particularly those form North America which also carry crayfish plague. The wise utilisation of native and alien crayfish is to be encouraged. However, where there is not a tradition for harvesting native crayfish they should still be considered not only as a valuable freshwater resource, but also from the aesthetic and biological diversity points of view.

More needs to be known about the distribution of both native and alien crayfish at regional, national and European levels. The distribution of crayfish is not static and a European data base should be set up into which individuals can enter records and access information so as to highlight threats to their native species. The situation needs to be regularly monitored. The introduction of alien crayfish into Europe may have seemed like a good idea at the time in order to improve stocks decimated by crayfish plague and other factors but it is turning out, at least from a conservationist's point of view, to be another example of the Frankenstein Effect.

Introduction

Native and alien crayfish in Europe: An introduction

David M. Holdich
School of Biological Sciences, University of Nottingham, Nottingham, UK
Francesca Gherardi
Department of Animal Biology and Genetics 'Leo Pardi', University of Florence, Florence, Italy

ABSTRACT

An overview of the problems and issues related to introduced crayfish in Europe is presented. The historical magnitude of the problem is made manifest. Sources of ready information are presented. The basic terms applied in the field, and the scientific and common names employed are outlined.

1 INTRODUCTION

Nobody can deny the fact that we owe much of our modern culture, particularly in the form of food, to introduced animals and plants (Pullin 1994). However, some ecologists are worried about the implications of this and are warning that we are in danger of homogenizing the world's biota by our activities (Culotta 1991, Lodge 1993). Elton (1958) in his classic book on the ecology of invasions was one of the first scientists to try and highlight the environmental problems being caused by animals and plants being moved outside their natural ranges by humans. He cited the ever increasing volume and speed of transport by air, land, and water as the root of the problem. He stated that we were living in a period of the world's history when the mixing of the biota from different parts of the world, either accidentally or on purpose, was setting up 'dislocations in nature' and causing 'huge changes in the natural population balance of the world'. Since 1958 the situation has got steadily worse and Williamson (1996) and Shigesada & Kawaski (1997) in their recent books on biological invasions have put a modern perspective on the problem. With increasing frequency articles with alarming titles such as 'Global change through invasion' (Lövell 1997) and 'Biological invasions as global environmental change' (Vitousek et al. 1996) are appearing, that highlight the global nature of the problem of introductions that become successfully established.

Tibbetts (1997) has pointed out, that, at the Earth Summit in Rio de Janeiro in 1992, 165 nations signed the United Nations Convention on Biodiversity which called on governments to prevent the introduction, control, or eradication of those alien species which threaten ecosystems, habitats or species. He goes on to say that the only things that have apparently been done since then were these. In 1995, agreement was reached on a plan of action with regard to marine and biological diversity – the Jakarta Mandate (which has no authority in international law). In 1996, parties to the biodiversity convention met in Trondheim, Norway, to discuss the issue of exotic species; however, no further guidelines were forthcoming.

Nowhere are introduced species having more of an impact than in the USA (Culotta 1991, Ruesink et al. 1995, Cohen & Carlton 1998). In 1997, 537 ecologists, agricultural officials and environmental experts signed a letter to Vice President Albert Gore expressing their concern about the impact of introduced species. They pointed out that the USA was in danger of losing the war against such invasive species and that their economic impacts were soaring. Apparently, attempts to stop the spread of introduced species in the USA is being hampered by the lack of a national policy and this is probably the case in many other countries too, including those in Europe. Consequently, to get international agreement on controlling exotics will be, as Tibbetts (1997) says 'a Sisyphean job.'

2 INTRODUCTIONS TO THE AQUATIC ENVIRONMENT

Welcomme (1988) has estimated that some 1354 introductions of 237 animal species (fish, crustaceans and amphibians) into 140 countries had been made into inland waters by the 1980s, most being made from 1950 to 1980. He estimated that 71 of these introductions have had a significant negative environmental impact and 25 of them are now regarded as pests, 47 have proved to be beneficial although with some negative effects, the rest being innocuous. He cites aquaculture as being the main reason for deliberate introductions (495 cases), followed by sport (191), improvement of wild stocks (165), ornaments (130) and control of other species (82). However, the reason for many introductions is unknown (312) or have occurred accidentally (139) through escapes from aquaria and fish farms, release of bait, or via the ballast water of ships.

The speed with which introduced species can become the dominant organism in an ecosystem is clearly illustrated by two ballast water introductions, i.e. the invasions of the Black Sea by the North American ctenophore, *Mnemiopsis leidyi* (Pearce 1995) and the North American Great Lakes by the European zebra mussel, *Dreissena polymorpha* (McMahon & Ram 1996). They also illustrate the fact that once a species is moved outside its home range and is not subject to controlling factors such as natural predators, then it may be very successful in a new environment. This is particularly so either if there is an unoccupied niche as with zebra mussels, i.e., there were no bivalve molluscs with byssus threads in the Great Lakes until the arrival of the zebra mussel, or if they are superior competitors to the native fauna. These facts are also very relevant to some freshwater crayfish species that have been introduced outside their home ranges (Holdich 1988).

3 FRESHWATER CRAYFISH

Crayfish are the largest, mobile invertebrates inhabiting freshwaters. There are thought to be in the region of 500 species worldwide. Where they occur they are often a keystone species and their elimination can have adverse environmental consequences. Alternatively, their addition to a freshwater ecosystem can also have adverse consequences as some species are able to build up their numbers very quickly and consume large amounts of animal and vegetable matter in the process (see Lindqvist & Huner and Nystrom this volume). In a number of countries, crayfish are considered a delicacy and they are farmed (astaciculture) and harvested from the wild, particularly in North America, China, Europe, and Australia. The total annual harvest from both sources may be in excess of 100,000 metric tons (mt) in some years (see Ackefors, this volume).

Details of the general biology of crayfish can be found in Holdich & Lowery (1988) and of their culture in Skurdal et al. (1989), Arrignon (1991), Lee & Wickins (1992), Holdich (1993) and Huner (1994, 1995). In addition, to the myriad of papers, particularly on the physiology of crayfish, to be found in standard journals, over 600 papers relating to all aspects of crayfish biology have been published in 'Freshwater Crayfish' the proceedings of the international symposia organised by the International Association of Astacology. Details of this organisation can be found on the IAA Homepage (http:/www.uku.fi/english/organisations/IAA/), which includes a bibliography and subject index to the papers in Freshwater Crayfish 1-10 and links to other valuable sources of information.

Ballast water introductions, as mentioned above, are an ever increasing problem (Moyle 1991), although one which has been with us for many decades. Indeed, the problems with alien crayfish in Europe are all thought to have stemmed from a ballast water introduction of crayfish carrying a fungal disease (crayfish plague – see Ackefors, Henttonen & Huner and Vogt this volume) into Italy from North America in the 1860s (Ninni 1865, Alderman 1996). This spread rapidly and led to the destruction of many populations of native crayfish species. Subsequently, crayfish species more immune to the disease were introduced from North America to supplement the remaining stocks (see Henttonen & Huner and Ackefors this volume). It is the spread of these often competitively superior species (see Lindqvist & Huner this volume) coupled with the reoccurring outbreaks of crayfish plague that have led to the current problems that are mainly the subject of this volume, although the introduction of some native species within Europe is also causing concern. The problems not only include the elimination of native European crayfish species, but also environmental damage due to burrowing and the alteration of trophic relationships (see Holdich and Nystrom this volume). In some countries, introduced crayfish have become pests once fully established and methods for eradicating them are being studied (see Holdich, Gydemo & Rogers this volume). However, on the positive side the farming and wild harvesting of the established alien species have brought some benefits (see Ackefors this volume).

4 TERMS

As the terms relating to crayfish introductions is not always the same, we have tried to clarify the situation for the purposes of this volume (see Table 1). The terms alien or exotic are usually used for any introduced organism that has not naturally occurred in historical times. As pointed out by Stubbs (1988), 'in historical times' is rarely pre-

Table 1. Terms relating to introductions (adapted from Stubbs 1988).

Term	Definition and comment
Imported	Brought into a country, contained
Introduction	The deliberate or accidental release of organisms into the wild where they have not occurred in historical times but *not* (yet) breeding
Reintroduction	The deliberate or accidental release of organisms into the wild in areas where they were indigenous in historical times
Restocking	The deliberate or accidental release of organisms into the wild in areas where they are already present
Translocation	The deliberate transfer of organisms for the purposes of introduction, reintroduction or restocking
Naturalized or established	The establishment of self-regenerating populations of any introduced organism in a free-living state in the wild
Native, indigenous, endemics or autochthonous	Native to and restricted to, a particular geographical area
Alien or exotic	Any introduced organism which has not naturally occurred in historical times
Pest	A species having a negative economic and/or environmental effect

cisely defined, but in Britain, at least, it is taken as approximately from the beginning of the Neolithic (ca. 5500 years ago). For the purposes of this volume, crayfish introductions in Europe only relate to what has happened in the last few hundred years or so (very little mention is made of crayfish in the literature prior to 1500), and these will cover both the translocation of native crayfish for stocking and restocking purposes, as well as the introduction of crayfish from other continents. As Stubbs (1988) has pointed out, ecologically there is no distinction between introducing a species from one country to another, than from one part of a country to another.

The term 'native' is used for organisms that have been native to, and restricted to a particular geographical area in historical times. For the purposes of this volume *Astacus astacus*, *Astacus leptodactylus*, *Astacus pachypus*, *Austropotamobius pallipes* and *Austropotamobius torrentium* are considered as native European crayfish (see Table 2). The five species occur to the west and north of the boundary between Europe and Asia as generally defined by modern geographers, i.e. from the Ural mountain range, the River Ural, the north-west part of the Caspian Sea, and then west along the Caucasus mountain range to the Black Sea, and have done so in historical times. The names, *Astacus leptodactylus* and *Astacus pachypus*, have been retained, although it is known that the taxonomy of these species has recently been revised (Starobogatov 1995). The fact that some crayfish species considered native to a particular country, e.g. *Astacus astacus* in Scandinavian countries and *Austropotamobius pallipes* in British countries, may have been introduced by humans relatively recently and are therefore 'alien' according to the definition above complicates the issue. However, it is not known how long these species have been in these geographical regions and they may have recolonised some parts naturally after the last Ice Age, which they are not thought to have survived (Albrecht 1982). Indeed there is also some fossil evidence from eastern England (Pleistocene deposits) (Bell 1920) and southern Finland (Littorina Period ca 3000-1000 BC) (Westman 1991), at least, that they may have been present much earlier. The fact that different countries in Europe

Table 2. Names used frequently in this volume (after Holdich 1992).

Scientific name	Authority and date	Common name	Origin
Aphanomyces astaci	(Schikora 1903)	crayfish plague	North America
Astacus astacus	(Linnaeus 1758)	noble	Europe
Astacus leptodactylus	Eschscholtz 1823	narrow-clawed	Europe/Near East
*Astacus pachypus**	Rathke 1837	thick-clawed	Europe/Near East
Austropotamobius pallipes	(Lereboullet 1858)	white-clawed	Europe
Austropotamobius torrentium	(Schrank 1803)	stone	Europe
Cherax destructor	Clark 1936	yabby	Australia
Cherax quadricarinatus	(von Martens 1868)	redclaw	Australia
Cherax tenuimanus	(Smith 1912)	marron	Australia
Orconectes limosus	(Rafinesque 1817)	spiny-cheek	North America
Orconectes rusticus	(Girard 1852)	rusty	North America
Pacifastacus leniusculus	(Dana 1852)	signal	North America
Procambarus clarkii	(Girard 1852)	red swamp	North America
Procambarus zonangulus	Hobbs & Hobbs 1990	white river	North America

*No common name has ever officially been given to *A. pachypus* – 'thick-clawed crayfish' is apparently used in some East European countries (Y. Machino pers. comm.) and will be adopted here.

consider the same species as being native for them, e.g., *A. pallipes* is native in England as well as in Italy, is not considered a problem, as we are considering Europe as a whole. *A. leptodactylus* probably originates from the Ponto-Caspian geographical area of Eastern Asia but also partly falls within the definition of Europe as given above. However, despite being a native European crayfish *A. leptodactylus* is considered by many to be an alien species in Northern and Western Europe (see Holdich this volume). It has many of the characteristics of the invasive North American species.

In addition to their natural colonization of new areas, there have been a number of different types of translocation involving native crayfish in Europe that have increased their range. First, there have been translocations of native crayfish to new localities, but within their area of natural distribution, e.g. neighbouring lakes and rivers. Second, the range of native crayfish within a country has been extended by translocating them to areas where they did not occur before, e.g. to more northerly parts of British and Scandinavian countries. Third, there have been translocations of one or more native crayfish species into the distributional area of another, e.g. *A. leptodactylus* and *A. astacus* in England, and *A. leptodactylus* in many European countries (see Holdich this volume).

5 AIMS

The aim of this volume is to bring together what is known about the impact that established populations of alien crayfish are having on the European freshwater environment, both from the positive and negative points of view, to highlight various aspects of their biology relevant to the topic, and to discuss what might be done about the problem. Of particular concern is the continuing decline of native crayfish stocks due to crayfish plague outbreaks and expanding alien crayfish populations. The series of papers represents a selection of those that were given at the workshop detailed in the Preface. This first section briefly reviews the problem of introductions, outlines the terminology relating to introductions, and introduces the volume. The second section of the volume includes a résumé of the history of crayfish introductions into Europe, an examination of what makes alien crayfish good colonisers, their impact – examined from both the negative and positive aspects, their trophic ecology, and the diseases they may transmit. The third section comprises a number of case studies, where the legislative, conservation and management steps taken by Italy, Belgium, France, Luxembourg, Switzerland, Germany, Spain, Portugal, Nordic and Baltic countries, the British Isles, and Belarus, to try and control the problem are dealt with. The fourth section starts with a review of possible methods for controlling alien crayfish. This is followed by an overview of the problem with a look into the future. The concluding chapter tries to bring together some of the thoughts outlined in the volume and from two round-table discussions held at the workshop. It asks the question – is the introduction of alien crayfish in Europe another example of the Frankenstein Effect (Moyle et al. 1986)? Also included in the final chapter is a summary of the known, present distribution of native and alien crayfish in Europe, which it is hoped will stimulate astacologists to confirm or deny and add to! It is imperative if we are to manage native and alien crayfish populations in Europe that we have a proper understanding of their distribution so that trends can be monitored and threats highlighted. In order to avoid some repetition, Table 2 lists those species, with their authorities and common names, which frequently appear in this volume. However, some repetition of the facts relating to European crayfish in a volume such as this is inevitable as the author(s) of each chapter has their own story to tell.

ACKNOWLEDGEMENT

The authors would like to thank Jostein Skurdal for his valuable comments on this chapter.

REFERENCES

Albrecht, H. 1982. Das System der europäischen Flusskrebse (Decapoda, Astacidae): Vorschlag und Begründung. *Mitt. hamb. zool. Mus. Inst.* 79: 187-210.

Alderman, D.J. 1996. Geographical spread of bacterial and fungal diseases of crustaceans. *Rev. Sci. Tech. Int. Epiz.* 15: 603-632.

Arrignon, J.C.V. 1991. *L'écrevisse et son elevage*. Paris: Lavoisier.

Bell, A. 1920. Notes on the later Tertiary Invertebrata. *Ann. Rep. Yorkshire Phil. Soc.* 1920: 1-21.

Cohen, A.N. & Carlton, J.T. 1998. Accelerating invasion in a highly invaded estuary. *Science* 279: 555-557.

Culotta, E. 1991. Biological immigrants under fire. *Science* 254: 1444-1447.

Elton, C.S. 1958. *The ecology of invasions by animals and plants*. London: Chapman & Hall.

Holdich, D.M. 1988. The dangers of introducing alien animals with particular reference to crayfish. *Freshwater Crayfish* 7: xv-xxx.

Holdich, D.M. 1992. Crayfish nomenclature and terminology: recommendations for uniformity. *Finn. Fish. Res.* 14: 149-155.

Holdich, D.M. 1993. Astaciculture: freshwater crayfish farming. *Aquat. Living Resour.* 6: 307-317.

Holdich, D.M. & Lowery, R.S. (eds) 1988. *Freshwater crayfish: biology, management and exploitation*. London: Croom Helm (now Chapman & Hall).

Huner, J. (ed.) 1994. *Freshwater crayfish aquaculture. in North America, Europe and Australia.* New York: The Haworth Press Inc.

Huner, J.V. 1995. An overview of the status of freshwater crawfish culture. *J. Shellf. Res*. 14: 539-543.

Lee, D. O'C. & Wickins, J.F. 1992. *Crustacean farming*. Blackwell: Oxford.

Lodge, D.M. 1993. Biological invasions: lessons for ecology. *TREE* 8(4): 133-137.

Lövell, G.L. 1997. Global change through invasion. *Nature* 388: 627-628.

McMahon, R.F. & Ram, J.L. (eds) 1996. The biology, ecology and physiology of zebra mussels. *Am. Zool*. 36: 239-384.

Moyle, P.B. 1991. Ballast water introductions. *Fisheries* 16(1): 4-6.

Moyle, P.B., Li, H.W. & Barton, B.A. 1986. The Frankenstein Effect: impact of introduced fishes on native fishes in North America. In R.H. Stroud (ed.), *Fish culture in fisheries management.* Bethesda, MD: American Fisheries Society.

Ninni, A.P. 1865. Sulla mortalita dei gamberi (*Astacus fluviatilis* L.) nel veneto e piu particolarmente nella provincia trevigiani, accompagnata da esemplari di gamberi ammalati e sani. *Atti Inst. Veneto* Ser. III, 10: 1203-1209.

Pearce, F. 1995. How the Soviet seas were lost. *New Scientist* 148(2003): 39-42.

Pullin, R.S.V. 1994. Exotic species and genetically modified organisms in aquaculture and enhanced fisheries: ICLARM's position. *Naga, ICLARM* 17(4): 19-24.

Ruesink, J.L., Parker, I.M., Groom, M.J. & Kareiva, P.M. 1995. Reducing the risks of nonindigenous species introductions. *Bioscience* 45(7): 465-477.

Shigesada, N. & Kawaski, K. 1997. *Biological invasions: theory and practice*. Oxford: OUP.

Skurdal, J., Westman, K. & Bergan, P.I. (eds) 1989. *Crayfish culture in Europe*. Report from the workshop on crayfish culture, 16-19 Nov. 1987, Trondheim, Norway.

Starobogatov, Ya.I. 1995. Taxonomy and geographical distribution of crayfishes of Asia and East Europe (Crustacea Decapoda Astacoidei). *Arthropoda Selecta* 4(3/4): 3-25.

Stubbs, D. 1988. *Towards an introductions policy. Conservation guidelines for the introduction and reintroduction of living organisms into the wild in Great Britain*. London: Wildlife Link.

Tibbetts, J. 1997. Exotic invasion. *Envir. Hlth. Persp*. 105: 590-593.

Vitousek, P.M., D'Antonio, C.M., Loope, L.L. & Westbrooks, R. 1996. Biological invasions as global environmental change. *Am. Sci*. 84: 468-478.

Welcomme, R.L. 1988. *International introductions of inland aquatic species*. FAO Fisheries Technical Paper 294. Rome: FAO.

Westman, K. 1991. The crayfish fishery in Finland – its past, present and future. *Finn. Fish. Res*. 12: 187-216.

Williamson, M. 1996. *Biological invasions*. London: Chapman & Hall.

Aspects of the biology of alien crayfish

The introduction of alien species of crayfish in Europe: A historical introduction

PAULA HENTTONEN
Department of Applied Zoology and Veterinary Medicine, University of Kuopio, Kuopio, Finland

JAY V. HUNER
Crawfish Research Center, University of Southwestern Louisiana, Lafayette, Louisiana, USA

ABSTRACT

The native, European, crayfish fauna belong to a single family, Astacidae, with two genera and five recognized species: *Astacus astacus*, *Astacus leptodactylus*, *Astacus pachypus*, *Austropotamobius pallipes,* and *Austropotamobius torrentium.* Habitat degradation and loss, overfishing and the crayfish plague fungus, *Aphanomyces astaci* Schikora, dramatically reduced and/or eliminated many populations of these crayfish. Several North American crayfish species were intentionally introduced into Europe: Two cambarids, *Orconectes limosus* was introduced in 1890s, and *Procambarus clarkii* in the 1970s, and one astacid, *Pacifastacus leniusculus,* in 1960s. Several Australian parastacid species, genus *Cherax*, have been or are being cultivated on a minor scale in Europe. *Orconectes limosus* stock in Europe was established by a successful stocking of 100 specimens from the Delaware River, Pennsylvania, USA, in a fish pond in Barnowko village, Germany. The first 60 *P. leniusculus* originated from California and were stocked in Sweden, initially for ecological studies of the species. In 1967-1969, great numbers of *P. leniusculus* were stocked in natural waterbodies in Finland and Sweden. Secondary introductions were later made with the Swedish signal crayfish to many countries in central Europe. *P. clarkii* was first introduced in 1973 from Louisiana to Spain with fewer that 40,000 individuals but has then spread through secondary introductions to many other south and central European countries. These introductions were often the result of modest numbers of crayfish imported live for food sales or pets. Successful stockings with alien crayfish have led to the situation that crayfish capable of resisting and carrying fungus plague are now permanently found in many of the waters that previously were inhabited by native crayfish. Alien crayfish, however, serve as food for many other animals and are of commercial value for man.

1 INTRODUCTION

1.1 *The motivation for introducing alien crayfish species into European waters*

Freshwater crayfish have been important components of European aquatic faunas ever since they invaded them and have become important commercial species for several centuries (Holdich & Lowery 1988). The native crayfish fauna is, however, impoverished. There is only one family, Astacidae, with two genera, *Astacus* and *Austropotamobius*. These include five recognized species including *Astacus astacus* in northwestern Europe and Scandinavia, *A. leptodactylus* in eastern and south-eastern Europe, *A. pachypus* in southeastern Europe, *Austropotamobius pallipes* in western and southwestern Europe, and *A. torrentium*, in central Europe (Lowery & Holdich 1988). However, strong crayfish populations that once supported important fisheries throughout Europe into the mid-1800s have collapsed as a consequence of the crayfish plague fungus, *Aphanomyces astaci* Schikora, habitat destruction, and pollution.

The need for introductions have been and are being promoted with the following arguments. The need for a plague-resistant stock to replace lost native populations, to occupy a vacant niche, for ornamental and aquacultural purposes, to control abundant water vegetation, and even to limit the existence of other species that are viewed as harmful. The reported reasons for historical introductions of alien stocks are not conclusive in the literature or among modern researchers.

Efforts to fill vacant benthic niches with crayfish that were both resistant to the crayfish plague and tolerant of degraded habitats were initiated toward the end of the 1800s (Holdich 1988) in western and central Europe where losses were greatest. Subsequent declines in native species in Scandinavia and southwestern Europe in the 1900s led to further interest in the introduction of additional crayfish species.

Crayfish introductions can be divided into a) primary introductions from USA and Australia to Europe, b) secondary introductions from European sites into other areas or countries in the continent, and c) natural distribution from the site of release. In this article, we will concentrate on primary introductions, although a few case of secondary introduction will be mentioned. Translocations of native crayfish are not included in this introduction.

1.2 *The alien crayfish species in Europe*

According to the literature (Kulmatycki 1935, translated by P. Smietana), the first introduction of the cambarid spiny-cheek crayfish (also called the striped American crayfish), *Orconectes limosus,* into Europe was conducted by Max von dem Borne in 1890. He stocked about 100 specimens of *O. limosus* from the Delaware River (Pennsylvania, USA) into a fishpond (of 0.1 ha surface area) in Barnowko village. This fishpond was connected to the Mysla River – a tributary of the Oder River (Kulmatycki 1935, translated by P. Smietana).

The second but unsuccessful introduction was made in France nearby Fecamp in 1895 with *O. limosus* originating from New York (Kossakowski 1966, translated by P. Smietana). A successful introduction in France was made with 2000 specimens of *O. limosus,* which were stocked in 1911-1913 in the River Cher (Kossakowski 1966, translated by P. Smietana). In Poland in 1911, Priest J. Cyr introduced *O. limosus* to the Lake Sieroslawskie, then known with the German name Juno Lake (Seligo 1911, translated by P. Smietana). Successful secondary introductions have also been made in later 1900s, e.g. in Austria (Spitzy 1973, Westman et al. 1992), in Germany (Schweng 1973) and in France (Laurent 1973).

Many years passed and Swedish authorities noted the biotic success but commercial failure of the *O. limosus* introductions and carefully surveyed additional North American crayfish species as candidates for introduction into Sweden (see the review of Svärdson 1995). Two species, the cambarid northern crayfish, *Orconectes virilis* and the astacid signal crayfish, *Pacifastacus leniusculus*, were brought to Sweden in 1960 for growth trials (Westman et al. 1992). The northern crayfish disappeared (Svärdson & Nilsson 1964, referred by Abrahamsson 1973a) and the signal crayfish was successfully introduced (Svärdson 1965) (note that French authorities had made an unsuccessful introduction of *O. virilis* into France in 1897, André 1960, referred by Abrahamsson 1973a, Westman et al. 1992).

Introductions of *P. leniusculus* in Sweden (Svärdson 1965, Abrahamsson 1973b, Brink 1977) and Finland (Westman 1973) are well documented. The first 60 *P. leniusculus* were imported to Sweden from the Natoma River, American River, Sacramento River drainages, California, in December 1959 and were released in 1960 (Svärdson 1965, 1995). After successful experiments carried out in different lakes and ponds, the Simontorp hatchery received a permit to import 100,000 specimens from Lake Tahoe. In 1969, 67 natural waters were stocked (Brink 1977).

The first introduction of *P. leniusculus* into Finland was made by the Finnish Game and Fisheries Research Institute with 3000 adult signal crayfish from Lake Hennessey and Lake Tahoe, California, in 1967-1968. The following year, a new batch of adult signal crayfish was imported from Lake Tahoe. At the same time, 3269 signal crayfish were stocked in eight small lakes in south and central Finland during the years 1967-1969. After that, in 1971-1972 secondary introductions into 22 lakes in south and central Finland were conducted with 10,000 newly hatched juveniles from Sweden. (Westman 1973).

Also Austria received the first signal crayfish shipment from California; this was organised in 1969 privately and not legally (Spitzy 1973) like in Sweden and Finland. After that, many secondary introductions of signal crayfish into other European countries have been conducted from Simontorp, Sweden, e.g. to Poland since 1971 (Kossakowski et al. 1978), to Germany in 1972 (Arrignon et al. 1997) and many times to Finland.

Soon after the Nordic introductions, Spanish authorities granted permission for trials on the suitability of two additional cambarid crayfishes, the red swamp crayfish, *Procambarus clarkii,* and the white river crayfish, *Procambarus zonangulus* (Hobbs & Hobbs) – then called *Procambarus acutus acutus* (Girard) – for introduction into Spanish waters (Habsburgo-Lorena 1978). The initial introduction from Louisiana State University in Baton Rouge, Louisiana and the Cities Services Oil Company crayfish farm near Monroe, was made in 1973 with the red swamp crayfish thriving and the white river crayfish disappearing. Over 500 kg of crayfish was involved in this project with well over 90% being *P. clarkii*. During the 1980s and 1990s, several Australian parastacid species of the genus *Cherax*, including the marron, *Cherax tenuimanus*, the yabby, *Cherax destructor,* and the redclaw, *Cherax quadricarinatus,* have been cultivated on a very small scale but so far there has been no reports of introductions into natural waters in Europe (Ackefors 1997, Perez et al. 1997, H. S. Habsburgo-Lorena pers. comm.).

2 SPREAD OF NORTH AMERICAN CRAYFISH SPECIES IN EUROPE

2.1 *Orconectes limosus*

The spiny-cheek crayfish is a hardy, aggressive, temperate-water crayfish species native to the north-eastern USA and south-eastern Canada. Well adapted for life in permanent lentic and lotic waterbodies, the spiny-cheek crayfish can reach sizes in excess of 40 g but often stunts at sizes under 15 g. Following Max von dem Borne's successful introduction of this species into German waters, it was intentionally stocked in France, Poland (Momot 1988), and Austria (Spitzy 1973, Momot 1988). Stockings and natural dispersal made the spiny-cheek crayfish the most important European crayfish from the naturalistic point of view by the end of the 1930s, but it has never achieved any real commercial significance despite its abundance (Holdich 1988, Momot 1988, Troschel & Dehus 1993). It is currently the dominant freshwater crayfish in most permanent water bodies in France, Germany, Czechia, Slovakia, and Poland, with its range including The Netherlands, Belgium, Switzerland, Spain, Italy, and Lithuania, and probably also Russia and Ukrania (Ackefors 1989). The spiny-cheek crayfish is clearly continuing to expand its natural range in Europe and may be expected to be present, eventually, in most continental waters.

2.2 *Procambarus clarkii*

The red swamp crayfish is a 'warm water', hardy species native to northeastern Mexico and the south central USA. Fully capable of reaching sizes in excess of 50 g in 3-5 months, this crayfish may mature at sizes as small as 10 g and adapt to very restricted, seasonal waters. As a result, the red swamp crayfish has become the dominant freshwater crayfish in the world during the 20th century. Over 85,000 tons is now commercially harvested each year from the USA, People's Republic of China, and the Iberian peninsula (Huner 1989, 1994, 1995).

Unlike the spiny-cheek and signal crayfish (however, see Holdich, Rogers & Reynolds this volume) and the native European species, the red swamp crayfish is a burrowing species adapted to life in seasonally-flooded wetlands where standing water may be absent for some months. As a result, the red swamp crayfish is well suited for life in natural and agricultural wetlands throughout south central Europe (Habsburgo-Lorena 1978, Huner 1994, Huner & Lindqvist 1995). For example, within 10 years of its initial introduction into Spain, the red swamp crayfish had become an important commercial species in that country with harvests reaching as much as 3 to 4000 tons in years of high rainfall. By 1979, it was well established in southern Portugal, probably as a result of natural dispersal from Spanish waters and, perhaps, intentional introductions. Commercial success apparently led to intentional, but illegal, introductions throughout Spain, France, and Italy during the period from the late 1970s into the late 1980s.

The habit of selling red swamp crayfish alive as a food item and as an aquarium and/or garden pond pet has surely accelerated successful introductions into natural waters throughout Europe. In addition, to the successful introductions previously referenced, the red swamp crayfish is now also known from The Netherlands, Switzerland, and Germany and is present in southern England (Holdich 1988, Huner 1989,

1994, Troschel & Dehus 1993). While Spain is certainly an important source of red swamp crayfish stocking material, it is abundantly clear that commerce in live crayfish from distant sources including the Far East, the USA and Kenya has been responsible for some, if not many, of the successful introductions of red swamp crayfish into England, The Netherlands, France, Germany and Switzerland.

The red swamp crayfish thrives in southern parts of Europe where warm, shallow wetland ecosystems, natural and agricultural, are widespread (Correia 1995, Guerra & Nino 1995). Elsewhere, in cooler areas of Europe, the crayfish becomes most abundant in isolated, small, permanent ponds and is less able to tolerate predaceous fishes typically found in larger water bodies (Troschel & Dehus 1993, Roqueplo et al. 1995, Delmastro & Laurent 1997, Huner pers. obs.). Efforts to limit and/or eradicate unwanted red swamp crayfish have been unsuccessful, wherever the species has been successfully introduced throughout the world (Hobbs et al. 1989, Huner 1989). Therefore, *Procambarus clarkii* can be expected to extend its range to much of Europe, in particular to warm and mild temperature areas.

2.3 *Pacifastacus leniusculus*

The signal crayfish is certainly the largest of the exotic North American crayfish species introduced successfully into European waters and, under ideal growth conditions able to achieve sizes in excess of 100 g (Lowery & Holdich 1988). The signal crayfish is native to permanent water bodies in cool temperate regions of the northwestern USA and southwestern Canada and is, therefore, well adapted for life in permanent lentic and lotic waters throughout Europe. As an astacid, this crayfish has a life cycle very similar to those of the native European crayfish species and quickly acclimated to areas where it has been introduced beginning, initially, in Sweden in 1960 (Abrahamsson 1973b, Svärdson 1995). Southern Sweden appears to be convenient for *P. leniusculus*. Several hundred tons of signal crayfish are harvested from natural Swedish waters (Huner 1989), while Finnish waters, where the species has been stocked now since 1967 (Westman 1973), generate no more than a few tons annually (O. V. Lindqvist pers. comm.). This suggests that the signal crayfish is unable to thrive under the extremely cold climatic conditions in Finland and in Northern Sweden.

The signal crayfish is now established throughout western and central Europe (Holdich 1988, Lowery & Holdich 1988). Reproductive populations have been reported from natural waters throughout continental Europe (see other papers in this volume). Most introductions took place during the 1980s and were the result of escapes (and intentional, but illegal releases) from aquacultural endeavours. Additional introductions will surely lead to the distribution of signal crayfish throughout Europe. Interestingly, with the exceptions of Sweden (Ackefors this volume) and England (Holdich, Rogers & Reynolds this volume), populations are not yet widespread and dense enough to support significant fisheries. Efforts were made to channel initial stocking material through quarantine facilities. Even so, the ready availability of live signal crayfish from US seafood dealers led to the arrival and stocking of specimens that were presumably carrying the crayfish plague as a chronic, sublethal infection. This was, nevertheless transmitted to non-resistant, native crayfish species, especially strong stocks of English white-clawed crayfish (Holdich & Reeve 1991, Holdich, Rogers & Reynolds this volume).

In Finland, signal crayfish stockings have been made with 'crayfish plague free juveniles' for many years. However, because the plague fungus exists somewhere in natural waters, signal crayfish sooner or later become infected and may thus spread the disease. Kirjavainen (1996) has stated that already tens of infected signal crayfish populations exist in Finland.

3 THREE AUSTRALIAN CRAYFISH SPECIES IN EUROPE

Although there are a number of species of *Cherax* in Australia, only three have shown any immediate potential for aquacultural endeavors (Mills et al. 1994). These include the large marron, *C. tenuimanus* (to 2 kg), the intermediate red claw, *C. quadricarinatus* (to 500 g), and the smaller yabby, *C. destructor* (to 100 g). Marron require fairly warm, Mediterranean conditions, permanent waters and high levels of dissolved oxygen. Red claw require warm, tropical conditions, permanent waters, and high levels of dissolved oxygen. Yabby, in contrast, are temperate species adapted to seasonally flooded wetlands, as a consequence of their burrowing habit, and tolerates very low levels of dissolved oxygen.

All three of these Australian *Cherax* species have been brought to Europe for aquacultural trials (Holdich 1993, Perez et al. 1997). Use of marron has apparently been abandoned. Red claw trials continue and require warm water systems, but the only significant commercial success to date appears to involve the sale of juveniles for aquarium pets although Ackefors (1997) notes a very limited amount of food-size red claw being cultivated in Spain, presumably in ponds. It is unclear if yabby are being currently cultivated in Europe. Based on the known habitat requirements for the marron, red claw and yabby, only the yabby appears to have any potential for successful introduction into any natural European waters and then only in southern Europe. However, all three species are presumed to lack resistance to the crayfish fungus plague (Mills et al. 1994). Therefore, it is questionable whether or not any of the species could become established on a continuing basis.

4 OBJECTIONS TO INTRODUCTIONS OF NON-NATIVE CRAYFISH SPECIES AND THE POTENTIAL FOR INTRODUCTIONS OF ADDITIONAL SPECIES

4.1 *Objections to the introductions of non-native crayfish species in Europe*

The initial European introductions of the spiny-cheek crayfish, red swamp crayfish, and signal crayfish were made by well intentioned authorities and entrepreneurs, who sought to restore freshwater crayfish numbers to levels existing before European crayfish populations were decimated by the crayfish plague and habitat destruction (Holdich 1988). They did not know or acknowledge that all three species were potential vectors for the crayfish plague if exposed to it (Huner 1995).

The spiny-cheek crayfish has been met with indifference from commercial and sports fishing interests because of its small size relative to the native *Astacus* species and success in eutrophic and often polluted waters. However, this 'small' crayfish is an acceptable human food especially in areas suffering from food shortages or economic problems, e.g. Polland (Kossakowski 1971). Environmental interests have objected to the spiny-cheek crayfish because it has displaced native European species when it has come in contact with them as a result of direct competition and, at least in some cases, from introduction of the crayfish plague.

Both red swamp crayfish and signal crayfish have been met by enthusiastic acceptance by most commercial and sports interests (Habsburgo-Lorena 1983a, b, Svärdson 1995) although the red swamp crayfish can, through its burrowing habit, damage water control structures. However, there has been general controversy, with some exceptions in the case of the red swamp crayfish about the environmental impacts of these two species (Delibes & Adrian 1987, Holdich 1988, Carral et al. 1993, Ruiz 1996, Gutiérrez-Yurrita et al. 1997). Both are aggressive and can serve as crayfish plague vectors to native European crayfish species. Because both species are keystone benthic organisms, they have caused significant changes in native floras and faunas but they have also become exceptionally good food resources for valuable wetland vertebrate species, including waterbirds and semi-aquatic mammals. In some countries stringent legislation has been imposed on the importation and movement of alien crayfish in order to protect the native crayfish and the freshwater environment, e.g. Great Britain, Ireland (Holdich & Reeve 1991, Holdich, Rogers & Reynolds this volume), Norway, and Estonia (Taugbøl & Skurdal 1997).

4.2 *The potential for introduction of additional, unwanted non-native crayfish species*

There are over 300 species of North American crayfish and 100 species of Australian crayfish (Hobbs 1988). These species occupy all types of habitats, ranging from wet meadows to wetlands, to small and large permanent water bodies. Many niches of the same kind are generally vacant in Europe and this has permitted the spiny-cheek crayfish, the red swamp crayfish and the signal crayfish to expand their populations very quickly throughout Europe. Any number of North American and Australian crayfish species can, therefore, be expected to be candidates for introductions into Europe, if afforded the opportunity to do so. The fact that 100 spiny-cheek crayfish were adequate to establish that species in Germany in 1890 (Kossakowski 1971) should give resource managers reason for concern, because all crayfish species may be transported for some days in damp, humid containers without standing water. Furthermore, the retail aquarium trade is based, in many instances, on novelty and a new crayfish species could be expected to be widely distributed in the pet trade as the red swamp crayfish has been and as the red claw is now being distributed. Additional species may, in fact, already be present in isolated locations not yet discovered by the authorities.

5 CONCLUSIONS

Introductions of alien crayfish in Europe have been both intentional and accidental.

Some of the intentional introductions have been well prepared and preceded by long-term research that included risk assessment. In many cases, the effects of alien introductions have been unforeseeable. Though some introduction have yielded commercial benefit to man, the same species in another natural environment may cause very different effects, depending on the vulnerability of the environment or of the native species.

We sincerely believe that all efforts must be made to preserve the fragile biodiversity of the native European crayfish fauna. Nations should have a crayfish strategy and introduction policy that is in harmony with the agreements the nation or authorities have signed (e.g. The Rio Convention). Research should focus on ecological, economic, and social benefits and costs that will be caused by introductions.

ACKNOWLEDGMENTS

The authors wish to acknowledge the intellectual contributions to this paper by Professor Ossi V. Lindqvist, Rector Magnificus, University of Kuopio, Kuopio, Finland, and to Dr Francesca Gherardi of the Dipartimento di Biologia Animale e Genetica, University of Florence, Florence, Italy, for inviting us to prepare this paper. We would like to thank Professor David Holdich, University of Nottingham, England, and Dr Jostein Skurdal for helpful remarks concerning a draft of this manuscript, and Dr Przemyslaw Smietana, University of Szczecin, Poland, for information concerning introductions in Poland.

REFERENCES

Abrahamsson, S. 1973a. Methods for restoration of crayfish waters in Europe. The development of an industry for production of young of *Pacifastacus leniusculus* Dana. *Freshwater Crayfish* 1:203-210.

Abrahamsson, S. 1973b. The crayfish *Astacus astacus* in Sweden and the introduction of the American crayfish *Pacifastacus leniusculus*. *Freshwater Crayfish* 1:27-40.

Ackefors, H. 1989. Intensification of European freshwater crayfish culture in Europe. *Special Session on Crayfish Culture on Aquaculture World Aquaculture Soc., Los Angeles, USA, Feb. 13, 1989.* Univ. Stockholm, Sweden.

Ackefors, H. 1997. Recent progress in European crayfish farming. Mscrpt. Univ. Stockholm, Sweden.

André, M. 1960. Les écrevisses francaises. Paris. (According to Abrahamsson 1973a).

Arrignon, J,. Gérard, P., Keller, M., Krier, A. & Stucki, T. Specific case histories of crayfish introductions: Belgium, France, Germany, Luxembourg, Switzerland. An international workshop The introduction of alien species of crayfish in Europe. Florence September 24-27, 1997. Abstract volume.

Brink, P. 1977. Developing crayfish populations. *Freshwater Crayfish* 3:211-228.

Carral, J.M., Celada, J.D., Gonzalez, J., Saez-Royuela, M., Gaudioso, V.R., Fernandez, R. & Lopez-Baisson, C. 1993. Wild freshwater crayfish populations in Spain: current status and perspectives. *Freshwater Crayfish* 9:158-162.

Correia, A.M. 1995. Population dynamics of *Procambarus clarkii* (Crustacea: Decapoda) in Portugal. *Freshwater Crayfish* 8:276-290.

Delamastro, G.B. & Laurent, P.J. 1997. Recent spreading of Nearctic crayfish species in Italy. *L'Astaciculteur de France* 50:2-8.

Delibes, M. & Adrian, I. 1987. Effects of crayfish introduction on otter *Lutra lutra* food in the Doñana National Park, SW Spain. *Biol. Conserv.* 42:153-159.

Guerra, J.L. & Nino, A.E. 1995. Ecology of red swamp crawfish (*Procambarus clarkii* Girard) in the central Meseta of Spain. *Freshwater Crayfish* 8:179-200.

Gutiérrez-Yurrita, P.J., Green, A. & Montes, C. 1997. Impact of crayfish *Procambarus clarkii* netting on vertebrates in the Guadalquivir marshes, Spain. *Crayfish News* 19(3):8-9.

Habsburgo-Lorena, H.S. 1978. Present situation of exotic species of crayfish introduced into Spanish continental waters. *Freshwater Crayfish* 4:175-184.

Habsburgo-Lorena, H.S. 1983a. Socioeconomic aspects of the crawfish industry in Spain. *Freshwater Crayfish* 5:552-554.

Habsburgo-Lorena, H.S. 1983b. Some observations on crawfish farming in Spain. *Freshwater Crayfish* 5:549-551.

Hobbs, H.H. Jr. 1988. Crayfish distribution, adaptive radiation and evolution 3. In D.M. Holdich & R.S. Lowery (eds), *Freshwater Crayfish. Biology, Management and Exploitation.* London & Sydney: Croom Helm/Portland, Oregon: Timber Press.

Hobbs, H.H. III, Jass, J. P. & Huner, J.V. 1989. A review of global crayfish introductions with particular emphasis on two North American species (Decapoda, Cambaridae). *Crustaceana* 56:299-316.

Holdich, D.M. 1988. The dangers of introducing alien animals with particular reference to crayfish. *Freshwater Crayfish* 7:15-30.

Holdich, D.M. 1993. A review of astaciculture: freshwater crayfish farming. *Aquat. Living Resour.* 6:307-317.

Holdich, D.M. & Lowery, R.S. 1988. Crayfish – an introduction 1. In D.M. Holdich & R.S. Lowery (eds), *Freshwater Crayfish. Biology, Management and Exploitation.* London & Sydney: Croom Helm/Portland, Oregon: Timber Press.

Holdich, D. M. & Reeve, I.D. 1991. Distribution of freshwater crayfish in the British Isles, with particular reference to crayfish plague, alien introductions and water quality. *Aquat. Conserv. Mar. Freshw. Ecosys.* 1:139-158.

Huner, J.V. 1989. Overview of international and domestic freshwater crawfish production. *J. Shellfish Res.* 8:259-266.

Huner, J.V. 1994 (ed.) *Freshwater Crayfish Aquaculture in North America, Europe and Australia. Families Astacidae, Cambaridae, and Parastacidae.* Binghamton, New York: Haworth.

Huner, J.V. 1995. An overview of the status of freshwater crawfish culture. *J. Shellfish Res.* 14:539-544.

Huner, J.V. & Lindqvist, O.V. 1995. Physiological adaptations of freshwater crayfishes that permit successful aquacultural endeavors. *Physiol. Zool.* 35:12-19.

Kirjavainen, J. 1996. Hämeen ravunviljelyopas. *Kala- ja riistahallinnon julkaisuja.* Nro 23/1996 (in Finnish).

Kossakowski, J. 1966. Raki (monograph). Panstwowe Wydawnictwo Rolnicze i Lesne: Warszawa. pp 23-25 (in Polish, translated by P. Smietana).

Kossakowski, J. 1971. Crayfish ('Raki'). Warsaw: Panstwowe Wydawnictwo Rolnicze i Lesne: Warszawa – Translation from Polish to English – US Dept. of Commerce – TT 70-55114: Washington, DC.

Kossakowski, J., Minch, M. & Kossakowski, G. 1978. The first introduction of the crayfish *Pacifastacus leniusculus* Dana into Polish waters. *Freshwater Crayfish* 4:195-196.

Kulmatycki, W. 1935. *Cambarus affinis* Say – rak amerykanski, nowy mieszkaniec Pomorza I Wielkopolski. Prezeglad Rybacki. Tom 10, 11:1-16 (in Polish, translated by P. Smietana).

Laurent, P. 1973. *Astacus* and *Cambarus* in France. *Freshwater Crayfish* 1:69-78.

Lowery, R.S. & Holdich, D.M. 1988. *Pacifastacus leniusculus* in North America and Europe, with Details of the Distribution of Introduced and Native Crayfish Species in Europe. In D.M. Holdich & R.S. Lowery (eds), *Freshwater Crayfish. Biology, Management and Exploitation.* London & Sydney: Croom Helm/Portland, Oregon: Timber Press.

Mills, B.J., Morrissy, N.M. & Huner, J.V. 1994. Cultivation of freshwater crayfishes in Australia. In J.V. Huner (ed.), *Freshwater Crayfish Aquaculture in North America, Europe, and Australia. Families Astacidae, Cambaridae, and Parastacidae*. Binghamton, New York: Haworth.

Momot, W.T. 1988. *Orconectes* in North America and elsewhere 10. In D.M. Holdich & R.S. Lowery (eds), *Freshwater Crayfish. Biology, Management and Exploitation*. London & Sydney: Croom Helm/Portland, Oregon: Timber Press.

Perez, J.R., Carral, J.M., Saez-Royuela, M., Munoz, C. & Sierra, A. 1997. Current status of astaciculture production and commercial situation of crayfish in Europe. Mscrpt. Univ. Leon, Spain.

Roqueplo, C., Laurent, P.J. & Neuev, A. 1995. *Procambarus clarkii* Girard (Louisiana red swamp crayfish) synthesis: problems caused by this species and attempts to control populations. *L'Astaciculteur de France* 44:2-14.

Ruiz, X. 1996. Crayfish food for Spanish waterbirds. *Crayfish News* 17(4):6-7.

Schweng, E. 1973. *Orconectes limosus* in Deutschland, insbesondere im Rheingebiet. *Freshwater Crayfish* 1:79-87.

Seligo, A. 1911. Mitteilungen des Westpreussischen Fisherei – Vereins. (translated by P. Smietana).

Spitzy, R. 1973. Crayfish in Austria, history and actual situation. *Freshwater Crayfish* 1:9-14.

Svärdson, G. 1965. The American crayfish *Pacifastacus leniusculus* (Dana), introduced into Sweden. *Rep. Inst. Freshw., Drottningholm*. 46:90-94.

Svärdson, G. 1995. The early history of signal crayfish introduction in Europe. *Freshwater Crayfish* 8:68-77.

Svärdson, G. & Nilsson, N.-A. 1964. Fiskeribiologi. Halmstad. (referred by Abrahamsson 1973a).

Taugbøl, T. & Skurdal, J. 1997. The future of native crayfish in Europe – how to make the best of a bad situation. An international workshop The introduction of alien species of crayfish in Europe. Florence September 24-27, 1997. Abstract volume.

Troschel, H.J. & Dehus, P. 1993. Distribution of crayfish species in the Federal Republic of Germany with special reference to *Austropotamobius pallipes*. *Freshwater Crayfish* 9:390-398.

Westman, K. 1973. The population of the crayfish *Astacus astacus* in Finland and the introduction of the American *Pacifastacus leniusculus* Dana. *Freshwater Crayfish* 1:41-55.

Westman, K., Ackefors, H. & Nylund, V. 1992. *Kräftor. Biologi, odling, fiske*. Kiviksgården: AB Ystad Centraltryckeri.

Life history characteristics of crayfish: What makes some of them good colonizers?

OSSI V. LINDQVIST
Department of Applied Zoology and Veterinary Medicine, University of Kuopio, Kuopio, Finland
JAY V. HUNER
Crawfish Research Center, University of Southwestern Louisiana, Lafayette, Louisiana, USA

ABSTRACT

The history of colonization of various animal and plant groups indicates a number of patterns. For instance, many European bird species have successfully colonized North America (but not vice versa), and nearly half of all North American insect pests are considered to be of exotic origin (Kim et al. 1993). Similarly, the European continent has been successfully invaded by deliberate introductions by several North American crayfish species. The notable example is *Procambarus clarkii,* which may now be considered nearly global by its current distribution. The most successful crayfish species to colonise are relatively r-selected species, with short life cycles and high fecundity. *Procambarus clarkii* is a relatively r-selected species, while the native *Astacus astacus*, for instance, is a relatively K-selected species, with a long life cycle and low fecundity. *A. astacus* is, thus, more competitive in mature ecosystems, while *P. clarkii* is able to colonize disturbed habitats and areas which have been modified by man. Modified aquatic habitats are often more tropical or temperate in terms of their thermal conditions. An additional factor influencing the success of North American crayfish in Europe often involves their simultaneous resistance and ability to transmit the crayfish fungus plague, *Aphanomyces astaci*, to highly susceptible native crayfish.

1 INTRODUCTION

The decline of native European crayfish species (family Astacidae) beginning in the 1860s and continuing to these days has been well documented (Ackefors & Lindqvist 1994). The deadly combination of the crayfish fungus plague, *Aphanomyces astaci*, presumably from North America, and habitat destruction and degradation led to elimination of native crayfish from many waters where their populations once thrived. The introduction of exotic, plague-resistant American species (family Cambaridae) began before 1900 with the cool-water species *Orconectes limosus* being successfully introduced with subsequent spread throughout western and central Europe. Although a very successful invader, *O. limosus* has never been widely accepted by consumers (Momot 1988). Demand for relatively rare native species, including *Astacus astacus*, *Austropotamobius pallipes* and *Austropotamobius torrentium,* was such that fisheries managers sought other North American species for introduction into vacant niches. Two large, aggressive species, the warm-water *Procambarus clarkii* (family Cambaridae) and the cool-water *Pacifastacus leniusculus* (family Astacidae) were widely introduced into Spain in the 1970s and Scandinavia in the 1960s and 1970s, respectively (Ackefors & Lindqvist 1994). Other papers in this volume document the subsequent northward and eastward dispersal of *P. clarkii* and the southward and eastward dispersal of *P. leniusculus*, almost entirely as a result of active human intervention.

In this paper, we address the attempt to explain why the three North American species have been able to rapidly invade nearly all suitable habitats into which they have been introduced in Europe. We also discuss the need and possibilities for restoration policies.

2 FACTORS EXPLAINING THE SUCCESS OF INVADING CRAYFISH SPECIES

2.1 *The crayfish fungus plague*

The crayfish fungus plague, *Aphanomyces astaci*, eliminated or sharply reduced native crayfish populations wherever it was introduced in Europe (Kossakowski 1971, Alderman & Polglase 1988). Re-introduction of crayfish, both intentional and natural, can lead to re-establishment of healthy populations over 5-10 years, but these may be again exposed to *A. astaci*, with subsequent population collapse (Huner et al. 1991). European crayfish are highly susceptible to *A. astaci,* whereas North American species are resistant and vectors for the malady (Unestam 1969, Unestam 1972, Söderhäll et al. 1988). Therefore, the highly successful introduction of *O. limosus* in the late 1800s ensured continuous exposure of native species to *A. astaci,* wherever this exotic crayfish encountered it. This situation was exacerbated with the subsequent introduction of *P. leniusculus* throughout southern Sweden and Finland in the 1960s and 1970s (Svärdson 1995) and subsequent introduction into England in the 1970s (Holdich, Rogers & Reynolds, this volume) and *P. clarkii* in Spain in the 1970s (Habsburgo-Lorena 1983) with spread into France and Italy in the 1980s (Arrignon 1997). Prior to the introductions of *P. clarkii* and *P. leniusculus*, *O. limosus* was not present in those regions to introduce or reinforce *A. astaci* infections of native crayfish.

There is now evidence, through analyses by molecular biology methods, that the introduced species can spread the plague fungus to native crayfish species in Europe (Huang et al. 1994, Lilley et al. 1997) and thus contribute to their demise.

2.2 *Man made changes*

Europe has experienced major works of canalisation of riverine habitats and impoundment of rivers and streams for hydroelectric purposes, flood control, and water supplies. Furthermore, air and water pollution further degraded riverine and lacustrine habitats (Westman 1985), and draining of mires in Fennoscandia has decreased the quality of natural waters. The native European crayfish are species that require high quality waters with large areas of littoral habitat well provided with hides, including cobble and woody debris (Kossakowski 1971).

In contrast to the native European crayfish, the North American exotic species, *O. limosus, P. clarkii*, and *P. leniusculus* are all tolerant of varying degrees of environmental extremes, especially elevated temperatures and low dissolved oxygen levels (Huner & Lindqvist 1995, Payne 1997). Although no crayfish species can develop high population levels in the presence of large numbers of predaceous fishes, more tolerant species readily occupy vacant protective habitat. Furthermore, both *O. limosus* and *P. clarkii* do excavate burrows in clay banks affording them with protective cover in the absence of other convenient shelters. *P. leniusculus* can also burrow extensively in suitable substrata (see Holdich this volume). Finally, because many fish that prey on crayfish are sight feeders and/or require the same high quality water that favours native crayfish, predator pressure may be reduced in degraded habitats.

2.3 *Life history strategies*

Invading species are those that come from elsewhere. Thus, the definition is geographical. Colonists, however, are species that keep moving into new territory, and their populations in one area need not be permanent. Therefore, the definition is ecological in nature and the species in question may be native or introduced (Williamson 1996). All three of the North American crayfish now widely distributed in Europe are clearly invading species and colonists as well. A review of the life history strategies of these crayfish may help to explain their success in both regards.

The relationships among habitat, ecological strategies, and population parameters have been termed r- and K-selection (MacArthur & Wilson 1967, Adams 1980). In unpredictable environments, species direct organic resources towards reproductive activities (r-selected species). In predictable environments, a species demonstrates increasing individual fitness through competitive ability (K-selected species).

The concepts of r- and K-selection are not absolute but represent a continuum. The native European crayfish all exhibit relatively stronger K-selected characteristics, including delayed maturity, reduced growth rates, low mortality rates, large body size and long life spans. The three invading North American crayfish all exhibit relatively r-selected characteristics, including early maturity, rapid growth rates, large numbers of offspring at a given parental size, maximum production of offspring at an early age, small body size, high mortality rates and short life spans. A 10 cm long *A. astacus* may have an ovarian egg count of 280, *P. leniusculus* 350 and *O. limosus* as high as 400 (Huner & Lindqvist 1991). Huner & Lindqvist (1991, 1995) further discuss these concepts at some length.

When each North American crayfish species was introduced initially into Europe, native crayfish may have been scarce or absent locally, and each area could be described as disturbed. These crayfish also did serve and do serve as vectors for *Aphanomyces astaci*, further enhancing their abilities to compete with the native crayfish that they encountered (cf. Lilley et al. 1997).

2.4 *Habitat considerations*

All three introduced crayfish in Europe are widely distributed in North America with *O. limosus* found in northeastern USA into southeastern Canada, *P. clarkii* throughout most of the USA, and *P. leniusculus* from central California (USA) into British Columbia (Canada) (Taylor et al. 1996). This clearly demonstrates that each species is a generalist rather than a specialist with respect to available habitats and 'predictably' suited to being successful away from its original distribution.

Both *O. limosus* and *P. leniusculus* have thrived in slow moving, silty permanent riverine and lacustrine habitats that would be described as degraded for native European crayfish (Crocker 1957, Meredith & Schwartz 1960, Kossakowski 1971, Shimizu & Goldman 1983, Momot 1988). *P. clarkii* is a very hardy, burrowing species that achieves highest populations in seasonal wetlands (Huner & Lindqvist 1995). Although *P. clarkii* populations can thrive well in permanent water bodies in the absence of intense predation by fishes (Huner & Barr 1994), and they do so in Europe (Arrignon 1997), the species occupies a relatively unoccupied freshwater crustacean niche in Europe, though historical studies are lacking especially concerning southern Europe. Therefore, the obvious success of *P. clarkii* in developing large populations in Europe, especially in Spain and Portugal, should come as no surprise. Furthermore, although often said to be a warm-water species, *P. clarkii* also thrives in colder, high latitude locations in the USA and in Japan, where the species has been introduced in this century (Huner and Barr 1994).

Procambarus clarkii has nearly become a cosmopolitan species as a consequence of successful introductions around the world outside Australia (Huner 1995). Unlike *O. limosus* and *P. leniusculus* and the European native *Astacus* spp. and *Austropotamobius* spp., *P. clarkii* has a unique life history characteristic that further enhances its ability to invade new habitats and remain there in the absence of an effective competitor. That is, *P. clarkii* populations include individuals that are incubating eggs or carrying young at any time during a year (Huner & Barr 1994). Therefore, whenever environmental conditions are favorable, there are young crayfish available to perpetuate the population. In this sense *P. clarkii* is ecologically a highly opportunistic species.

Even in the absence of the crayfish plague fungus, *Aphanomyces astaci*, it is questionable whether or not the native K-selected European crayfish could successfully compete with the aggressive, North American crayfish. Through his behavioral studies, Söderbäck (1991, 1992) demonstrated very persuasively that the native *A. astacus* is unlikely to be able to compete successfully with the exotic *P. leniusculus* within the same habitat. Likewise, Holdich & Domaniewski (1995) and Holdich et al. (1995) have shown that *A. pallipes* is at a competitive disadvantage when in mixed populations with plague-free *P. leniusculus* and are eventually eliminated. In North America, translocation of the r-selected *Orconectes rusticus* westward in the north-central USA/south-central Canada has resulted in a rapid expansion of its populations at the expense of the K-selected native *Orconectes virilis* (Momot 1991, Taylor et al. 1996).

3 EXPANSION OF PRESENT AND NEW SPECIES

3.1 *Extension of the distribution of existing alien crayfish*

The history of North American crayfish in Europe shows very clearly that there are only two factors that can limit the expansion of their ranges. These are physical and climatic barriers. Although introductions of *P. leniusculus* have been made repeatedly from lower to middle Finland since the 1960s, few populations have been established and then only in southern Finland (Ackefors & Lindqvist 1994). No *Orconectes* spp. population has been yet reported from Norway, Sweden, or Finland even though *O. virilis* was intentionally introduced into Sweden in the 1960s (Svärdson 1995) and *O. limosus* has been widely available for unsanctioned introductions since the turn of the century. *P. clarkii* is reported in reproductive populations in the cooler Netherlands, Germany, Italy, and Switzerland (cf. this volume) but large, expanding populations as seen in warmer Portugal, Spain, and France have not yet developed.

The value of physical barriers in inhibiting expansion of exotic crayfish in Europe is questionable. *O. limosus* has gradually expanded its distribution from Germany to much of Europe since the end of the last century (Ackefors & Lindqvist 1994). *P. leniusculus* and *P. clarkii* are well established in many areas. This expansion took place over two to three decades, far too quickly to have been accomplished without the hand of man being involved. These introductions have come about primarily as a consequence of commerce in living crayfish for food and the ornamental pet trade.

3.2 *Other possible introductions into Europe*

Freshwater crayfish are widely esteemed for food in Europe with the largest crayfish commanding the highest prices (see Ackefors, this volume). Therefore, it is not surprising that there is considerable interest in the aquaculture of several large Parastacidae species, genus *Cherax*, from Australia in Europe. Only one, *Cherax destructor*, seems to have any real potential for becoming established in the warmer regions of southern Europe. However, all parastacid crayfish are considered to be highly susceptible to the plague fungus, suggesting that no *Cherax* sp. could persist anywhere in Europe in anything but relatively small populations outside of aquaculture systems. Huner & Lindqvist (1995) should be consulted for more information about the characteristics of *Cherax* spp.

4 PREDICTORS OF COLONIZING ABILITY

What is then the ecological predictability of the colonising ability of freshwater crayfish species? Despite the apparent usefulness of the r-K continuum, we can still see the situation post facto only. There are no studies or observations of possible applicability of the tens rule with the crayfish (Williamson 1996), although Holdich (this volume) discusses this concept with regard to crayfish introductions in Great Britain.

The chance effect may also have some role. There are over 300 species of freshwater crayfish in North America (Taylor et al. 1996), compared to only five in Europe. Thus we may predict that the 'traffic' is from North America to Europe, but not vice versa. Many North American species have very small, localized distributions but there are 15-20 species of the genera *Cambarus*, *Cambarellus*, *Fallicambarus*, *Orconectes,* and *Procambarus* that may be described as generalists with extensive distributions. Some of these species grow large enough to be desirable foods and others might be unique enough, such as the dwarf *Cambarellus* spp., to enter Europe through the ornamental pet trade. Furthermore, several of the *Cambarus* spp. and *Fallicambarus* spp. are primary or secondary burrowers (Huner & Barr 1994) and occupy niches in North America totally unoccupied in Europe. Therefore, there is still potential for further colonization of exotic crayfish into Europe.

Furthermore, commercial importance often helps the spreading of crayfish species, usually through regulated (or often also unregulated) human activities. Cultured species are often both r-strategists and have a wide-range physiological tolerance (Huner & Lindqvist 1995). Finally, earlier success in the colonizing ability may still be the best predictor of a species' success in new environments and localities.

5 CONCLUSIONS

We believe that it is highly desirable to sustain Europe's unique native crayfish fauna. Encouragement of aquaculture programs and establishment of localized, soundly managed natural populations of native European crayfish are strategies to ensure that European astacid biodiversity could be maintained. Lindqvist (1988) proposed a series of management steps for the maintenance and restoration of native crayfish stocks in Europe. These actions should be planned and executed throughout Europe by both environmental and fisheries agencies, since fragmented policies in different countries may produce only limited and often short-term results only. The reasons for a European-wide program in this respect have been elaborated above.

Although idealistic, we feel that aggressive public information campaigns can also limit the extension of the ranges of the exotic North American crayfish presently established in Europe and prevent introductions of new species. A moratorium on the importation and commerce on all living exotic crayfish seems also advisable.

REFERENCES

Ackefors, H. & Lindqvist, O.V. 1994. Cultivation of freshwater crayfish in Europe. In J.V. Huner (ed.), *Freshwater crayfish aquaculture in North America, Europe, and Australia. Families Astacidae, Cambaridae, and Parastacidae*: pp. 157-216. Binghamton, New York: Hayworth Press.

Adams, P.B. 1980. Life history patterns in marine fishes and their consequences for fisheries management. *Fishery Bulletin* 78(1):1-12

Alderman, D.J. & Polglase, J.L. 1988. Pathogens, parasites, and commensals. In D.M. Holdich & R.S. Lowery (eds), *Freshwater crayfish: biology, management and exploitation*: pp.167-212. London & Sydney: Croom Helm.

Arrignon, J.C.V. 1997. The status of foreign crayfish in France. *Freshwater Crayfish* 11:665-670.

Crocker, D.W. 1957. The crayfish of New York State. *New York State Museum and Science Service. Bull.* 355:1-97.

Habsburgo-Lorena, A.S. 1983. Some observations on crawfish farming in Spain. *Freshwater Crayfish* 5:549-551.

Holdich, D.M. & Domaniewski, J.C.J. (1995). Studies on a mixed population of the crayfish *Austropotamobius pallipes* and *Pacifastacus leniusculus* in England. *Freshwater Crayfish* 10:37-45.

Holdich, D.M., Rogers, W.D., Reader, J.P. & Harliouglu, M. (1995). Interactions between three species of freshwater crayfish (*Austropotamobius pallipes*, *Astacus leptodactylus* and *Pacifastacus leniusculus*). *Freshwater Crayfish* 10:46-56.

Huang, T.-S., Cerenius, L. & Söderhäll, K. 1994. Analysis of genetic diversity in the crayfish plague fungus, *Aphanomyces astaci*, strain from the freshwater crayfish *Procambarus clarkii. Mycol. Res.* 99(5):574-578.

Huner, J.V. 1995. An overview of the status of freshwater crayfish culture. *J. Shellfish Res.* 14:539-543.

Huner, J.V. & Barr, J.E. 1994. *Red swamp crawfish, biology and exploitation.* Lousiana State University, Baton Rouge, Lousiana, USA.

Huner, J.V. & Lindqvist, O.V. 1991. Special problems in freshwater crayfish egg production. *Crustacean Issues* 7:235-246.

Huner, J.V. & Lindqvist, O.V. 1995. Physiological adaptations of freshwater crayfish that permit successful aquacultural ventures. *Amer. Zool.* 35:12-19.

Huner, J.V., Henttonen, P. & Lindqvist, O.V. 1991. Observations on noble crayfish, *Astacus astacus* Linnaeus, (Decapoda, Astacidae), populations in central Finland – Management implications. *J. Shellfish. Res.* 10:187-193.

Kim, K.C. & McPheron, B.A. 1993. Insect pests and evolution. In K.C Kim & B.A. McPheron (eds), *Evolution of insect pests: patterns of variation*: pp. 3-25. New York: John Wiley & Sons.

Kossakowski, J. 1971. *Crayfish* ('Raki'). Washington, DC., US Dept. of Commerce. Translation TT 70-55114.

Lilley, J.H., Cerenius, L. & Söderhäll, K. 1997. RAPD evidence for the origin of crayfish plague outbreaks in Britain. *Aquaculture* 157:181-185.

Lindqvist, O.V. 1988. Restoration of native European crayfish stocks. *Freshwater Crayfish* 7:6-12.

MacArthur, R.H. & Wilson, E.O. 1967. *The theory of island biogeography.* Princeton, USA: Princeton University Press.

Meredith, W.G. & Schwartz, F.J. 1960. *Maryland crayfish.* Maryland Dept. Res. Ed. Educational Series No. 46.

Momot, W.T. 1988. *Orconectes* in North America and elsewhere. In D.M Holdich. & R.R. Lowery (eds), *Freshwater crayfish: biology, management and exploitation*: pp. 262-282. London & Sydney: Croom Helm.

Momot, W.T. 1991. Potential and exploitation of freshwater crayfish in coolwater systems: management guidelines and issues. *Fisheries* 16(5):14-21.

Payne, J.F. 1997. Adaptive success within the cambarid life cycle. *Freshwater Crayfish* 11:1-12.

Shimizu, S. & Goldman C. 1983. *Pacifastacus leniusculus* (Dana) production in the Sacramento River. *Freshwater Crayfish* 5:210-228.

Söderbäck, B. 1991. Interspecific dominance relationship and aggressive interactions in the freshwater crayfish *Astacus astacus* (L.) and *Pacifastacus leniusculus* (Dana). *Can. J. Zool.* 69: 1321-1325.

Söderbäck, B. 1992. Predator avoidance and vulnerability of two co-occurring crayfish species, *Astacus astacus* (L.) and *Pacifastacus leniusculus* (Dana). *Ann. Zool. Fennici* 29: 253- 259.

Söderhäll, K., Johansson, M.W. & Smith, V.J. 1988. Internal defence mechanisms. In D.M. Holdich & R.R. Lowery (eds.), *Freshwater crayfish: biology, management and exploitation*: pp. 213-235. London & Sydney: Croom Helm.

Svärdson, G. 1995. Early history of signal crayfish introduction into Europe. *Freshwater Crayfish* 8: 68-77.

Taylor, C.A., Warren, Jr, M.L., Fitzpatrick, Jr, J.F., Hobbs III, H.H., Jezerinac, R.F., Pflieger, W.L. & Robinson, H.W. 1996. Conservation status of crayfish in the United States and Canada. *Fisheries* 21(4): 25-38.

Unestam, T. 1969. Resistance to the crayfish plague in some American, Japanese, and European crayfish. *Rept. Inst. Freshwater Res., Drottingholm* 49:202-209.

Unestam, T. 1972. On the host range and origin of the crayfish plague fungus. *Rept. Inst. Freshwater Res., Drottingholm* 52:192-198.

Westman, K. 1985. Effects of habitat modification on freshwater crayfish. In J.S. Alabaster (ed.), *Habitat modification and freshwater fisheries. Proceedings of a Symposium on Stock Enhancement in Management of Freshwater Fisheries*. Budapest, Hungary, May-June 1982. *EIFAC TECHNICAL Paper* 42:21-22.

Williamson, M. 1996. *Biological Invasions*. London: Chapman & Hall, 244 pp.

The negative effects of established crayfish introductions

DAVID M. HOLDICH
School of Biological Sciences, The University of Nottingham, Nottingham, UK

ABSTRACT

Superimposed on the natural distribution of freshwater crayfish is one orchestrated by humans, mainly from deliberate introductions to satisfy demand for this food, although sometimes accidentally via fishing bait and other means. While some crayfish introductions had positive effects when they became established, in the majority of cases they also had negative effects, which often outweighed the positive ones. These negative effects include displacement of native crayfish species, transfer of disease, consumption of fish eggs, reduction of fish stocks, consumption of large amounts of invertebrates and macrophytes, upsetting production in rice fields, and displacement of amphibians. In addition, the burrowing activities of some species can cause physical damage to irrigation structures and banks of rivers and lakes. All these factors cause environmental degradation of freshwater habitats. Unfortunately, such changes cannot be easily rectified as without extreme measures it is almost impossible to remove established alien crayfish populations. The main species causing concern are *Pacifastacus leniusculus* in California and Europe, *Orconectes limosus* in Europe, *Orconectes rusticus* in North America, *Procambarus clarkii* in Africa, California, Europe and many other parts of the world, *Astacus leptodactylus* in Europe, and *Cherax destructor* in Africa and Australia. All these crayfish are invasive and build up their numbers very quickly. Not all problems relate to trans-continental introductions but inter-state introductions can be equally damaging when native crayfish are moved outside their home ranges as is happening within North America, Australia, and Europe. Legislation is in place in many countries to try and curtail the spread of alien crayfish, but this is often ineffective against those wishing to propagate a valuable crayfish species. Educating governments and the public about the negative impacts of crayfish introductions can play an important role in this respect. Degradation of the freshwater environments is set to continue unless means can be found to control, if not eradicate, nuisance populations of alien crayfish.

1 INTRODUCTION

There are approximately 500 known species of freshwater crayfish – 100+ in the southern hemisphere, five in Europe, four in Asia, with the rest in North and Central America (Hobbs Jr. 1988, Fitzpatrick 1995, Taylor et al. 1996, C.M. Austin personal communication). Only a small proportion of these are known to cause environmental problems, and this is usually when they are moved outside their natural geographical range (Holdich 1988).

Crayfish have been moved outside their natural ranges either naturally (migrations, floods and continental drift), accidentally (in ballast water, via canals, as bait, by escapes from holding facilities, by predators and humans), or deliberately by humans (for aquaculture and stocking, as aquarium pets, for food, for snail and weed control, and because of inappropriate disposal). Most of the problems associated with crayfish introductions are due to our desire to eat them, but it must be remembered that nowadays crayfish are a luxury food, very few populations relying on them for a subsistence food (see Goddard 1988). Many countries have derived benefit from the restocking or stocking of their waters with alien crayfish and in some countries such crayfish are cultivated by extensive and semi-intensive methods, although only juveniles are usually produced by intensive methods (Holdich 1993, Huner 1994, Ackefors this volume). It is virtually impossible to retain crayfish in holding facilities, as they are the masters of escape, and introductions often become spread much further in the environment than was originally intended because of this.

In this review, the environmental impact of established introductions will be discussed, whether they be derived from a deliberate or accidental act.

2 DISTRIBUTION

Figure 1 illustrates the 'natural distribution' of the families of freshwater crayfish and, except for Western Europe, probably represents the situation as it was in the early 19th century. It is clear that Africa and much of Asia and South America do not have indigenous crayfish. Hobbs Jr. (1988) and Scholtz (1995) discuss the reasons for this uneven distribution in more detail. The situation is complicated in Western Europe as some native species (particularly *Astacus astacus* and *Austropotamobius pallipes*) were almost certainly eliminated from the northerly parts of their ranges during the last Ice Age (Albrecht 1983). Their current distribution probably owes much to human intervention, particularly in Scandinavia and the British Isles (Hobbs III et al. 1989, Ackefors & Lindqvist 1994, Holdich & Rogers 1997, Holdich, Rogers & Reynolds this volume). In addition, the occurrence of *Astacus leptodactylus* in Western Europe has been greatly assisted by humans (see below).

Superimposed on the global 'natural' distribution of crayfish is one orchestrated by humans, which has involved the translocation of various North American species to every continent with the exception of Australasia. On a global scale, the introduction of the North American crayfish, *Procambarus clarkii*, into Africa, Asia, Europe and South America this century (Hobbs III et al. 1989), whilst benefiting the local economies, has also led, or is likely to lead to, environmental problems. Likewise, the introduction of another North American crayfish, *Pacifastacus leniusculus*, from California to Europe in the 1960s has been beneficial to some countries, particularly Sweden and Finland (Svärdson 1995, Westman 1995), but in others such as England there have been negative effects (Holdich, Rogers & Reynolds this volume). The in-

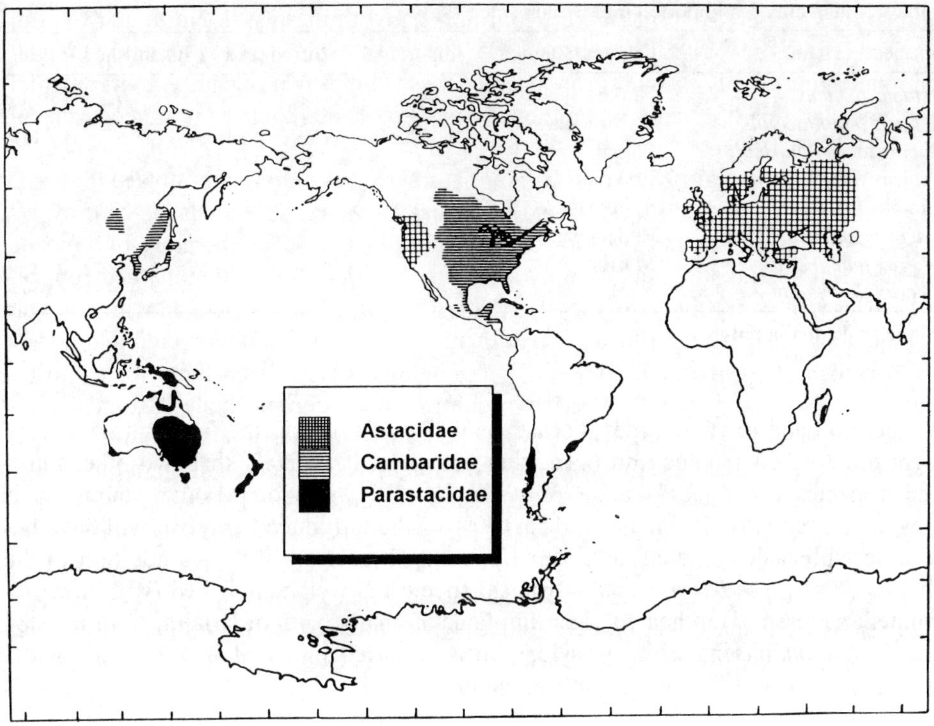

Figure 1. Distribution of crayfish families. (Modified from Horton H. Hobbs Jr 1988).

troduction of Australian *Cherax* species into Europe, Asia, South Africa, and North, Central, and South America may well lead to problems in the future. For example, Romero (1997) reports on the recent upsurge in interest in farming *C. quadricarinatus* in Ecuador after attempts to farm *P. clarkii* met with little success.

3 THE THREE TENS RULE

In relation to introductions, the Three Tens Rule of Williamson (1996) is a useful concept to consider. This states that approximately 10% of imported species become introduced, 10% of introduced species become established, and 10% of those established become pests. The full sequence being: imported ⇒ escaping ⇒ introduced ⇒ establishing ⇒ established ⇒ becoming a pest ⇒ pest. The terminology used above is defined in Holdich & Gherardi (this volume).

There are many exceptions to this rule (Williamson 1996) and freshwater crayfish are a case in point. For example, of the eight alien crayfish species imported into Britain since 1970 five have become introduced, three of these have become established in the wild (i.e. *A. astacus*, *A. leptodactylus*, and *P. leniusculus*) and are now officially classified as pests (Table 1) (Holdich & Rogers 1997, Holdich, Rogers & Reynolds this volume). Their classification as pests is based on the fact that they have the potential to damage stocks of the single native crayfish species, *A. pallipes*, by

Table 1. Alien crayfish imported into Britain since 1970.

Species	Common name	Imported	Introduced	Established	Pest
Astacus astacus	Noble	–	–	–	–
Astacus leptodactylus	Narrow-clawed	–	–	–	–
Pacifastacus leniusculus	Signal	–	–	–	–
Procambarus clarkii	Red swamp	–	–	?	
Orconectes limosus	Spiny cheek	–	–	?	
Cherax quadricarinatus	Redclaw	–	*	*	
Cherax tenuimanus	Marron	–			
Cherax destructor	Yabby	–			

* In heated aquaria only.

competition and in some cases disease, as well as damaging the freshwater environment due to their trophic and burrowing activities. It is likely that two other introduced species, i.e. *P. clarkii* and *Orconectes limosus*, will soon become established, if they are not already. If this is so, then 100% of the introduced crayfish will have become established – a wide deviation from the Three Tens Rule! In addition to the above, *Cherax quadricarinatus*, a tropical species, is well established (with the government's blessing!) in heated aquarium tanks in many parts of Britain. *C. destructor* and *C. tenuimanus* have on occasions been imported into Britain for resale or the aquarium trade, but they have not become introduced.

4 ENVIRONMENTAL IMPACT

The environmental impact caused by an established introduction can be either posiive, neutral, or negative. However, what might be a positive feature from one person's point of view, e.g. a crayfish farmer, might be a negative one from another's, e.g. a conservationist. Figure 2 illustrates some of the positive and negative impacts that might result from an introduction becoming established in a freshwater environment with (e.g. Hill et al. 1993, Holdich & Domaniewski 1995, Söderbäck 1995, Taylor & Redmer 1996) and without (e.g. Harper et al. 1990, Gamradt & Kats 1996, Mikkola 1996) an established native crayfish fauna. In addition, some of the possible solutions to the negative impacts are listed, some of which are discussed below and elsewhere in this volume.

Huner & Lindqvist (1995) and Lindqvist & Huner (this volume) have dealt with the characteristics that make certain species of crayfish successful colonizers and suitable species to cultivate. A short life cycle, high fecundity, and resistance to crayfish plague are characteristic of an r-selected species such as *P. clarkii*. All of the alien crayfish that have become established in Europe have at least some of these characteristics, while the native species are largely of the K-selected type, i.e. long-lived, low fecundity, and susceptibility to crayfish plague, e.g. *A. pallipes*. The one exception amongst Eurasian crayfish is perhaps *A. leptodactylus*, which has high fecundity and the ability to build up its numbers very quickly in suitable conditions (see below), although it is susceptible to crayfish plague. Among the introduced North American species, *P. leniusculus* is exceptional in its longevity, individuals over ten

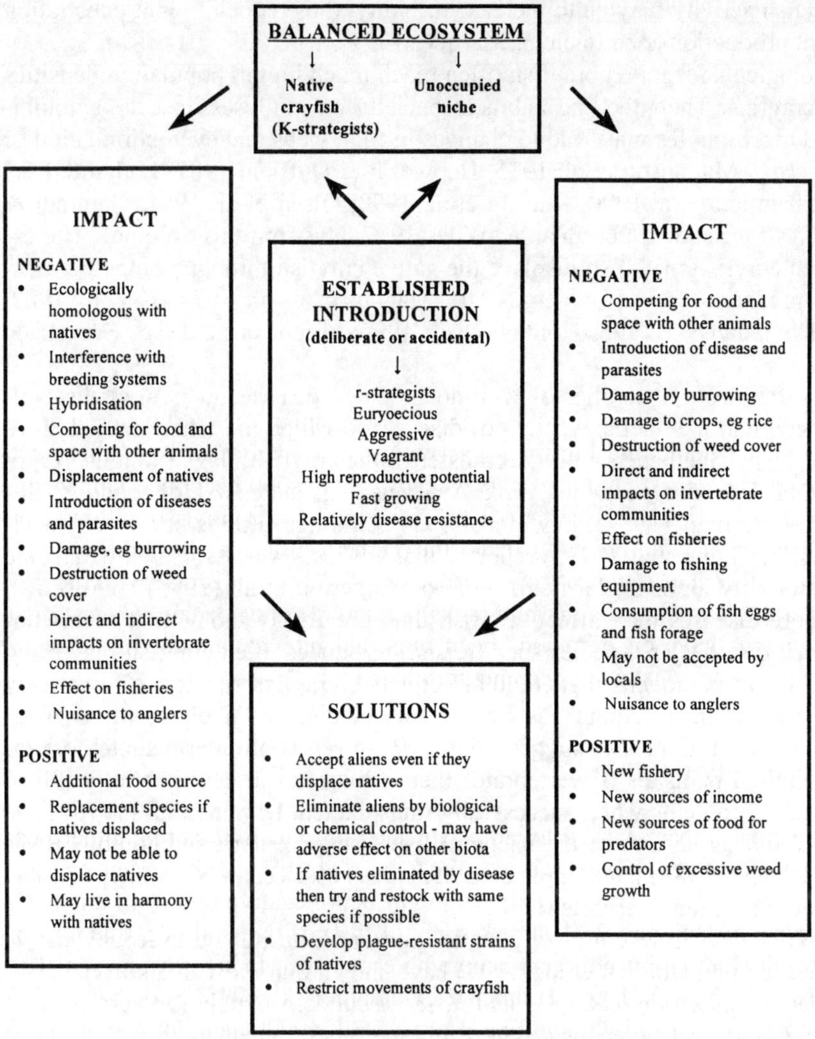

Figure 2. Summary diagram to illustrate the negative and positive impacts of an established introduction of alien crayfish on an ecosystem inhabited by native crayfish and one without native crayfish. (Modified from Holdich 1988).

years old having been recorded by the lipofuschin technique (Belchier et al. 1998). (For a discussion of the value of the terms r and K see Williamson 1996).

The invasive capabilities of r-selected crayfish species can result in both physical and biological environmental degradation.

On the physical side, the activities of burrowing species such as *P. clarkii*, *P. leniusculus*, and *Cherax destructor*, can be a problem in agricultural and recreational areas, e.g. rice plantations, lawns, golf courses, levees, dams, dykes, and also in rivers and lakes where they may destabilize the banks (e.g. Huner 1988, Guan 1994, Holdich et al. 1995b, Anastácio & Marques 1997, Fonseca et al. 1997). An indirect effect of crayfish activity may be to increase turbidity, thus reducing light penetration and thus plant production (Anastácio & Marques 1997).

On the biological side, alien crayfish often reach much higher population densities than native crayfish. Their feeding habits, especially on molluscs, insects, amphibians, fish, and macrophytes may lead to changes in food webs and even elimination of some species (e.g. Magnuson et al. 1975, Hepworth & Duffield 1987, Lodge & Lorman 1987, Chambers et al.1990, Hanson et al. 1990, Olsen et al. 1991, Gamradt & Kats 1996, Nyström et al.1996, Guan & Wiles 1997, Nyström this volume). The established alien crayfish may also displace the native crayfish through competitive exclusion and predation (e.g. Bovbjerg 1970, Magnuson et al. 1975, Capelli 1982, Holdich & Domaniewski 1995, Holdich et al. 1995a, Light et al. 1995, Söderbäch 1995).

Crayfish suffer from a number of well-documented diseases (see Vogt this volume). Alien crayfish may act as vectors of diseases to which the other crayfish have no immunity, as has happened with the transfer of the crayfish plague fungus, *Aphanomyces astaci*, into Europe from North America (e.g. Diéguez-Uribeondo & Söderhäll 1993, Alderman 1996, Lilley et al. 1997). Although little is known about viruses in crayfish, Thune et al. (1991) reported that vibriosis was associated with some large-scale mortality amongst farmed *P. clarkii*. Edgerton et al. (1995) have shown that many more diseases may affect crayfish than previously thought. It seems unlikely that such diseases are species specific and introduced alien crayfish may bring with them a host of pathogens that could affect native crayfish species. Alternatively crayfish may act as intermediate hosts or vectors of diseases likely to affect other animals. Hobbs III et al. (1989) suggest that as *P. clarkii* is an intermediate host for numerous helminth parasites of vertebrates then it may create new public health or veterinary health problems when successfully transplanted. Dunlin et al. (1976) have suggested crayfish as vectors of the bacterial fish disease, enteric redmouth (ERM), and Halder & Ahne (1988) have implicated *A. astacus* as a vector of another fish disease, infectious pancreatic necrosis (IPN).

Alien crayfish may also colonize aquatic environments not occupied by native crayfish. For example, Holdich et al. (1997) have shown that individuals from established populations of *A. leptodactylus* and *P. leniusculus* in Britain are very tolerant of saline conditions and may invade the estuarine environment as they have done within their home ranges (Cherkasina 1975, McGriff 1983). If they become established then they may compete with the local estuarine fauna for resources and upset the trophic stability.

5 CASE STUDIES

5.1 *Introduction*

As mentioned above relatively few species of crayfish cause environmental problems and, at present, the problems are not global but are mainly confined to Europe, North America, Australia, and Africa. The situation in these regions will be examined further to illustrate some of the negative effects of established introductions.

5.2 *Europe*

Native crayfish in Europe are under threat from many fronts, including pollution, habitat modification, disease and alien crayfish (Westman 1985). In historical terms, there have been five important 'crayfish' events in Europe since the mid-1800s that have led to the current problems.

5.2.1 *Crayfish plague*

First, the introduction of crayfish plague from North America into Italy in the mid-1800s and its subsequent spread throughout Europe (Alderman 1996, Ackefors this volume, Hentonnen & Huner this volume, Vogt this volume) was without doubt the most important event.

Crayfish plague has had a devastating effect on European crayfish populations and outbreaks are still occurring in the 1990s (see Westman 1995, Diéguez-Uribeondo et al. 1997, Machino & Diéguez-Uribeondo 1998, and many papers in this volume). North American crayfish act as vectors of the disease, consequently their introduction and establishment into most European countries to boost stocks of crayfish has only exasperated the situation as they act as a reservoir of the fungus. The disease does not require its host in order to spread as the spores can become attached to damp surfaces and be transported in this manner, as is thought to have happened with the crayfish plague outbreak in central Ireland in 1986 (Reynolds 1988). The disease can spread very quickly through a native population as evidenced by its rapid movement down some British rivers (Alderman & Polglase 1988, Alderman 1993).

5.2.2 *Orconectes limosus*

Second, the introduction and establishment of *O. limosus* from North America into Germany in the 1890s and its subsequent spread to much of east, central and Western Europe (Lowery & Holdich 1988, Momot 1988) has had grave consequences.

This widespread species has displaced native crayfish in Europe (Laurent 1988) and acts as a vector of crayfish plague (Vey et al. 1983). Although absent from Scandinavia, *O. limosus* was considered for introduction into Sweden in the 1950s, but the idea was rejected (Svärdson 1995). However, *O. virilis* (Hagen) was introduced but never became established (Svärdson 1995). The recent finding of populations of *O. limosus* in England could pose an additional threat to the future survival of *A. pallipes* (Holdich, Rogers & Reynolds this volume), as may its recent discovery in Belarus (Alekhnovich et al. this volume) affect the native *A. astacus*, as it has done in many European countries.

5.2.3 *Pacifastacus leniusculus*

Third, the introduction and establishment of *P. leniusculus* into Sweden from North America in the 1960s (Abrahamsson 1973, Brink 1983, Svärdson 1995) followed by its subsequent introduction into many European countries from Sweden and North America by the 1980s (Lowery & Holdich 1988) was important.

This species is very similar in many ways to *A. astacus* and consequently has proved popular as a substitute for it since the decline of that species due to crayfish plague (Ackefors & Lindqvist 1994, Skurdal et al. this volume). Compared to *P. clarkii, P. leniusculus* is relatively slow-growing (Lowery 1988) but is ideally suited to cool-temperate waters (Ackefors this volume, Lindqvist & Huner this volume). It is a vector of crayfish plague (Smith & Söderhäll 1986, Alderman et al. 1990, Diéguez-Uribeondo et al. 1997) and has been implicated in many recent outbreaks.

Not all populations of *P. leniusculus* harbor crayfish plague and mixed populations with native European crayfish can develop. However, in the majority of cases, *P. leniusculus e*liminates the native species through predation and competitive exclusion (Holdich & Domaniewski 1995, Söderbäch 1995), although in some cases long-term survival has been shown to be possible (Westman et al. 1993).

Although there is no documented evidence that *P. leniusculus* is a burrowing species in North America or mainland Europe, in Britain it burrows extensively in suitable substrata (Guan 1994) and can cause considerable damage to lake and river banks. It can reach higher densities than native crayfish (Guan & Wiles 1996). Consequently, it exerts a greater effect on the other biota, particularly other crayfish species (Cukerzis 1988, Holdich et al. 1995a, Söderbäch 1995, Holdich, Rogers & Reynolds this volume), benthic fish (Guan & Wiles 1996), molluscs (Warner et al. 1995, Nyström this volume), and macrophytes (Nyström et al.1996, Nyström & Strand 1996).

The impact that this species is having on the freshwater environment in Britain has led to stringent legislation being introduced to try and restrict its spread (Holdich & Rogers 1997, Rogers & Holdich 1997, Holdich, Rogers & Reynolds this volume). Escapes and deliberate implants have led to many wild populations developing and the harvest from these, plus those of *A. leptodactylus* (see below) are probably greater than from that derived from cultivation (Rogers & Holdich 1995).

There have been positive and negative aspects to the introduction of *P. leniusculus* into Europe. On the positive side, its establishment has allowed countries whose stocks of native crayfish were devastated by crayfish plague to develop an alternative species for human consumption (Ackefors this volume). On the negative side it is becoming clear that wild populations of *P. leniusculus* are capable of considerable physical and biological damage to the freshwater environment. It may prove necessary to try and find ways of eradicating such populations, as is currently being explored in Britain (see Holdich, Gydemo & Rogers this volume).

5.2.4 *Procambarus clarkii*

Fourth, the introduction and establishment of *P. clarkii* into Spain from North America in the 1970s (Huner 1988, Diéguez-Uribeondo et al. 1997, Gutiérrez-Yurrita et al. this volume) and its subsequent spread throughout that country is a cause for concern. This was followed by its introduction and establishment in Portugal (Gutiérrez-Yurrita et al. this volume), and then Cyprus (Hobbs III et al.1989), England (Holdich, Rogers & Reynolds this volume), France (Arrignon et al. this volume), Germany (Dehus et al. this volume), Italy (Gherardi et al. this volume), Majorca and The Netherlands (Hobbs III et al.1989) and, most recently, Switzerland (Stucki 1997, Stucki & Staub this volume).

Procambarus clarkii is a relatively large, prolific, aggressive, burrowing crayfish (Hobbs III et al. 1989, Lindqvist & Huner this volume). Hobbs III et al. (1989) point out that the majority of introductions of this species have had negative consequences and that they have had little or no economic value.

Undoubtedly, in Spain, at least, there have been benefits to its introduction with a export industry based on the developing wild populations quickly becoming established (Ackefors & Lindqvist 1994, Ackefors 1998 and this volume). However, *P. clarkii* is a proven vector of crayfish plague in Europe (Diéguez-Uribeondo & Söderhäll 1993). In Spain and Portugal, crayfish plague has severely reduced the range of *A. pallipes* (Bernardo et al. 1997, Diéguez-Uribeondo et al. 1997), as well as having effects on the aquatic ecosystem as a whole (Delibes & Adrian 1987).

As a burrowing species and a very rapid colonizer, it can cause considerable environmental damage, particularly to earthen dykes, levees, and water control structures in irrigated agricultural systems (Huner 1988). In Europe, it has so far caused most problems in Spain and Portugal within their rice growing areas (Anastácio & Marques 1997, Diéguez-Uribeondo et al. 1997, Fonseca et al. 1997). However, considerable concern has also been expressed about the rapid spread of this species in other parts of Europe (Laurent et al. 1991, Arrignon 1997 and this volume, Stucki 1997, Gherardi et al. this volume). In Spain, rice farmers have used organophosphate pesticides to try and eliminate the crayfish from their fields because of the damage they do, sometimes with disastrous consequences for bird life (MacKenzie 1986). Crayfish can be harvested by the public in Spain, and thus additional damage can be caused to the crops (Habsburgo-Lorena 1983).

5.2.5 *Astacus leptodactylus*

Fifth, the westerly spread of *A. leptodactylus* deserves comment. Although a native of Turkey and the Near East, *A. leptodactylus* has become widely established in Europe, although not in the Iberian Peninsula and Nordic countries. At one time some 8000 tons of *A. leptodactylus* were being harvested from Turkish lakes and large quantities were being exported to Europe (Köksal 1988), before over fishing and crayfish plague in the 1980s reduced production by 85% (Ackefors & Lindqvist 1994, Ackefors 1998).

In some cases, *A. leptodactylus* has spread naturally, but for the most part it has been introduced by man to supplement stocks of other European crayfish eliminated by crayfish plague. For example, it became commercially important in the former USSR after *A. astacus* stocks were affected by crayfish plague (Cukerzis 1988). Huxley (1879) stated that '... the invading *A. leptodactylus* is everywhere overcoming and driving out *A. nobilis* in the struggle for existence, apparently in virtue of its more rapid multiplication.' He was referring to the displacement of *A. astacus* in the White Sea region of Russia and the fact that *A. leptodactylus* had probably reached this region via canals connecting its rivers to the River Volga. There is no reason to believe that the same thing has not happened elsewhere in Europe, e.g. Switzerland (Stucki & Staub this volume). Albrecht (1983) has highlighted the importance of canals in the movement of crayfish in Europe. Cukerzis (1968, 1988) has also reported on the interactions between and the displacement of *A. astacus* by *A. leptodactylus* in Lithuania, although in some cases the two species lived in the same water body, sometimes hybridizing.

Despite its widespread occurrence in Europe, there have been very few reports of *A. leptodactylus* causing environmental degradation, except in England (Holdich, Rogers & Reynolds this volume). In that country, wild populations have reached densities high enough to be harvested (Rogers & Holdich 1995), but at the same time the high numbers are causing considerable problems for anglers and eradication methods have been implemented in one case (Holdich et al. 1995b). High population densities are also likely to be having significant detrimental impacts on the native biota, although this has yet to be quantified.

As mentioned above *A. leptodactylus* has more in common with some of the North American species, except for the fact that it is susceptible to crayfish plague. The main effect of this species seems to be the displacement of native species such as *A. astacus* due to its invasive capabilities and its fast population growth.

5.3 *Africa*

There are no indigenous freshwater crayfish in Africa (Hobbs Jr 1988). However, from the 1970s onwards, various North American and Australian species have been introduced, particularly *P. clarkii.* This crayfish has been introduced into Egypt, Kenya, South Africa, and Zambia (Henttonen & Huner 1997), and Sudan, although its fate there is unknown (Mikkola 1996). According to Mikkola (1996), those crayfish that now occur in African fresh waters have been introduced without the knowledge and the permission of the relevant authorities.

One country that is particularly concerned about the introduction of crayfish is South Africa, where a number of species, particularly *Cherax destructor* from Australia, are being targeted for aquaculture. The introduced species are considered to be a potential threat to the fragile aquatic ecosystems that so far have only been home to freshwater crabs (Mikkola 1996). *P. clarkii* has already been found in the wild in South Africa but is more commonly known from Kenya, where it has become widespread in all the major drainage systems since its introduction in 1966, and Zambia where it is farmed along with *C. destructor* and *C. quadricarinatus* (Mikkola 1996). Although alien crayfish have not been reported from the wild in Zambia, they, particularly the burrowing species, have the potential to become serious pests if they do.

Procambarus clarkii was introduced into Kenya between 1966 and 1970 to broaden the range of the commercial fishery in the lakes and dams (Mikkola 1996). It was introduced into Lake Naivasha in 1971 (Lowery & Mendes 1977) and a fishery was opened in 1975 (Mikkola 1996). Many problems occurred over this and other introductions in Kenya for social and economic reasons, e.g. local people did not eat crayfish, nor did they have the economic or technical means to catch or export them properly (Mikkola 1996). Some 40 tons were being harvested annually in the 1980s, mainly for tourist consumption (Harper et al. 1990). However, negative environmental effects have been reported, such as interference with the gill net fishery for bass and cichlids by damaging nets and fish (Lowery & Mendes 1977), out competing tilapia for food (Brummett & Alon 1994), and destruction (assisted by introduced coypu) of submerged macrophytes (Harper et al. 1990). Attempts have been made to control the snails which are the intermediate hosts of the organism causing schistosomiasis (*Bilharzia*) in Kenya by introducing *P. clarkii* as a predator (Hofkin et al. 1991). This may encourage an even greater spread of this crayfish in Africa with its attendant environmental problems. For example, it is becoming a problem in the Nile delta where it is interfering with the tilapia fishery by eating fish eggs, and also by burrowing into irrigation dams (R. Hartnoll pers. comm.).

Mikkola (1996) lists possible solutions to the problems in Africa, including the setting up of protected areas in which the introduction of alien species is prevented, better control of ornamental crayfish imports, harmonization of conservation and fisheries policies between countries, and the inclusion of crayfish in national fishery policies.

5.4 *Australia*

Although Australia does not allow the import of live crayfish, there are concerns about the translocation of indigenous *Cherax* species from one state to another for aquacultural, culinary and recreational purposes (Horwitz 1990). Such movements may result in the displacement of native species, hybridization and habitat alteration. Both *C. destructor* and *C. tenuimanus* have had their ranges artificially extended to most suitable parts of Australia and *C. quadricarinatus* is set to follow suite, although some states do have laws preventing translocation, e.g. *C. tenuimanus* into Northern Territory and Victoria. Although crayfish plague is not known from Australia, it is known that Australian crayfish are susceptible (Unestam 1975) and an illegal implant of a plague-carrying species such as *P. clarkii* could have devastating effects as it has done in Europe (this species is currently on sale in the Philippines and may reach the rice fields, R.S.V. Pullin pers. comm.). Despite concerns about the impact of translocations within Australia there is a growing export market for *Cherax* species (Mills et al. 1994) and it could soon become globally widespread.

5.5 *North America (USA and Canada)*

Despite the large number of crayfish species present in North America, there is concern about the future of 48% of them mainly due to human activity and the threat from alien species, including North American crayfish translocated outside their native range (Taylor et al. 1996). The introduction of alien crayfish species for commercial culture has occurred in the western, southwestern and southeastern USA (Hobbs III et al. 1989) and 'bait-bucket introductions' have led to range expansions of several species, particularly *Orconectes rusticus* (Taylor et al. 1996).

A number of species of *Orconectes* have been moved outside of their home range in North America and have caused displacement of other crayfish species (Magnuson et al. 1975, Hobbs III et al. 1989, Lodge & Hill 1994), but *O. rusticus* is causing the most concern. This species is native to the Ohio river drainage system (Momot 1997), but it has spread, usually aided by humans, north and west to many other states and also to Canada (Hobbs III et al. 1989). It is well known for its negative environmental impact, and legislation has been passed in two states banning its possession (Taylor et al. 1996). *O. rusticus* has been implicated in causing environmental damage through competitive exclusion of other crayfish species, predation on fish eggs, reduction in fish stocks and consumption of large amounts of macrophytes and invertebrates (see Magnuson et al. 1975, Hobbs III et al. 1989). Many papers emanating from the laboratories of D.M. Lodge and others have documented the spread of this species and its negative impact (Capelli 1982, Butler & Stein 1985, Lodge et al. 1985, 1986, Lodge & Lorman 1987, Weber & Lodge 1990, Olsen et al. 1991, DiDonato & Lodge 1993, Hill et al. 1993, Hill & Lodge 1994, 1995, Lodge & Hill 1994, Lodge et al. 1994, 1995, Taylor & Redmer 1996, Momot 1997). *O. rusticus* appears to be a superior competitor to all other species of crayfish it encounters and its ability to quickly build up dense populations means that it has significant detrimental impacts on the native biota when it becomes established outside its home range.

One of the species causing concern in Europe, *P. leniusculus*, is a native of Oregon. It was introduced to California in 1912 (Goldman 1973) but not apparently to any other states, then or since (Hobbs III et al. 1989). It now supports a commercial fishery in the Sacramento-San Joaquin delta (McGriff 1983) and has been introduced widely elsewhere, probably as bait for anglers (Goldman & Rundquist 1977, Eng & Daniels 1982). Elser et al. (1994) have shown that a recent introduction of this crayfish into a California lake had a marked impact on the littoral zone habitat, although no negative effects were reported from Lake Tahoe and Lake Donner after the introduction of this species (Goldman & Rundquist 1977). *P. leniusculus* has been implicated in the extinction of *P. nigrescens* in the San Francisco area (Bouchard 1977) and the decline of *P. fortis*, a native of Northeastern California (Light et al. 1995).

Another species causing problems in Europe, *Procambarus clarkii*, is a native to the south central USA and northeastern Mexico, but it has been introduced into at least 15 other states, although few studies have documented its impact (Hobbs III et al. 1989), unlike in Europe (see above). Feminella & Resh (1986) have shown that introduced *P. clarkii* can be an effective regulator of sago pond-weed, a habitat for larval mosquitoes in California. However, Gamradt & Kats (1996) found that it is having a major impact on California newts, despite their antipredator defenses, i.e. tetrodotoxin in both adults and eggs. *P. clarkii* is very common in California rice fields and attempts have been made to eradicate them (Chang & Lange 1967) rather than harvest them (Sommer & Goldman 1983), because of the damage they do to irrigation structures and the rice crop.

6 CONCLUSIONS

The foregoing review clearly shows that when crayfish are moved outside their home range and become established elsewhere, they often become a nuisance and have a negative impact on the freshwater environment. The following statement by Hobbs III et al. (1989) sums up very effectively the negative aspects of crayfish introductions:

'...once an 'exotic' crayfish is introduced into a lake or stream, it may impose considerable environmental stress on the system and, in all too many instances, irreparable shifts in species diversity occur. Every effort should be taken to prevent unsolicited transplantations of any crayfish outside of its native range. Resource managers should realize that successful crayfish transfers for aquacultural purposes will almost certainly lead to the establishment of breeding populations in the natural environment.'

This sentiment was also expressed in a resolution on transplantations accepted by delegates at the 7th International Association of Astacology Symposium in Lausanne in 1987 (see p. xxvi, Freshwater Crayfish 7, 1989).

As mentioned in the introduction to this volume, signatories to the United Nations Convention on Biodiversity at the 1992 Earth Summit in Rio de Janeiro called on governments to prevent the introduction of, control, or eradicate those alien species which threaten ecosystems, habitats, or species (Tibbetts 1997). However, very little is being done to meet these objectives. Holdich, Gydemo and Rogers (this volume) have reviewed the methods that could be used to control nuisance crayfish populations, but found none that could eradicate them without affecting other arthropods, except crayfish plague for susceptible species! Until methods specific to crayfish are developed then we are going to have to make the best of a bad situation in Europe and rely on educating governments and the public on the dangers of allowing introductions to be made in the first place. We can be thankful that Europe has not been invaded by *Orconectes rusticus* – yet!

REFERENCES

Abrahamsson, S.A. 1973. The crayfish *Astacus astacus* in Sweden and the introduction of the American crayfish *Pacifastacus leniusculus*. *Freshwater Crayfish* 1:27-40.

Ackefors, H. 1998. The culture and capture crayfish fisheries in Europe. *World Aquaculture* 29(2):18-24, 64-67.

Ackefors, H. & Lindqvist, O.V. 1994. Cultivation of freshwater crayfish in Europe. In J.V. Huner (ed.), *Freshwater crayfish aquaculture in North America, Europe and Australia*. New York: The Haworth Press, Inc.

Albrecht, H. 1983. Besiedlungsgeschichte und ursprhnglich holozäne Verbreitung der europischen Flusskrebse. *Spixiana* 6:61-77

Alderman, D.J. 1993. Crayfish plague in Britain, the first twelve years. *Freshwater Crayfish* 9:266-272.

Alderman, D.J. 1996. Geographical spread of bacterial and fungal disease of crustaceans. *Rev. Sci. Tech. Int. Epiz.* 15:603-632.

Alderman, D.J. & Polglase, J.L. 1988. Pathogens, parasites and commensals. In D.M. Holdich & R.S. Lowery (eds*), Freshwater crayfish: biology, management and exploitation:* 167-212 pp. London: Croom Helm (Chapman & Hall).

Alderman, D.J., Holdich, D.M. & Reeve, I.D. 1990. Signal crayfish as vectors in crayfish plague in Britain. *Aquaculture* 86:3-6.

Anastácio, P.M. & Marques, J.C. 1997. Crayfish, *Procambarus clarkii*, effects on initial stages of rice growth in the lower Mondego River valley (Portugal). *Freshwater Crayfish* 11:608-617.

Arrignon, J.C.V. 1997. Status of foreign crayfish in France. *Freshwater Crayfish* 11:665-670.

Belchier, M., Edsman, L., Sheehy, M.R.J. & Shelton, P.M.J. 1998. Estimating age and growth in long-lived temperate freshwater crayfish using lipofuschin. *Freshwater Biology* 39:439-446.

Bernardo, J.M., Ilhéu, M. & Costa, A.M. 1997. Distribution, population structure and conservation of *Austropotamobius pallipes* in Portugal. *Bull. Fr. Pêche Piscic.* 347:617-624.

Bouchard, R.W. 1977. Distribution, ecology and systematic status of five poorly known Western North American crayfishes (Decapoda: Astacidae and Cambaridae). *Freshwater Crayfish* 3:409-421.

Bovbjerg, R.V. 1970. Ecological isolation and competitive exclusion in two crayfish (*Orconectes virilis* and *Orconectes immunis*). *Ecology* 51:226-236.

Brummet, R.E. & Alon, N.C. 1994. Polyculture of Nile tilapia (*Oreochromis niloticus*) and Australian red claw crayfish (*Cherax quadricarinatus*) in earthen ponds. *Aquaculture* 122:47-54.

Butler IV, M.J. & Stein, R.A. 1985. An analysis of the mechanisms governing species replacements in crayfish. *Oecologia* 66:168-177.

Capelli, G.M. 1982. Displacement of northern Wisconsin crayfish by *Orconectes rusticus* (Girard). *Limnol. Oceanogr*. 27:741-745.

Chambers, P.A., Hanson, J.M., Burke, J.M. & Prepas, E.E. 1990. The impact of the crayfish *Orconectes virilis* on aquatic macrophytes. *Freshwater Biology* 24:81-91.

Chang, V.C.S. & Lange, W.H. 1967. Laboratory and field evaluation of selected pesticides for control of the red crayfish in California rice fields. *J. Econ. Entomol.* 60:473-477.

Cherkasina, N.Ya. 1975. Distribution and biology of crayfishes of genus *Astacus* (Crustacea, Decapoda, Astacidae) in Turkmen waters of the Caspian Sea. *Freshwater Crayfish* 2:553-555.

Cukerzis, J. 1968. Interspecific relations between *Astacus astacus* L. and *A. leptodactylus* Esch. *Ekologia Polska*, Ser. A. 16:629-636.

Cukerzis, J. 1988. Astacus astacus in Europe. In D.M. Holdich & R.S. Lowery (eds), *Freshwater crayfish: biology, management and exploitation*: 309-340 pp. London: Croom Helm (Chapman & Hall).

Delibes, M. & Adrian, I. 1987. Effects of crayfish introduction on otter *Lutra lutra* food in the Doñana National Park, SW Spain. *Biological Conservation* 42:153-159.

DiDonato, G.T. & Lodge, D.M. 1993. Species replacements among *Orconectes* crayfishes in Wisconsin lakes: the role of predation by fish. *Can. J. Fish. Aquat. Sci.* 50:1484-1488.

Diéguez-Uribeondo, J. & Söderhäll, K. 1993. *Procambarus clarkii* Girard as a vector for the crayfish plague fungus, *Aphanomyces astaci* Schikora. *Aquacult. Fish. Manag.* 24:761-765.

Diéguez-Uribeondo, J., Temiño, C. & Múzquiz, J.L. 1997. The crayfish plague fungus (*Aphanomyces astaci*) in Spain. Bull. *Fr. Pêche Piscic.* 347:753-763.

Dunlin, M., Huddleston T., Larson, R.E. & Klontz, G.W. 1976. Enteric redmouth disease. *Univ. Idaho Forest. Wildl. Range Exp. Stat. Bull.* 8:1-15.

Edgerton, B., Owens, L., Harris, L., Thomas, A. & Wingfield, M. 1995. A health survey of farmed redclaw crayfish, *Cherax quadricarinatus* (von Martens), in tropical Australia. *Freshwater Crayfish* 10:322-338.

Elser, J.J., Junge, C. & Goldman, C.R. 1994. Population structure and ecological effects of the crayfish *Pacifastacus leniusculus* in Castle Lake, California. *Great Basin Naturalist 54*(2):162-169.

Eng, L.L. & Daniels, R.A. 1982. Life history, distribution, and status of *Pacifastacus fortis* (Decapoda: Astacidae). *Calif. Fish Game* 68:197-212.

Feminella, J.W. & Resh, V.H. 1986. Effects of crayfish grazing on mosquito habitat at Coyote Hills marsh. *Proc. Calif. Mosquito Vector Control Assoc.* 54:101-104.

Fitzpatrick, J.F. Jr. 1995. The Eurasian Far-eastern crawfishes: a preliminary review. *Freshwater Crayfish* 8:1-11.

Fonseca, J.C., Marques, J.C. & Madeira, V.M.C. 1997. Oxygen uptake inhibition in *Procambarus clarkii*, red swamp crayfish by biodegradable surfactants: an ecotechnological approach for population control in rice fields. *Freshwater Crayfish* 11:235-242.

Gamradt, S.C. & Kats, L.B. 1996. Effect of introduced crayfish and mosquito fish on California newts. *Conservation Biology* 10:1155-1162.

Goddard, J.S. 1988. Food and feeding. In D.M. Holdich & R.S. Lowery (eds), *Freshwater crayfish: biology, management and exploitation*: pp. 145-166. London: Croom Helm (Chapman & Hall).

Goldman, C.R. 1973. Ecology and physiology of the California crayfish *Pacifastacus leniusculus* (Dana) in relation to its suitability for introduction into European waters. *Freshwater Crayfish* 1:106-120.

Goldman, C.R. & Rundquist, J.C. 1977. A comparative ecological study of the California crayfish, *Pacifastacus leniusculus* (Dana), from two subalpine lakes (Lake Tahoe and Lake Donner). *Freshwater Crayfish* 3:51-80.

Guan, R. 1994. Burrowing behaviour of signal crayfish, *Pacifastacus leniusculus* (Dana), in the River Great Ouse, England. *Freshwater Forum* 4:155-168.

Guan, R. & Wiles, P.R. 1996. Growth, density and biomass of crayfish, *Pacifastacus leniusculus*, in a British lowland river. *Aquat. Living Resourc.* 9:265-272.

Guan, R. & Wiles, P.R. 1997. Ecological impact of introduced crayfish on benthic fishes in a British lowland river. *Conservation Biology* 11:641-647.

Habsburgo-Lorena, A.S. 1983. Socioeconomic aspects of the crawfish industry in Spain. *Freshwater Crayfish* 5:552-554.

Halder, M. & Ahne, W. 1988. Freshwater crayfish *Astacus astacus* – a vector for infectious pancreatic necrosis virus (IPVN). *Diseases of Aquatic Organisms* 4:205-209.

Hanson, J.M., Chambers, P.A. & Prepas, E.E. 1990. Selective foraging by the crayfish *Orconectes virilis* and its impact on macroinvertebrates. *Freshwater Biology* 24:69-80.

Harper, D.M., Mavuti, K.M. & Muchiri, S.M. 1990. Ecology and management of Lake Naivasha, Kenya, in relation to climatic change, alien species' introductions, and agricultural development. *Environmental Conservation* 17:328-336.

Henttonen, P. & Huner, J.V. 1997. Where is *Procambarus clarkii*? *Crayfish News* 19(4):5, 8.

Hepworth, D.K. & Duffield, D.J. 1987. Interactions between an exotic crayfish and stocked rainbow trout in Newcastle Reservoir, *Utah. North Amer. J. Fish. Manag.* 7:554-561.

Hill, A.M. & Lodge, D.M. 1994. Diel changes in resource demand: competition and predation in species replacement among crayfishes. *Ecology* 75:2118-2126.

Hill, A.M. & Lodge, D.M. 1995. Multi-trophic-level impact of sublethal interactions between bass and omnivorous crayfish. *J. N. Am. Benthol. Soc.* 14:306-314.

Hill, A.M., Sinars, D.M. & Lodge, D.M. 1993. Invasion of an occupied niche by the crayfish *Orconectes rusticus*: potential importance of growth and mortality. *Oecologia* 94:303-306.

Hobbs, H.H. Jr 1988. Crayfish distribution, adaptive radiation and evolution. In D.M. Holdich & R.S. Lowery (eds), *Freshwater crayfish: biology, management and exploitation*: 52-82 pp. London: Croom Helm (Chapman & Hall).

Hobbs, H.H. III, Jass, J.P. &. Huner, J.V 1989. A review of global crayfish introductions with particular emphasis on two North American species (Decapoda, Cambaridae). *Crustaceana* 56:300-316.

Hofkin, B.V., Mkoji, G.M., Koech, D.K. & Loker, E.S. 1991. Control of schistosome-transmitting snails in Kenya by the North American crayfish *Procambarus clarkii*. *Amer. J. Trop. Med. Hyg.* 45:339-344.

Holdich, D.M. 1988. The dangers of introducing alien animals, with particular reference to crayfish. *Freshwater Crayfish* 7:xv-xxx.

Holdich, D.M. 1993. A global review of astaciculture – freshwater crayfish farming. *Aquat. Living Resour.* 6:307-317.

Holdich, D.M. & Domaniewski, J.C.J. 1995. Studies on a mixed population of the crayfish, *Austropotamobius pallipes* and *Pacifastacus leniusculus*. *Freshwater Crayfish* 10:37-45

Holdich, D.M. & Rogers, W.D. 1997. The white-clawed crayfish, *Austropotamobius pallipes*, in Great Britain and Ireland with particular reference to its conservation in Great Britain. *Bull. Fr. Pêche Piscic.* 347:597-616.

Holdich, D.M., Harlioglu, M.M. & Firkins, I. 1997. Salinity adaptations of crayfish in British waters with particular reference *to Austropotamobius pallipes, Astacus leptodactylus* and *Pacifastacus leniusculus. Est. Cstl Sh. Sci.* 44:147-154.

Holdich, D.M., Reader, J.P., Rogers, W.D. & Harlioglu, M. 1995a. Interactions between three species of crayfish (*Austropotamobius pallipes, Astacus leptodactylus* and *Pacifastacus leniusculus*). *Freshwater Crayfish* 10:46-56.

Holdich, D.M., Rogers, W.D. & Reader, J.P. 1995b. Crayfish conservation. Project Record 378/10/N&Y. Bristol: National Rivers Authority.

Horwitz, P. 1990. The translocation of freshwater crayfish in Australia: potential impact, the need for control and global relevance. *Biological Conservation* 54:291-305.

Huner, J.V. 1988. *Procambarus* in North America and elsewhere. In D.M. Holdich & R.S. Lowery (eds), *Freshwater crayfish: biology, management and exploitation*: pp. 239-261. London: Croom Helm (Chapman & Hall).

Huner, J.V. (ed.) 1994. Freshwater crayfish aquaculture in North America, Europe and Australia. New York: The Haworth Press, Inc.

Huner, J.V. & Lindqvist, O.V. 1995. Physiological adaptations of freshwater crayfishes that permit successful aquacultural enterprises. *Amer. Zool.* 35:12-19.

Huxley, T.H. 1879. The crayfish. An introduction to the study of zoology. London: Keegan Paul.

Köksal, G. 1988. *Astacus leptodactylus* in Europe. In D.M. Holdich & R.S. Lowery (eds), *Freshwater crayfish: biology, management and exploitation*: pp. 365-400. London: Croom Helm (Chapman & Hall).

Laurent, P.J. 1988. *Austropotamobius pallipes* and *A. torrentium*, with observations on their interactions with other species in Europe. In D.M. Holdich & R.S. Lowery (eds), *Freshwater cray-*

fish: biology, management and exploitation: pp. 341-364. London: Croom Helm (Chapman & Hall).

Laurent, P.J., Leloirn, H. & Neveu, A. 1991. Remarques sur l'acclimation en France de *Procambarus clarkii* (Decapoda, Cambaridae). *Bull. mens. Soc. linn. Lyon* 60(5):166-173.

Light, T., Erman, D.C., Myrick, C. & Clarke, J. 1995. Decline of the Shasta crayfish (*Pacifastacus fortis* Faxon) of Northeastern California. *Conservation Biology* 9:1567-1577.

Lilley, J.H., Cerenius, L. & Söderhäll, K. 1997. RAPD evidence for the origin of crayfish plague outbreaks in Britain. *Aquaculture* 157:181-185.

Lodge, D.M. & Hill, A.M. 1994. Factors governing species composition, population size, and productivity of cool-water crayfishes. *Nordic J. Freshw. Res*. 69:111-136.

Lodge, D.M. & Lorman, J.G.1987. Reductions in submersed macrophyte biomass and species richness by the crayfish *Orconectes rusticus*. *Can. J. Fish. Aquat. Sci.* 44:591-597.

Lodge, D.M., Beckel, A.L. & Magnuson, J.J. 1985. Lake-bottom tyrant. *Natural History* 94(8):32-37.

Lodge, D.M., Kershner, M.W. & Aloi, J.E. 1994. Effects of an omnivorous crayfish (*Orconectes rusticus*) on a freshwater littoral food web. *Ecology* 75:1265-1281.

Lodge, D.M., Kratz, T.K. & Capelli, G.M. 1986. Long-term dynamics of three crayfish species in Trout lake, Wisconsin. *Can. J. Fish. Aquat. Sci.* 43:993-998.

Lowery, R.S. 1988. Growth, moulting and reproduction. In D.M. Holdich & R.S. Lowery (eds), *Freshwater crayfish: biology, management and exploitation*: pp. 83-113. London: Croom Helm (Chapman & Hall).

Lowery, R.S. & Holdich, D.M. 1988. *Pacifastacus leniusculus* in North America and Europe, with details of the distribution of introduced and native crayfish species in Europe. In D.M. Holdich & R.S. Lowery (eds), *Freshwater crayfish: biology, management and exploitation*: pp. 309-340. London: Croom Helm (Chapman & Hall).

Lowery, R.S. & Mendes, A.J. 1977. The biology of *Procambarus clarkii* in Lake Naivasha, Kenya; with a note on its distribution. *Freshwater Crayfish* 3:203-210.

Machino, Y. & Diéguez-Uribeondo, J. 1998. Un cas de peste des écrevisses en France dans le bassin de la Seine. *L'Astaciculteur de France* 54:2-11.

MacKenzie, D. 1986. Crayfish pesticide decimates Spanish birds. *New Scientist* 112 (1530):24.

Magnuson, J.J., Capelli, G.M., Lorman, J.G. & Stein, R.A. 1975. Consideration of crayfish for macrophyte control. In P.L. Bezonik & J.L. Fox (eds), *The Proceedings of a Symposium on Water Quality Management Through Biological Control*. Gainesville: University of Florida.

McGriff, D. 1983. The commercial fishery for *Pacifastacus leniusculus* Dana in the Sacramento-San Joaquin delta. *Freshwater Crayfish* 5:403-417.

Mikkola, H. 1996. Alien freshwater crustacean and indigenous mollusc species with aquaculture potential in eastern and southern Africa. *South Afr. J. Aquat. Sci.* 22(1-2):90-99.

Mills, B.J., Morrissy, N.M. & Huner, J.V. 1994. Cultivation of freshwater crayfishes in Australia. J.V. Huner (ed.), *Freshwater crayfish aquaculture in North America, Europe and Australia*. New York: The Haworth Press, Inc.

Momot, W.T. 1988. *Orconectes* in North America and elsewhere. In D.M. Holdich & R.S. Lowery (eds), *Freshwater crayfish: biology, management and exploitation. London*: Croom Helm (Chapman & Hall).

Momot, W.T. 1997. History of the range extension of *Orconectes rusticus* into northwestern Ontario and Lake Superior. *Freshwater Crayfish* 11:61-72.

Nyström, P. & Strand, J.A. 1996. Grazing by a native and an exotic crayfish on aquatic macrophytes. *Freshwater Biology* 36:673-682.

Nyström, P., Brönmark, C. & Granéli, W. 1996. Patterns in benthic food webs: a role for omnivorous crayfish? *Freshwater Biology* 36:631-646.

Olsen, T.M., Lodge, D.M., Capelli, G.M. & Houlihan, R.J. 1991. Mechanisms of impact of an introduced crayfish (*Orconectes rusticus*) on littoral congeners, snails and macrophytes. *Can. J. Fish. Aquat. Sci.* 48:1853-1861.

Reynolds, J.D. 1988. Crayfish extinctions and crayfish plague in Ireland. *Biological Conservation* 45:279-285.

Romero, X.M. 1997. Redclaw crayfish aquaculture in Ecuador: the new boom. *NAGA, The ICLARM Quarterly* 20(1):18-21.

Rogers, W.D. & Holdich, D.M. 1995. Crayfish production in Britain. *Freshwater Crayfish* 10:583-596.

Rogers, W.D. & Holdich, D.M. 1997. New legislation to conserve the native crayfish in Britain – will it work? *Freshwater Crayfish* 11:619-626.

Scholtz, G. 1995. Ursprung und Evolution der Flusskrebse (Crustacea, Astacidae*). Sber. Ges. Naturf. Freunde Berlin* 34:93-115.

Söderbäck, B. 1995. Replacement of the native crayfish *Astacus astacus* by the introduced species *Pacifastacus leniusculus* in a Swedish lake: possible causes and mechanisms. *Freshwater Biology* 33:291-304.

Smith, V.J. & Söderhäll, K. 1986. Crayfish pathology: an overview. *Freshwater Crayfish* 6:199-211.

Sommer, T.R. & Goldman, C.R. 1983. The crayfish *Procambarus clarkii* from California ricefields: ecology, problems, and potential for harvest. *Freshwater Crayfish* 5:418-428.

Stucki, T.P. 1997. Three American crayfish species in Switzerland. *Freshwater Crayfish* 11:130-133.

Svärdson, G. 1995. The early history of signal crayfish introduction into Europe. *Freshwater Crayfish* 8:68-77.

Taylor, C.A. & Redmer, M. 1996. Dispersal of the crayfish *Orconectes rusticus* in Illinois, with notes on species displacement and habitat preference. *J. Crust. Bio*l. 16:547-551.

Taylor, C.A., Warren Jr, M.L., Fitzpatrick Jr, J.F., Hobbs III, H.H., Jezerinac, R.F., Pflieger, W.L. & Robinson, H.W. 1996. Conservation status of crayfishes of the United States and Canada. *Fisheries* 21(4):25-38.

Thune, R.L., Hawke, J.P., & Siebling, R.J. 1991. Vibriosis in the red swamp crawfish. *J. Aquat. Anim. Hlth.* 3:188-191.

Tibbetts, J. 1997. Exotic invasion. *Envir. Hlth. Persp.* 105:590-593.

Unestam, T. 1975. Defense reactions in and susceptibility of Australian and New Guinean freshwater crayfish to European crayfish plague fungus. *Aust. J. Exp. Med. Biol. Sci.* 53:349-359.

Vey, A., Söderhäll, K. & Ajaxon, R. 1983. Susceptibility of *Orconectes limosus* Raff. to the crayfish plague, *Aphanomyces astaci* Schikora. *Freshwater Crayfish* 5:284-291.

Warner, G.F., Wood, J.C & Orr-Ewing, R.H. 1995. Signal crayfish (*Pacifastacus leniusculus*) feeding on pond snails: optimal foraging? *Freshwater Crayfish* 8:352-359.

Weber, L.M. & Lodge, D.M. 1990. Periphytic food and predatory crayfish: relative roles in determining snail distribution. *Oecologia* 82:33-39.

Westman, K. 1985. Effects of habitat modification on freshwater crayfish, In J.S. Alabaster (ed.) *Habitat modification and freshwater fisheries*. London: Butterworths.

Westman, K. 1995. Introduction of alien crayfish in the development of crayfish fisheries; experience with signal crayfish (*Pacifastacus leniusc*ulus (Dana)) in Finland and the impact on the native noble crayfish (*Astacus astacus* (L.)). *Freshwater Crayfish* 10:1-17.

Westman, K., Savolainen, R. & Pursiainen, M. 1993. A comparative study of the growth and moulting of the noble crayfish, *Astacus astacus* (L.), and the signal crayfish, *Pacifastacus leniusc*ulus (Dana), in a small forest lake in southern Finland. *Freshwater Crayfish* 9:451-465.

Williamson, M. 1996. *Biological invasions*. London: Chapman & Hall.

The positive effects of established crayfish introductions in Europe

HANS ACKEFORS
Department of Zoology, Stockholm University, Stockholm, Sweden

ABSTRACT

Most scientists agree that the introduction of exotic species in aquatic environments is a risky business that can cause several problems. Diseases and parasites may be introduced into the new environment, the introduced species may be accompanied by less desirable species, the introduced species may out compete the indigenous fauna, the food web may change to a less favourable status, the introduced species is usually impossible to eradicate. The long-term gain, if any, is very difficult to foresee and it is said that the real judgement of an introduction cannot be made until 100 years have passed. However, there are good examples of species which seem to have acclimatized to the new environment without any ecological disadvantages, e.g. rainbow trout and some carp species. Successful introductions of freshwater crayfish have also been made in the past. *Pacifastacus leniusculus* was introduced during the period 1895-1916 into Lake Tahoe in California. Specimens from this population were brought into Sweden in Europe in the 1960s and later from Sweden into several other European countries. In some countries, the species has established viable populations. In the 1970s, *Procambarus clarkii* was introduced into Spain from Louisiana, USA. This latter species now dominates the harvest of crayfish in Europe. The objective of this paper is to evaluate the commercial, socio-economic, recreational and biological values of the introduced crayfish species in Europe. The positive impact of introduction of alien crayfish in Europe can be summarised as: 1) Rapid restoration of traditional habitats, e.g. fishing for crayfish in Sweden; 2) Economic benefit for local crayfishermen, e.g. Spanish netsmen; 3) Diversification of agriculture to include astaciculture, e.g. crayfish farmers in Britain and in Sweden; 4) Developmental research into astaciculture, e.g. various universities and institutes including commercial companies, e.g. in Sweden; 5) Creating large artificial waterbodies; 6) Restocking of lakes with plague-resistant species, e.g. Finland and Sweden; 7) Preventing lakes from being overgrown by water plants, e.g. Finland and Sweden; 8) Socio-economic advantages for rural people; and 9) Increased trade between countries inside Europe as well as between European countries and countries outside Europe.

1 INTRODUCTION

With the introduction of exotic species in aquatic environments, diseases and parasites may be introduced. The introduced species may be accompanied by less desirable species. The introduced species may out compete the indigenous fauna, the food web may change to a less favorable status The introduced species is usually impossible to eradicate. The long-term gain, if any, is very difficult to foresee, and it is said that the real judgement of an introduction can not be made until 100 years have passed. However, there are examples of species that seem to have acclimatized to the new environment without any ecological disadvantages, e.g. rainbow trout (*Oncorhynchus mykiss*) and carp species.

Successful introductions of crayfish have also been made in the past, where the economic benefits exceeded the disadvantages as far as we can see at present. *Pacifastacus leniusculus* was introduced during the period 1895-1916 into Lake Tahoe in California (Goldman 1973). In 1969, the adult population was estimated to be around 55 million specimens. It was also introduced into Sacramento River where there is now a substantial fishery for it (McGriff 1983).

Originally, the number of crayfish species was only five in Europe, viz. *Astacus astacus*, *A. leptodactylus*, *A. pachypus*, *Austropotamobius pallipes* and *A. torrentium*.

Since the crayfish plague fungus (*Aphanomyces astaci*) spread to Europe in 1860 from USA with American crayfish species (Ackefors 1989) (Fig. 1) the populations

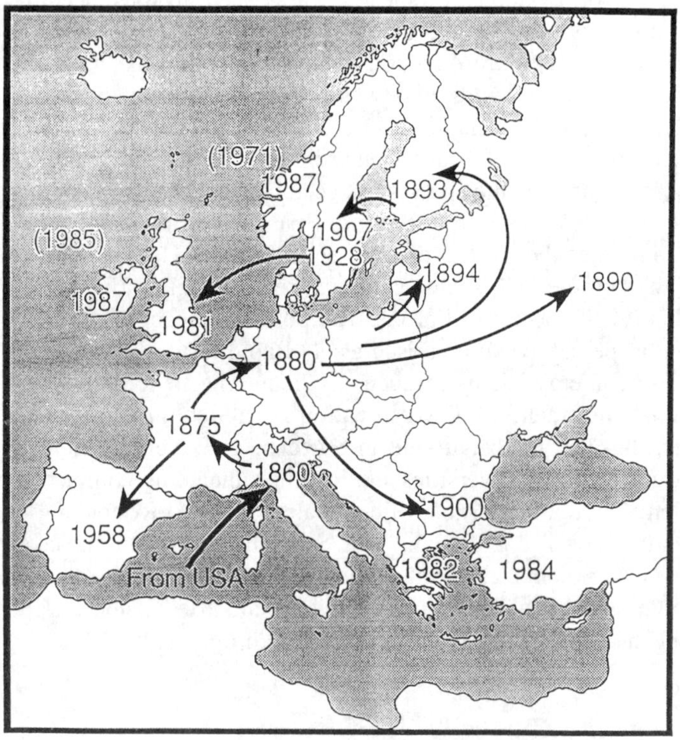

Figure 1. The spread of the crayfish fungus plague, *Aphanomyces astaci*, in Europe. Figures in brackets indicate an outbreak of the plague that was stopped and eradicated (after Ackefors 1989)

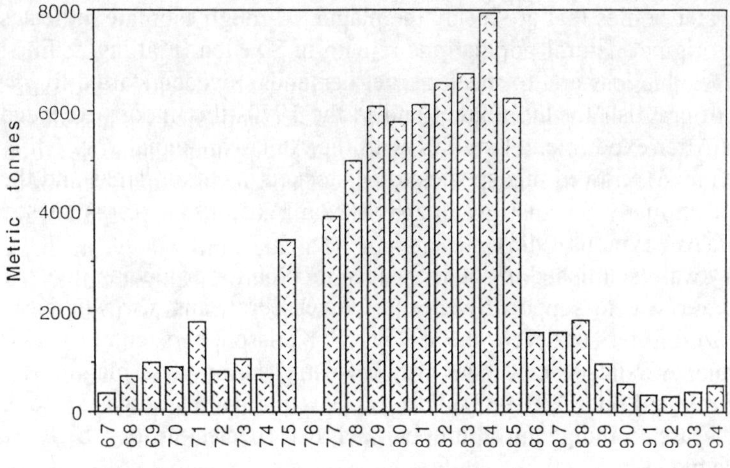

Figure 2. The harvest of narrowed-clawed crayfish, *Astacus leptodactylus*, in Turkey 1967-1994 (P. Baggot, personal communication).

of indigenous species have gradually decreased and nowadays the main part of the harvest in Europe is introduced American species, mainly resistant to the plague under optimal conditions.

One of the most important crayfish populations in Europe used to be *A. leptodactylus* in Turkey. During the peak year 1984, no less than 8000 metric tons were harvested. In 1984-1985, however, the plague arrived in Turkey and now the annual harvest is less than 500 tons (Fig. 2).

The spiny-cheek crayfish, *Orconectes limosus,* was introduced into Germany in 1890 to replace eradicated indigenous crayfish populations. In the 1960s, the stocking of signal crayfish, *P. leniusculus,* started in Sweden (Fürst 1977) and in 1973, the red swamp crayfish, *Procambarus clarkii,* was introduced into Spain (Habsburgo-Lorena 1979, 1983a, b). In 1983, an Australian species, the yabby, *Cherax destructor,* was introduced into Spain (Gutiérrez-Yurrita 1997). This species has also been introduced into Italy (D. Lanari, personal communication) as well as the redclaw crayfish, *C. quadricarinatus.*

The objective of this paper is to evaluate the positive impact of introduced crayfish species in Europe with regard to commercial, socio-economic, recreational, and biological values. For this reason the author has reviewed the current literature and included the results from a recent investigation on the present status of the European crayfish species described in three papers (Ackefors 1997a, b, 1998).

2 SIGNAL CRAYFISH

Among introduced species the signal crayfish and the red swamp crayfish are most important from the commercial point of view. The signal crayfish, which is considered as a cold water species, was introduced into Sweden in the 1960s to compensate for the loss of the indigenous species, the noble crayfish (Svärdson 1995). Already in 1907, the plague fungus reached Sweden with crayfish from Finland (Westman et al. 1992, Ackefors & Lindqvist 1994).

Crayfish plague has continuously spread to new areas and there are still reports every year of new water bodies that are hit by the plague. A rough estimate indicates that only 5% of the original natural populations remain in Sweden (Fjälling & Fürst 1988). To make up for this loss and to meet market demands, Sweden started to import large amounts of crayfish for the market and in the 1970s the import exceeded 2000 tons annually. After experimental studies with the American signal crayfish in the 1960s, the government allowed this species to be stocked in certain areas and the species is now found in many Swedish waters (Svärdson 1965, 1995, Abrahamsson 1973a, b, Fürst 1977, Anonymous 1993).

A private hatchery was established in south Sweden in 1969 for producing juvenile signal crayfish. The aim was to support Swedes and other Europeans who wanted to stock the species in wild waters or to rear the species (S. Karlsson, pers. comm.). This initiative became controversial, since signal crayfish carry the crayfish plague as a latent infection (Persson & Söderhäll 1983). The species can thus function as a vector for spreading this disease to indigenous European crayfish species, all of which are susceptible to the plague.

The signal crayfish is a cold-water species and it matures under good conditions when 2-3 year old. The optimal temperature for growth seems to be 15-20°C. Mating takes place in autumn when the water temperature decreases to below 10°C. In Sweden, at least, females carry the eggs under the abdomen on the pleopods from November until June the following year. The number of eggs varies with the size of the females from 200 to 400. Molting is frequent during the first summer. The hatchlings remain with their mother for a couple of weeks only and then the latter starts to predate on her own juveniles. Shelters are therefore very important for juveniles as well as for young and adult specimens.

Sweden was thus the first country to introduce signal crayfish and the above mentioned hatchery has delivered juvenile signal crayfish (stage 2-3 juveniles) to many countries. According to the results from a questionnaire in 1995, directed to scientists and administrators in all European countries, the species now occurs in cultivation in many European countries (Table 1) (Ackefors, 1997a). The results are compared with information from Huner (1988), describing the overall occurrence in wild waters in Europe.

It is seen from Table 1 that signal crayfish occur in 24 countries in Europe, when the two sources are combined. Signal crayfish is now the most important species in crayfish cultivation in Europe besides the red swamp crayfish (Table 2) (Ackefors, 1997a, 1998). The yield is 51 and 62 tons, respectively. According to Gutiérrez-Yurrita (pers. comm.) the Spanish culture of red swamp crayfish is originally caught wild but cleansed in 'farms'.

It is obvious that cultivation of signal crayfish has started in many countries. However, only few of them have a rather high production. About 50 tons are cultivated, which is about 32% of the overall European harvest of all species in cultivation (Table 3). Sweden is ranked as number one with about 40 tons followed by the UK with 7 tons. Minor quantities are reported from Finland, France and Spain. The wild harvest is seven times higher in Europe (Ackefors 1998). About 355 tons are caught, which is rather moderate compared to the yield of harvested red swamp crayfish. Also in this context Sweden is the leading country with an estimated yield of 247-282 tons followed by France with 50 tons, while Denmark, Finland and U.K. have reported smaller catches (Ackefors 1997a).

Table 1. The occurrence of signal crayfish in Europe in cultivation in 1995 according to Ackefors 1997a and the overall occurrence of the species in wild waters (x) according to Huner (1988).

Country	Huner 1988	Ackefors 1997
Austria	x	C
Belgium	x	
Bulgaria	x	C
Denmark	x	
Czeckia	x	
Cyprus	x	
Finland	x	C
France	x	C
Germany	x	W
Greece	x	
Hungary	x	C (experimental)
Holland	x	
Italy	x	C
Latvia*		W
Lithuania*		C
Luxembourg	x	
Poland	x	C
Portugal	x	C
Russia*		C
Spain	x	C
Sweden	x	C
Switzerland	x	
UK	x	C
USSR*	x	
Yugoslavia	x	
SUM	22	14

C = cultivated, W = wild only, *Latvia, Lithuania, and Russia were parts of the former USSR in 1988.

Table 2. The occurrence (x) of the red swamp crayfish in Europe according to Huner (1988, 1996), Rogers & Holdich (1995), Schmid (1996), Keller (1996).

Country	Occurrence
Austria	x?
Belgium	x
Cyprus	x
France	x
Germany	x
Italy	x
Portugal	x
Spain	x
Switzerland	x
UK	x
Number of countries	10

*The species does not occur in Sweden which was indicated in Huner (1988).

Table 3. The total yield of crayfish in Europe in 1994 (Ackefors, 1997a, 1998). For capture fishery some data from 1993 have been included. Figures for culture in Spain are according to Pérez et al. (1997).

Harvest	*A.a.*	*A.l.*	*A.p.*	*O.l.*	*P.l.*	*P.c.*	*C.d.*	Total
Metric tons								
Capture	287	1252	3	15	355	2340	0	4252
Culture	27.2	12.5		+	50.8	62	3.7	156
Total	314.2	1265	3	15	408	2402	3.7	4408
Per cent								
Capture	6.8	29.4	0.1	0.3	8.4	55.0	0	100
Culture	17.3	8.0	0	+	32.4	39.7	2.4	99.8
Total %	7.1	28.7	0.1	0.3	9.3	54.6	0.1	100.2

For USSR figures from 1968 have been used (Köksal 1988). *A.a.= Astacus astacus , A..l. = A. leptodactylus , A.p.= Austropotamobius pallipes , O.l. = Orconectus limosus ,P.l. = Pacifastacus leniusculus, P.c. = Procambarus clarkii, C.d.= Cherax destructor.*

3 RED SWAMP CRAYFISH

The red swamp crayfish is a warm-water species from southern USA. It has been successfully introduced into Spain and Portugal (Habsburgo-Lorena 1979, Ramos & Pereira 1981, De Bikuna et al. 1989, Gutiérrez-Yurrita 1997).

All native species of Europe, including the introduced *P. leniusculus,* belong to the small family Astacidae. The introduced *P. clarkii,* however, belongs to the larger family Cambaridae and is able to adapt itself to temporarily dry conditions. Such species occupy all kinds of habitats, including subterranean situations, wet meadows, seasonally flooded swamps and marshes and permanent lakes and streams (Huner 1988). All individuals, mature and immature, burrow if the habitat dries up, regardless of the time of the year.

The reproductive cycle in Cambaridae is quite different from that of the European species. *P. clarkii* exhibits a cyclic dimorphism of sexually active and sexually inactive forms alternating during the life cycle. After the production of the young, a unique metamorphosis takes place in both the mature males and females. They begin to eat voraciously and within two to three weeks they moult. After completing the moult, they again assume the immature appearance (Huner & Barr 1994).

The growth rate of *P. clarkii* is very rapid under optimal conditions. A generation can be produced in four and one-half months. Egg production can be completed within six weeks. Incubation and maternal attachment take three to four weeks. Maturation can take place in as little as eight weeks. Optimal temperatures are 21-27°C and growth is severely inhibited by temperatures below 10-12°C.

Procambarus clarkii matures at a size range of 60-125 mm. The growth increment (total length) is about 6 mm at each molt. At 100 mm a female produces as many as 500 eggs, while smaller females produce only 100 eggs. The eggs are much smaller (0.4 mm) than those produced by species within the family Astacidae. The newly hatched crayfish must undergo two moults before they can fend for themselves, and

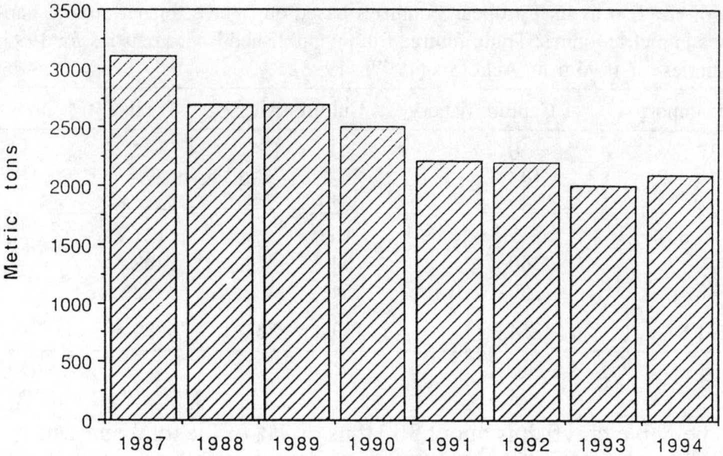

Figure 3. Spanish harvest of red swamp crayfish, *Procambarus clarkii*, 1987-1994 (FAO 1996).

they therefore remain with their mother a longer time than the European species. The young will remain in the burrow for 6-12 weeks and females may eat both hatchlings and eggs.

Feasibility of crayfish farming with regard to local ecological conditions has been described by Gaudé (1986). Soil salinity is the major limiting factor for expansion in certain districts. However, aquaculture of *P. clarkii* is now practised in Spain, although the harvests from wild waters are larger.

From an estimated harvest of 3000 tons in 1982 in Spain, the yield was reduced to 700 tons in 1983 (Habsburgo-Lorena 1986). Maximum yield of red swamp crayfish was about 5000 tons during 1986 (Gutiérrez-Yurrita 1997). According to the official FAO (1996) statistics, the capture fishery since 1987 and onwards have fluctuated between 2000 and 3000 tons (Fig. 3) (please, note the differences between official statistics and the amounts reported by Gutiérrez-Yurrita 1997). The species is very sensitive to long periods of drought and in 1981 and 1982 the annual harvest was only 100 tons.

Although this species is considered subtropical, it is spreading to the north in Europe, to countries with a temperate climate. At present, the species occurs in some ten countries, of which five to six can be considered as temperate (Table 2).

4 TOTAL HARVEST OF INTRODUCED AND NATIVE SPECIES IN EUROPE

In 1994 the introduced species from USA, the signal crayfish and the red swamp crayfish made up 64% of the total harvest. The latter species amounted to 53.8% and the former to 9.4%. In Table 3, the total yield is reported in metric tons and recalculated to percentage for all crayfish capture and culture in Europe. Red swamp crayfish is the dominating species in Europe today and the harvest is at least 2000-3000 tons annually, some years up to 5000 tons. The second species in rank is the narrow-clawed crayfish mainly harvested in the eastern European countries. The amount is more than 1000 tons, in 1994 about 29% of the total harvest. In the western countries, the signal crayfish is ranked as number two with a total yield of 500 tons or roughly

Table 4. Consumption of crayfish in six European countries based on figures for net import, capture and culture fisheries in metric tonnes. Trade figures from unpublished FAO statistics for 1995. Capture and culture fisheries for 1994 from Ackefors (1997b, 1998).

Country	Net import	Capture fishery	Culture fishery	Total
Belgium	132	0	0	132
Finland	119	105	2	226
France	354	57	2.6	413.6
Germany	98.8	12.3	10.0	121.1
Spain	0	2340	64.8	2408.8
Sweden	2588	350	54	2992
Total	3291.8	2864.3	133.4	6289.5

9%. The harvest of the noble crayfish is about 300 tons or 7% of the total harvest.

It is thus clear that, without the introduced species, the European harvest would only be about 1500 tons (in western Europe only about 300-400 tons), annually. If one just considers the western countries where trade statistics are available (Table 4) and combine the net import with the harvest from capture and culture fishery, it is seen that the yield in western Europe for all species is about 3000 tons. The net import to Europe is about 3300 tons. A rough figure for the consumption is thus 6300 tons. This might be a minimum figure, as the main part of artisanal fishery is not included nor the production from small crayfish rearing units.

5 DISCUSSION

5.1 *General background*

There is no doubt that introduced species in the historic perspective have played an important role in Europe to enhance the economy and to raise the nutritional status of people. Cereals such as wheat and rye originate from the Middle East and Asia Minor. Corn and potatoes have been introduced from South America. Examples are manifold with regard to both plants and animals (see Holdich & Gherardi this volume).

That there are drawbacks is also obvious. Indigenous plants and animals have been eradicated due to the introduction of exotic species. In modern time, this is evident, e.g. when the Europeans introduced new animals on the American continent. The American bison, bighorn sheep. and pronghorn antelope were all displaced on the American Plains by cattle (Bixby 1992). As the prairies were broken, many indigenous plants were lost, replaced by exotic forage grasses and grains for livestock feed.

However, species introductions should be recognized as an important component of human-induced global change. The introduced species make up 26-40% of all plants in isolated regions according to Vitousek et al. (in Lövei 1997).

The literature on introductions of all inland aquatic species was reviewed by Welcome (1988). In his paper, 1354 introductions of 237 species in 140 countries were recorded. Since 1850, the pattern of introductions has shown a steady increase until 1960 when 41 introductions were made in one year. Since then, the rate has declined. This general trend is not typical for the crustacean group that increased particularly during the 1980s. In that decade, crustacean introductions made up 23.9% of all species groups introduced.

Introduced species intended for aquaculture are more numerous than introductions for other purposes (Welcome 1988). Since the 1970s, introductions made for this purpose have accounted for well over half of all introductions made. During the 1980s both crustaceans and species of tilapia made up 21% of those species introduced for aquaculture, i.e. nearly half of the introduced species (Welcome 1988). Sport fishing has provided the second major motive for introduction with a relatively constant number of species per decade. From 1950 to 1979, the introductions made for improvement of capture fisheries in lakes, reservoirs and rivers increased in number, when they overtook sport fishing as the second most important motive. The use of species for control of unwanted organisms has a long history, but most of the introductions made for this purpose were in the period 1950-1979.

5.2 *Introduced alien crustacean species in Europe*

As this paper is aimed at discussing mainly the positive side of the introductions of alien crayfish in Europe, the gains in various environments will mainly be dealt with. This approach is different from the one in a paper published by Holdich (1988), where both negative and positive aspects of introduced crayfish species are discussed.

The stocking of alien species, especially with regard to the signal crayfish, has been made in three types of water to be discussed below under Sections 5.3-5.5. The significance of such introductions is briefly dealt with in Section 5.6. Concerning other species, it is of special interest to look at the red swamp crayfish in Spain (Section 5.7).

5.3 *Waters where native European species still occur*

The native species are primarily K-strategists, while the introduced American species in Europe are primarily r-strategists (McArthur & Wilson 1967, see Lindqvist & Huner this volume). This means that the fecundity is higher for the American species. In addition, some of them such as the signal crayfish are more aggressive (Söderbäck 1991). The signal crayfish will ultimately out compete the noble crayfish as has happened in Swedish waters (Fjälling & Fürst 1988, Svärdson et al. 1991). This is thus a negative outcome for the indigenous species and it seems ecologically impossible that the two species can live in harmony. A similar situation exists in the UK where the introduced signal crayfish is out competing the native, white-clawed crayfish (see Holdich, Rogers & Reynolds this volume).

5.4 *Waters where crayfish have never occurred*

Due to the unsafe situation for native species in a region where the crayfish plague occurs, most people seem to prefer signal crayfish for stocking in ponds. This is quite obvious when looking at the number of licences granted for crayfish culture in Sweden. Between 1984 and 1994, 500 licences were granted for production units of nearly 600 tons (Ackefors 1997b), about 400 licences referring to the planned rearing of signal crayfish (Fig. 4). The harvest in 1995 followed this distribution with about

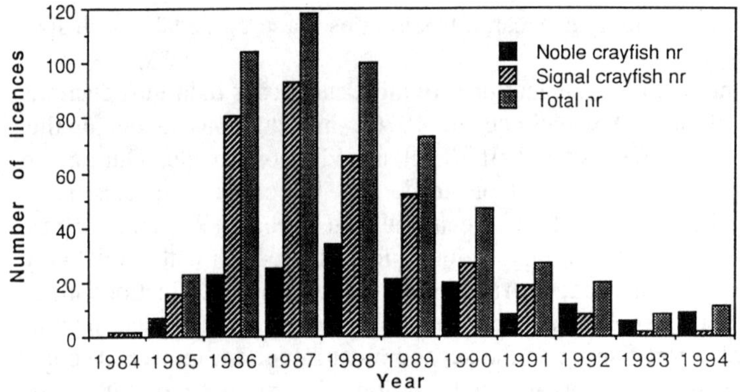

Figure 4. The number of licences for rearing noble (*Astacus astacus*) and signal crayfish (*Pacifastacus leniusculus*) species 1984-1994 in Sweden (Ackefors 1997b).

80% from signal crayfish or around 42 tons. Even if this is well below the legal production 'ceiling', the economic gain is not to be disregarded. Calculating with a price of about 150 Swedish crones (SEK) per kg, this means a gross income of about 6.2 million SEK to people living in rural areas. On the other hand, the price to the farmers nowadays is double for the indigenous species, the noble crayfish.

From other countries, there are examples of successful introductions of signal crayfish that have favoured sport fishing as well (Laurent & Vey 1986). For example, in France (in Lake Divonne) juveniles from Simontorp hatchery were introduced in July 1973 and in July 1976 into a gravel-pit lake created in the 1960s. The annual catch in the 1980s was estimated to 4.5 tons. This is an example of a good investment for recreational fishery.

5.5 *Plague stricken waters where introductions have been made to replace eradicated European species*

Here Sweden provides a good example. Many waters both small and large, where the plague had completely eradicated the native populations, have been stocked with signal crayfish. The introductions have mostly been successful and populations are now giving large yields. In 1992, the harvest of feral signal crayfish and noble crayfish in Sweden were assessed. Signal crayfish gave five times higher yield than the one of noble crayfish, 247-282 tons vs. 46-58 tons (Anonymous 1993).

5.6 *Ecological, economical and socio-economic aspects of introductions*

To begin with, it can be stated that it is possible to stock American crayfish species in some waters where crayfish have never occurred. European crayfish species can also be successfully replaced by American species in plague stricken waters. For example, the signal crayfish, which is a homologue of the noble crayfish, can fit into the ecosystem and adapt itself to the niche earlier occupied by the noble crayfish. It is thus possible to increase the biological diversity, which in itself may be regarded as desirable.

A more conspicuous effect of such introductions is seen when studying the role of crayfish in the ecosystem. When the noble crayfish was eradicated, many water bodies became overgrown with water plants (weeds) and small ponds could be choked by the growth of plants. The grazing effect of signal crayfish has opened up water bodies and prevented lakes becoming overgrown. They also contribute to recycle energy, and organic and inorganic material, and thus increase the biological productivity with regard to other organisms.

The economic yield from many lakes in Sweden decreased substantially when noble crayfish caught the plague and disappeared. Incomes from noble crayfish fishery sometimes exceeded the income from agriculture in certain areas. This was the case with the Swedish Lake Erken, in the county of Uppland. A replacement with signal crayfish has nowadays partially compensated for the loss.

The income from signal crayfish fishery is thus important for rural people, and in many areas this is a valuable additional income to agriculture or to other activities. There are also examples of crayfish farmers who have initiated new types of business around their cultures, with tourist content. All of this is of special significance in remote areas to help keeping the countryside alive. The socio-economic aspects are thus important to consider.

5.7 *Red swamp crayfish*

Procambarus clarkii was first introduced into Spain in 1973, with two shipments from Louisiana (Habsburgo-Lorena 1979). The introduction has created a flourishing crayfish industry in Spain, which already by 1980 had an overall monetary turnover of 7.6 million US dollars (Habsburgo-Lorena 1983a).

The *P. clarkii* crayfish industry has revitalised the local economy in certain districts of Spain. Several hundred fishermen make their living year around from this fishery, which actually might be considered as an extensive form of aquaculture. The main trapping seasons are the months of March-May and September-October. However, an antagonism between farmers and fishermen exists, because fishermen are allowed to fish in the farmers' rice fields, the walls of which are damaged by the burrowing crayfish.

6 CONCLUSIONS

The introduced American crayfish species have contributed to an increasing commercial value of culture and capture fishery of crayfish in Europe. A conservative estimate indicates that fishermen and aquaculturists earn about 13 million US dollars annually. Four hundred tons of signal crayfish can give an income of 8 million US dollars (20 US dollars per kg) and 4000 tons of red swamp crayfish about 8 million US dollars (2 US dollar per kg). According to Pérez et al. (1997), live crayfish fetch 8 US dollars/kg in Spain, but the author has used a lower price as an average price for the total catch. Considering the overall monetary turnover in the fishing industry, the sum should at least be doubled.

The socio-economic value of crayfish culture or fishery for rural people is difficult to estimate in monetary terms. However, it is an important issue to keep in mind. Additional income for rural farmers may be decisive for their possibilities to remain in a non-urban area and it may create new employment as well.

Many authors have pointed out the recreational value of catching crayfish. Sport fishing of crayfish is as important as that of fish in some areas.

The ecological and biological values of introducing exotic crayfish in lakes with no indigenous crayfish species are evident. The grazing effect of crayfish helps to keep water bodies clear from excessive growth of water plants and their occurrence indirectly leads to an increase of other organisms, because of their contributions to a turn-over of mainly biological material.

REFERENCES

Abrahamsson, S. 1973a. Methods for restoration of crayfish waters in Europe. *Freshwater Crayfish* 1:203-210.

Abrahamsson, S. 1973b. The crayfish *Astacus astacus* in Sweden and the introduction of the American crayfish *Pacifastacus leniusculus*. *Freshwater Crayfish* 1:27-40.

Ackefors, H. 1989. Intensification of European freshwater crayfish culture in Europe. *Special session on crayfish aquaculture.* World Aquaculture Society, Los Angeles, USA, February 13, 1989.

Ackefors, H. 1997a. Recent progress in European crayfish farming (submitted).

Ackefors, H. 1997b. The development of crayfish culture in Sweden during the last decade. *Freshwater Crayfish* 11:627-654.

Ackefors, H. 1998. The culture and capture crayfish fisheries in Europe. *World Aquaculture* 29(2):18-24, 64-67.

Ackefors, H. & Lindqvist, O. 1994. Cultivation of freshwater crayfishes in Europe. In Huner, J.V. (ed.) *Freshwater crayfish aquaculture in North America, Europe, and Australia. Families Astacidae, Cambaridae, and Parastacidae*: pp 157-216. Binghatonson, New York: Haworth Press.

Anonymous 1993. Möjlighterna att öka flodkräftbestånd i svenska vatten (Possibilities to enhance the populations of noble crayfish in Swedish waters). *Inst. of Fresh. Res. Drottningholm* 2:1-66.

Bixby, D.E. 1992. Species introductions and transfers in agriculture. Proceedings of the Conference and Workshop 'Introduction and transfers of marine species, achieving a balance between economic development and resource protection', pg. 198.

De Bikuna, B.G., Docampo, L. & Asensio, R. 1989. Distribution et autoecologie de l'écrevisse a pattes blanches, *Austropotamobius pallipes* (Ler.) a Bizkaia (Pays Basques, Espagne). *Annal. Limnol.* 25(3):219-229.

FAO 1996. *Aquaculture production statistics 1985-1994.* FIDI/C815 (Rev.8).

Fjälling, A. & Fürst, M. 1988. Signalkräftan i Sverige 1969-1984. (English summary: The introduction of the crayfish *Pacifastacus leniusculus* into Swedish waters). *Inf. Sötvattenslab., Drottningholm* (8), 29 pp.

Fürst, M. 1977. Introduction of *Pacifastacus leniusculus* (Dana) into Sweden: methods, results and management. *Freshwater Crayfish* 3:229-247.

Gaudé, A.P. 1986. Ecology and production of Louisiana red swamp crayfish *Procambarus clarkii* in southern Spain. *Freshwater Crayfish* 6:11-130.

Goldman, C.R. 1973. Ecology and physiology of the California crayfish *Pacifastacus leniusculus* (Dana) in relation to its suitability for introduction into European waters. *Freshwater Crayfish* 1:106-120.

Gutiérrez-Yurrita, P.J. 1997. The status of crayfish populations in Spain and Portugal. Abstract Volume of the International Workshop 'The introductions of alien species of crayfish in Europe. How to make the best of a bad situation?', pp.7-8. Florence, September 24-27, 1997.

Habsburgo-Lorena, A.S. 1979. Crayfish situation in Spain. *Crayfish News* 3(2)1-2, July 1979.

Habsburgo-Lorena, A.S. 1983a. Socioeconomic aspects of the crawfish industry in Spain. *Freshwater Crayfish* 5: 552-554.

Habsburgo-Lorena, A.S. 1983b. Some observation on crawfish farming in Spain. *Freshwater Crayfish* 5:549-551.

Habsburgo-Lorena, A.S. 1986. The status of the *Procambarus clarkii* population in Spain. *Freshwatrer Crayfish* 6:131-133

Holdich, D. 1988. The dangers of introducing alien animals with particular reference to crayfish. *Freshwater Crayfish* VII, XV-XXX.

Holdich, D.M. & Lowery, R. (eds.) 1988. *Freshwater crayfish: biology, management and exploitation.* London: Croom Helm.

Huner, J.V. 1988. Status of crayfish transplantations. *Freshwater Crayfish* 7:29-34.

Huner, J.V. 1996. Comments on the occurrence of *Procambarus clarkii* in Belgium. *Crayfish News* 18(2):5.

Huner, J.V. & Barr, J.E. 1994. *Red swamp crawfish, biology and exploitation.* Baton Rouge, Louisiana, USA: Louisiana State University Press.

Keller, M. 1996. Crayfish situation in Germany, Switzerland and Austria. *Crayfish News* 18(2):4.

Köksal, G. 1988. *Astacus leptodactylus* in Europe. In: Holdich, D.M. & Lowery, R. (eds) 1988. *Freshwater crayfish: biology, management and exploitation*: pp. 365-400. Croom Helm, London.

Laurent, P.J. & Vey, A. 1986. The acclimation of *Pacifastacus leniusculus* in Lake Divonne. *Freshwater Crayfish* 6:146-155.

Lövei , G.L. 1997. Global change through invasion. *Nature* 388:627-628

McArthur, R.H. & Wilson, E.O. 1967. *The theory of island biography* . Princeton, USA: Princeton University Press, 203 pp.

McGriff, D. 1983. The commercial fishery for *Pacifastacus leniusculus* Dana in the Sacramento-San Joaquin delta. *Freshwater Crayfish* 5:403-417.

Pérez, J.R., Carral, J.M., Celada, J.D., Sáez-Royuela, M., Munoz, C. & Sierra, A.S. 1997. Current status of astaciculture production of crayfish in Europe. *Aquaculture Europe* 22(1):6-13.

Persson, M. & Söderhäll, K. 1983. *Pacifastacus leniusculus* Dana and its resistance to the parasitic fungus *Aphanomyces astacis* . *Freshwater Crayfish* 5:292-298.

Ramos, M.A. & Pereira, T.M.G. 1981. Un novo Astacidae para a fauna Portuguesa: *Procambarus clarkii* (Girad, 1852). *Biol. Inst. Nacional Invest. Pesca Lisboa,* 6 Jul-Aug., pp. 37-41.

Rogers, W.D. & Holdich, D.M. 1995. Crayfish production in Britain. *Freshwater Crayfish* 10:585-596.

Schmid, P. 1996. *Procambarus clarkii* in Switzerland. *Crayfish News* 18 (2):5

Svärdson, G. 1965. The American crayfish *Pacifastacus leniusculus* (Dana), introduced into Sweden. *Rep. Inst. Freshwater Res., Drottningholm* 46:90-94.

Svärdson, G. 1995. The early history of signal crayfish introduction into Europe. *Freshwater Crayfish* 7:68-77.

Svärdson , G., Fürst, M. & Fjälling, A. 1991. Population resilience of *Pacifastacus leniusculus* in Sweden. *Finn. Fish. Res.* 12:165-177.

Söderbäck, B., 1991. Interspecific dominance and aggressive interaction in the freshwater crayfishes *Astacus astacus (*L.) and *Pacifastacus leniusculus* (Dana). *Can. J. Zool.* 69:1321-1325.

Welcome, R.L.1988. International introductions of inland aquatic species. *FAO Fisheries Technical Paper* 294, 318 pp.

Westman, K., Ackefors, H. & Nylund, V. 1992. *Kräftor- Biologi, Odling, Fiske* (Crayfish- Biology, Cultivation, Fishing). Kiviksgårdens förlag, 152 pp.

Ecological impact of introduced and native crayfish on freshwater communities: European perspectives

PER NYSTRÖM

Limnology, Department of Ecology, University of lund, Ecology Building, Lund, Sweden

ABSTRACT

Interest in the culture of freshwater crayfish throughout Europe has increased during the last few decades. However, the effects of both introduced and native crayfish species on the structure and diversity of freshwater communities has received little attention. In this paper, I review the impact of crayfish on macrophytes, algae, invertebrates, amphibians and fish; also I compare the impact of introduced and native crayfish species. By feeding selectively on submerged species and seedling stages of emergent macrophytes, abundant crayfish populations reduce the biomass and species richness of such macrophytes. The impact of crayfish on algae is complex and less predictable: macro-algae may be grazed by crayfish, whereas micro-algae may increase when crayfish reduce the number of invertebrate grazers. Among the invertebrates, thin-shelled snails are particularly vulnerable to crayfish predation, whereas swimming invertebrates are affected less. While crayfish consume amphibian eggs, larvae, and froglets, the impact, however, of crayfish on amphibian eggs deposited when water temperature is low is probably negligible. Amphibian larvae may outswim crayfish, whereas less mobile froglets will be vulnerable to predation. Some amphibians such as bufonids may be unpalatable to crayfish (e.g. *Pacifastacus leniusculus* and *Astacus astacus*), *Procambarus clarkii* being a possible exception. The loss of vegetation in crayfish-rich environment has several indirect effects on biodiversity and may reduce the value of a habitat for invertebrates, amphibians and trout fry. Bird species dependent on macrophytes and invertebrates as food may be negatively affected, whereas birds utilizing crayfish as food may benefit from their introduction. Most crayfish species have similar preferences for macrophytes and invertebrates, but due to greater consumption and more rapid population development the introduced species may have stronger impact on macrophyte and invertebrate biomass. Moreover, these introduced species appear to have adaptations that differ from native species. *Procambarus clarkii* may, for example, colonize ephemeral habitats that are uninhabitable to native crayfish.

1 INTRODUCTION

The current rate of species introductions is increasing considerably. The invasion of new species also may threaten specific ecological communities and global diversity (Coblenz 1990, Lövei 1997). Exotic species negatively impact native species, through predation on native species and habitat changes from disturbance by the introduced species (Lodge 1993). Throughout the world, there are numerous examples of non-native crayfish species being stocked (Hobbs et al. 1989). Introduced crayfish species may have negative effects on native crayfish by transferring lethal diseases (e.g. the crayfish plague; Alderman & Polglase 1988), and by competitive exclusion (Butler & Stein 1985, Mather & Stein 1993, Garvey et al. 1994, Söderbäck 1995, Holdich this volume). Introduced crayfish also may have negative effects on native amphibians and fishes (Gamradt & Kats 1996, Guan & Wiles 1997). During the last few decades the introduction and cultivation of both native and introduced crayfish species throughout Europe has increased (e.g. Westman et al. 1989, Pérez et al. 1997). Today, *Procambarus clarkii* contributes more than 90% of the total crayfish production in Europe (Pérez et al. 1997, Ackefors this volume). Native crayfish populations of Europe have decreased dramatically, primarily due to the spread of the crayfish plague (Hentonnen & Huner and Holdich this volume). Because of the high demand for crayfish as food for humans, many plague stricken waters have been restocked with introduced crayfish species, especially *Pacifastacus leniusculus*, and most European countries have at least one introduced species (Westman et al. 1989). In Sweden for example, several lakes and streams containing previously abundant native *Astacus astacus* now have dense populations of the introduced *P. leniusculus*. Several Swedish ponds, previously devoid of crayfish have now been stocked with *P. leniusculus* (Anonymous 1993). Because of their high commercial value and conservation significance, the stocking of *A. astacus* will likely increase. Also there are examples of native European species being stocked in areas outside their normal distribution (Qvenild et al. 1989, Holdich et al. 1995). In order to meet the high demand of crayfish for human consumption, and to preserve native crayfish species, the future scenario in European crayfish production may be to increase the spread of native species, as well as stocking of introduced species in habitats where production of native species is impossible (Pérez et al. 1997). In many countries, waters are created and restored with the intention of farming crayfish. However, the effects of both introduced and native crayfish species on the structure and diversity of freshwater communities has received surprisingly little attention. Will there be a future conflict between the need for increased crayfish production and the maintenance of biodiversity in freshwater ecosystems? Should we encourage the stocking of native species into as many new habitats as possible? Do introduced crayfish species have the same impact on the ecosystem as native crayfish species? In order to predict the impact of different crayfish species (both natives and exotics) in Europe, we must understand the role of these species in aquatic ecosystems (i.e. lakes, rivers and ponds). Crayfish are important consumers in many aquatic food webs and may dominate the invertebrate biomass in lakes and streams (Momot et al. 1978, Momot 1995). They are omnivores and may act as predators, herbivores, and detritivores, but they are also important prey for many organisms (Hogger 1988, Foster & Slater 1995). Thus, crayfish often act as key organisms in many food webs (Momot et al. 1978, Hart 1992, Creed 1994,

Momot 1995, Nyström et al. 1996). Subsequently, the loss of a crayfish population or an introduction of crayfish to a non-native habitat may significantly affect the aquatic environment. The purpose of this paper is to review the relative impact of native and introduced crayfish species on other trophic levels, particularly on macrophytes, algae, macroinvertebrates, amphibians, and fish. The aim also is to discuss possible mechanisms responsible for the impact of crayfish on aquatic food webs.

1.1 *Physiological characteristics of introduced and native crayfish in Europe*

Both abiotic and biotic factors influence the population size and distribution of freshwater crayfish species (Lodge & Hill 1994). The introduced American crayfish species *Procambarus clarkii*, *Pacifastacus leniusculus*, and *Orconectes limosus* are all plague resistant species, with a higher fecundity than most crayfish native to Europe (Lowery 1988). In temperate regions, *P. leniusculus* grows faster than most native species, whereas *P. clarkii* is the fastest growing species in warmer regions (Lowery 1988, Firkins & Holdich 1993, Westman et al. 1993, Kirjavainen & Westman 1994, Jonsson 1995, Tulonen et al. 1995). Temperature tolerance may affect crayfish survival, reproduction, and growth rate (e.g. Lodge & Hill 1994). *Pacifastacus leniusculus* has both a greater overall temperature tolerance and faster growth rate compared to *Astacus leptodactylus* and *Austropotamobius pallipes*. Although *P. clarkii* is well adapted to warmer climates it also may survive in temperate regions (Carral et al. 1993, Huner & Lindqvist 1995). Unlike other crayfish species present in Europe, *P. clarkii* can survive in burrows in desiccated ponds for several months. Burrowing species such as *P. clarkii*, also are better adapted for survival in oxygen depleted waters (McMahon & Hankinson 1993). Consequently, *P. clarkii* may invade habitats (e.g. ephemeral ponds) where native crayfish usually do not occur.

Orconectes limosus may reach very high densities in Europe. Its growth rate is comparable to *P. leniusculus* and *A. leptodactylus*, although its overall final size is smaller than these species. Unlike native species it seems tolerant of deoxygenated, eutrophic or polluted waters (Momot 1988, Troschel & Dehus 1993). Hence, crayfish species introduced into Europe differ in physiological adaptations from that of native species. They can be regarded as r-selected species, with relatively short life cycles, high fecundity, high growth rates, and adaptations to changing environments (Lowery & Holdich 1988, Huner & Lindqvist 1995, Lindqvist & Huner this volume). Many of these characteristics help spread of these species into less favorable habitats. Also introduced crayfish often replace native species (Söderbäck 1995). Where these introduced species colonize, their rapid population development and feeding activities may have significant effects on the composition of freshwater food webs.

2 MACROPHYTES

2.1 *The role of macrophytes in freshwater ecosystems*

Aquatic macrophytes play an important role in freshwater ecosystems, by influencing water chemistry (e.g. water oxygenation) and biotic interactions within the ecosystem (for review see Carpenter & Lodge 1986). Macrophytes are the primary source of littoral detritus in freshwater (Wetzel 1983), an important food source for many invertebrates, including crayfish (France 1996). Macrophytes also may influence the distribution of vertebrates and invertebrates. In lakes and streams, areas devoid of macrophytes have a lower biomass and invertebrate richness compared to those areas with macrophytes (Gregg & Rose 1985, Blindow et al. 1993, Hargeby et al. 1994). Why are macrophyte stands important for invertebrate diversity and biomass? Macrophytes are consumed by many herbivores (see reviews in Lodge 1991 and Newman 1991) and function as a substrate for epiphytic algae, an important food source for many grazers (Steinman 1996). Dense macrophyte stands may affect predator-prey interactions. Macrophyte beds may reduce fish predation on invertebrates (Crowder & Cooper 1982, Diehl 1988), predation by predatory invertebrates on other invertebrates (Thompson 1987), but also may reduce predation by aquatic insects (Babitt & Jordan 1996, Babitt & Tanner 1997) and crayfish on amphibian tadpoles (Axelsson et al. 1997). Moreover, many waterfowl feed on macrophytes and associated invertebrates. For example, during the expansion of submerged macrophytes in a shallow eutrophic lake in southern Sweden, the number of resting and breeding waterfowl increased (Hargeby et al. 1994). Thus, any reduction in macrophyte biomass and species richness would be expected to have negative effects on many invertebrate and vertebrate species.

2.2 *Crayfish impact on macrophytes*

Several crayfish species graze on aquatic macrophytes and may significantly reduce macrophyte biomass (Table 1). The reduction of macrophyte biomass may not be due only to direct consumption, since crayfish also may destroy macrophytes by non-

Table 1. Studies showing a negative impact of different crayfish species on aquatic macrophytes.

Species	Study	Country	Reference
Astacus astacus	C	Sweden	Abrahamsson 1966
Austropotamobius pallipes	C	Ireland	Matthews & Reynolds 1992
Austropotamobius pallipes	F	Ireland	Matthews et al. 1993
Orconectes causeyi	C	USA	Dean 1969
Orconectes nais	C	USA	Rickett 1974
Orconectes rusticus	C	USA	Capelli 1982
Orconectes rusticus	F	USA	Lodge & Lorman 1987
Orconectes rusticus	F	USA	Lodge et al. 1994
Orconectes spp.	C	USA	Magnusson et al. 1975
Orconectes spp.	F	USA	Hazlett et al. 1992
Orconectes virilis	E	Canada	Chambers et al. 1990
Pacifastacus leniusculus	F	USA	Flint & Goldman 1975
Pacifastacus leniusculus	C	UK	Hogger 1986
Pacifastacus leniusculus	C	Sweden	Nyström et al. 1996
Pacifastacus leniusculus	E	Sweden	Axelsson et al. 1997
Procambarus clarkii	F	USA	Feminella & Resh 1989
Procambarus clarkii	C	Spain	Gutiérrez-Yurrita 1997

C = correlative, E = experimental, F = field experiment.

consumptive cutting of the stems (Dean 1969, Lorman & Magnuson 1978, Lodge & Lorman 1987, Chambers et al. 1990, Lodge et al. 1994, Nyström & Strand 1996). By consuming macrophytes, crayfish may also affect water transparency. In some habitats where crayfish have reduced macrophytes, water turbidity increased due to plankton blooms and resuspended particles (Dean 1969, Rickett 1974, P.J. Gutiérrez-Yurrita, personal communication). By feeding selectively, crayfish may reduce the number of macrophyte species (Lodge & Lorman 1987, Lodge et al. 1994, Nyström et al. 1996). Lodge et al. (1998a) proposed a conceptual model for the feeding preferences of generalist herbivores (e.g. crayfish), suggesting that macrophyte preferences should be similar for different crayfish species. For a macrophyte to be preferentially consumed it must: 1) Have a structure that makes it possible for the herbivore to consume it, 2) Lack chemical deterrents, and 3) Be nutritious. Concurring with this, crayfish usually impact submerged macrophytes (easy to handle), whereas the robust established emergent taxa, such as *Typha* and *Scirpus*, are resistant (Dean 1969, Chambers et al. 1991, Lodge 1991, Nyström & Strand 1996). Consumption of seeds and seedlings may, however affect the establishment of otherwise unpreferred robust emergent species (Nyström & Strand 1996). Several studies have revealed that macro-algae of the genus *Chara* are particularly preferred by crayfish (Chambers et al. 1991, Saffran & Barton 1993, Nyström & Strand 1996, Gutiérrez-Yurrita 1997, Cronin 1998) Why is Charophyta preferable to other submerged macrophytes? Providing that a macrophyte is structurally suitable, a preferred plant should be highly digestible (i.e. nutritious, but not chemically defended). Most fresh plants contain carbon, whereas assimilation of carbon is low, plant nitrogen is highly assimilated by crayfish (Hessen & Skurdal 1988). Ilhéu & Bernardo (1995) compared the digestibility and preference for plant material for *P. clarkii* in Portugal. They found that the digestibility of different food items ranged from 42.4% to 98.5%, with the highest values for animal muscle (fish) and the lowest for fresh macrophytes (*Ceratophyllum*). Digestibility was relatively high for filamentous algae (94.6%), while detritus had an intermediate digestibility (62.4%). Food items with high digestibility (i.e. high in protein but low in fibre) were also the most frequently consumed. Brown et al. (1990) compared the digestibility of aquatic macrophytes by *O. virilis* and found similar low digestibility values for most macrophytes, but the digestibility was almost 90% for *Chara*. *Chara* was not only low in fiber but also in protein, and had a high ash content (almost 48%). Although *Chara* species are low in fiber, they are encrusted with calcium (therefore the high ash content), and may therefore be relatively indigestible to many herbivores with low calcium demands. Crayfish, however, require calcium during molting and they can absorb calcium directly from the gut (Hessen et al. 1991) and thus, *Chara* species may therefore be well digested by crayfish.

Several field observations indicate that crayfish negatively impact *Chara* (Abrahamsson 1966, Rickett 1974, Matthews & Reynolds 1992). Abrahamsson (1966) observed that dense populations of *A. astacus* had eliminated the submerged vegetation; however, following the crayfish plague and the subsequent eradication of *A. astacus*, the bottoms of the ponds were rapidly covered by *Chara*. *Pacifastacus leniusculus* was introduced in the same ponds in 1970, and today the ponds lack submerged vegetation (P. Nyström personal observation).

Although studies suggest that introduced crayfish species may have similar preferences for macrophytes as native crayfish (Olsen et al. 1991, Nyström & Strand 1996), the impact on macrophyte biomass may differ. Olsen et al. (1991) compared the effect of the introduced *Orconectes rusticus* and the natives *O. virilis* and *O. propinquus* on aquatic macrophytes. The amount of consumed macrophytes on a per biomass basis revealed that *O. rusticus* had higher consumption rates than the native species. Nyström & Strand (1996) showed that *P. leniusculus* consumed significantly more *Chara* over a wide range of temperatures than the native *A. astacus*. In order to predict the impact of crayfish on macrophyte biomass and species richness, crayfish preference and consumption rates must be considered. Also requiring consideration is the ability of the macrophyte to compensate for losses due to grazing, and the plants' sensitivity to losses of tissue. A plant is probably most sensitive to grazing during the seedling stage when there is little energy stored in the plant available to rebuild lost tissue (Nyström & Strand 1996). Macrophytes, such as *Elodea canadensis* and *Chara*, may disperse as fragments and should therefore be less sensitive to grazing and cutting compared to macrophytes that do not have the capability to establish themselves vegetatively. Nyström et al. (in press) compared the effects of *P. leniusculus* and *A. astacus* on a benthic community in outdoor pools. They found that *Chara* was significantly reduced, but not *Elodea*. Crayfish probably preferred *Chara*, but *Elodea* also grew three times faster than *Chara*. In this experiment *P. leniusculus* also had a stronger impact on macrophyte biomass than *A. astacus*.

In conclusion, most crayfish have strong negative effects on the biomass and species richness of aquatic macrophytes. Crayfish feed selectively on submerged species, especially on Charophyta. By consuming seeds and seedlings, robust emergent species also may be reduced. In Sweden, new ponds created for the culture of crayfish are often stocked before the macrophytes have become established. As *P. leniusculus* populations develop rapidly, very few macrophyte species establish in these culture ponds (P. Nyström pers. obs.). The underlying mechanisms behind the crayfish/macrophyte system may involve direct consumption as well as non-consumptive destruction. The impact should therefore be most significant on nutritious submerged life stages that are unable to establish themselves by fragments, and that have a slow growth rate. Because several species of *Chara* are threatened (Blindow 1995) and crayfish have a strong preference and impact on these species, the stocking of crayfish may pose a threat to the distribution of these species. Fast-growing introduced crayfish species have similar preferences for macrophytes as native species, but due to greater consumption and more rapid population development, the introduced species may have stronger impact on macrophyte biomass. Reduction of macrophyte biomass and consumption of detritus by crayfish may affect the ontogeny of shallow lakes and ponds. During eutrophication, many shallow lakes develop emergent vegetation, which eventually result in swamp, marsh and terrestrial conditions (Wetzel 1983). The consumption of macrophytes and detritus by crayfish may prolong this succession.

3 ALGAE

Benthic algae are an important food source for numerous consumers in aquatic ecosystems and macro-algae also may provide an important habitat for invertebrates (e.g. Lamberti 1996). Algae may also be an important energy source for crayfish (France 1996) and are frequently found in crayfish stomach contents (Lund 1944, Abrahamsson 1966, Hessen & Skurdal 1986, Lirås et al. 1998). Available data suggest that crayfish grazing may negatively impact the distribution of macro-algae (e.g. *Mougeotia* and *Cladophora*), whereas their direct impact on micro-algae is probably minor. Compared to snails, which are efficient grazers on micro-algae (Brönmark 1989, Barnese et al. 1990), crayfish are not particularly well adapted to grazing on tightly attached micro-algae (Nyström et al. in press). However, as crayfish impact macrophytes, the surface area available for micro-algae colonization may decrease (Lodge et al. 1994). France et al. (1991) and France & Welbourn (1992) studied the relationship between littoral grazers and macro-algae in Ontario lakes and found a negative correlation between the development of macro-algae and crayfish (*Orconectes* spp. and *Cambarus* spp.) abundance, whereas there was no relationship between the abundance of small grazers (including snails) and macro-algae. In streams, crayfish may significantly reduce the biomass of *Cladophora* (Hart 1992, Creed 1994). This in turn had strong effects on the distribution of micro-algae and invertebrate taxa. There are several examples where crayfish have positive effects on periphyton, but also examples were periphyton are unaffected or may decline in biomass. Positive effects on the periphyton often occur when crayfish reduce the numbers of important periphyton grazers, such as snails (Weber & Lodge 1990, Lodge et al. 1994, Charlebois & Lamberti 1996, Nyström et al. in press) (Fig. 1c), or when the

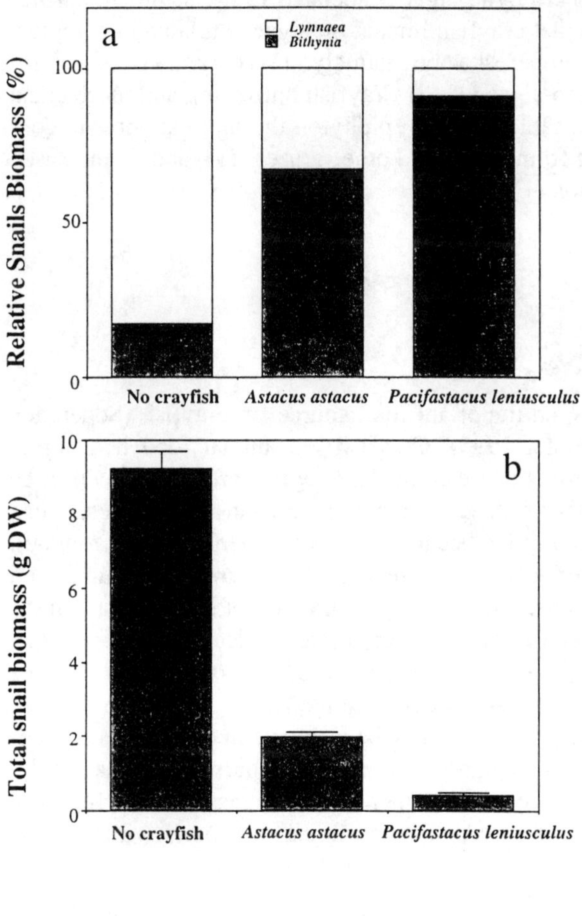

Figure 1. Relative biomass of thin-shelled *Lymnaea stagnalis* and *L. peregra* to a) The hard-shelled *Bithynia tentaculata* and b) Total biomass in outdoor pools (each with an area of 4.5 m^2) without crayfish, with *Astacus astacus*, or *Pacifastacus leniusculus*. Error bars denote 1SE (N = 5). Data from Nyström et al. (in press). (Figure 1b reproduced with permission from Oikos).

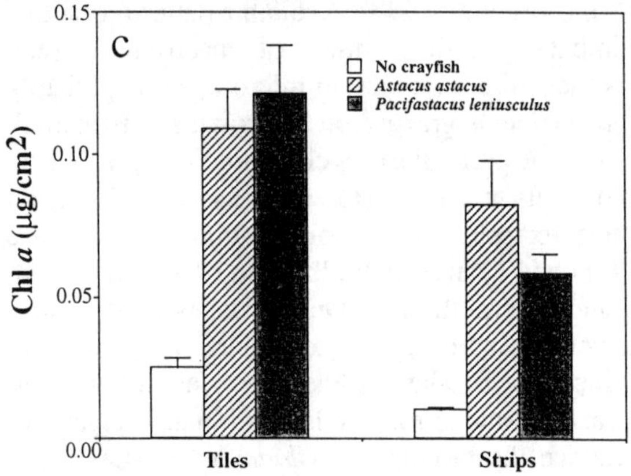

Figure 1. Continued. c) Relative biomass of thin-shelled *Lymnaea stagnalis* and *L. peregra* to peryphytic algae (measured as chlorophyll a on glazed tiles: placed on sediment surface, and on plastic strips). (Figure 1c reproduced with permission from Oikos).

excretion of nutrients by crayfish stimulates primary production (Flint & Goldman 1975, McCormick 1990). Although *P. leniusculus* reduced the snail biomass in ponds in southern Sweden, periphyton biomass (measured as chlorophyll a) growing on stones did not increase. This was probably due to crayfish grazing (Nyström et al. 1996). Because crayfish feeding and burrowing activities may stir the bottom layer (Momot 1995), loosely attached periphytic algae associated to the sediment surface may be affected. Thus, it appears that crayfish impact on algae is relatively complex, and they may affect algae in a number of ways, namely: 1) The grazing of macro-algae (–) facilitates growth of micro-algae (+), 2) Crayfish burrowing and mechanical disturbance (–), 3) Reduced colonisable area for epiphyton through macrophyte consumption (–), 4) Reduced grazing from snails and other grazers (+), and 5) Increased levels of nutrients through excretion (+).

4 INVERTEBRATES

4.1 *Crayfish impact on invertebrates*

Freshwater macroinvertebrates are an important food source for crayfish (Söderbäck et al. 1988, Hill et al. 1993, Momot 1995). Crayfish fed animal food had higher growth rates than animals fed detritus or fresh plants. Crayfish are also known to be cannibalistic. They predate on other species of crayfish and introduced species can have a dramatic impact on native crayfish because of this (see Holdich this volume). Ilhéu & Bernardo (1993) studied the food preferences of *P. clarkii* in Portugal and found that detrital plant material constituted the largest fraction of gut contents. In the laboratory, however, *P. clarkii* preferred macroinvertebrates to detritus. *Procambarus acutus* preferred *Physa* snails to the aquatic macrophyte *Elodea* (Covich 1977). The impact of a number of species of crayfish on aquatic invertebrates has been studied. Correlative studies in ponds suggest that there may be a change in the species composition of macroinvertebrates towards swimming and sediment burrowing taxa not dependent on macrophytes in crayfish-rich environments (Abrahamsson 1966, Nyström et al. 1996). Abrahamsson (1966) observed that the fauna in a pond with a dense population of *A. astacus* was characterized by the relatively large number of swimming species, but after an outbreak of crayfish plague, molluscs, leeches, and tadpoles increased. Today, these ponds have a dense population of *P. leniusculus* and an invertebrate species composition similar to the one when *A. astacus* was present (P. Nyström personal observation). The number of macrophyte-associated invertebrate taxa declined significantly in ponds where *P. leniusculus* was abundant (Nyström et al. 1996). Likewise, *O. rusticus* reduced the abundance and taxa richness of invertebrates in a Michigan stream (Charlebois & Lamberti 1996). In laboratory tests, *P. clarkii* preferred invertebrates with weak escape reactions (Odonata, Ephemeroptera, and snails), and had a low predation efficiency on live mosquito fish (*Gambusia affinis*) (Ilhéu & Bernardo 1993). In Swedish ponds suitable for farming *P. leniusculus*, the density has been estimated to approximate 10 adults/m^2. In these ponds, there are very few invertebrate taxa except for some sediment living oligochaetes and mobile taxa such as Ephemeroptera and Heteroptera (Nyström & Granéli 1996). In the long term, most invertebrate taxa will decline in crayfish-rich environments due to habitat changes (loss of macrophytes and macro-algae) and crayfish predation. In the short term, the direct impact of crayfish is expected to be significant on less mobile non-burrowing invertebrate taxa such as snails (Lodge et al. 1994, Perry et al. 1997).

4.1.1 *Impact on snails*

Numerous studies have shown that crayfish have a strong negative impact on different snail species (Table 2). Crayfish are selective in their choice of snails, and thin-shelled species are preferred over thick-shelled species. Why do crayfish prefer thin-shelled snails? Although crayfish are able to consume hard-shelled snails, these are probably difficult for them to handle (Alexander & Covich 1991a, b, Nyström & Pérez 1998), especially for smaller animals (Alexander & Covich 1991a, Warner et al. 1995). The choice of snail may be explained by the shell-thickness (size). Handling time of thick shells increases exponentially, making large snails less profitable (Nyström & Pérez 1998). According to Nyström & Pérez (1998) the handling time

Table 2. Studies showing a negative impact of different crayfish species on freshwater snails.

Species	Study	Country	Reference
Astacus astacus	C	Sweden	Abrahamsson 1966
Austropotamobius pallipes	C	Ireland	Matthews & Reynolds 1992
Orconectes nais	C	USA	Rickett 1974
Orconectes spp.	C	USA	Lodge et al. 1987
Orconectes rusticus	F	USA	Lodge et al. 1994
Orconectes rusticus	E, F, C	USA	Lodge et al. 1998b
Orconectes rusticus	F	USA	Perry et al. 1997
Orconectes spp.	C	USA	Weber & Lodge 1990
Orconectes virilis	C	USA	Crowl & Schnell 1990
Orconectes virilis	E	Canada	Hanson et al. 1990
Pacifastacus leniusculus	C	Sweden	Nyström et al. 1996
Pacifastacus leniusculus	E	Sweden	Nyström & Pérez 1998
Procambarus clarkii	C	Spain	Gutiérrez-Yurrita 1997
Procambarus simulans	E, F	USA	Alexander & Covich 1991a, b

C = correlative, E = experimental, F = field experiment.

for adult *P. leniusculus* feeding on *Lymnaea stagnalis* was less than a minute for small snails (mean size: 11.0 mm) but increased to more than 27 minutes for large snails (mean size: 37.4 mm) (Fig. 2). When crayfish were offered a choice of snails they also preferred the smaller snails. Although the choice of smaller snails may be explained from an energetic point of view, crayfish may simultaneously minimize time spent feeding in order to reduce the risk of predation, or attack by other crayfish. In nature, however, the environment is often heterogeneous, and they may not always be able to choose snails of an optimal size, and snails that are encountered are probably consumed independent of their size (Nyström & Pérez 1998). Some vulnerable snail species co-existing with predatory American crayfish have evolved anti-predator behavior. These snails (e.g. *Physella virgata* and *Planorbella trivolvis*) crawl above the water line (Alexander & Covich 1991a, b, Covich et al. 1994) and increase growth rate at the expense of reproduction (Crowl & Covich 1990). Although some snail species may reduce the predation risk by crawling out of the water or by having thicker shells, crayfish have a negative impact on most snail species in lakes, ponds, and streams (Crowl & Schnell 1990, Lodge et al. 1994, Lodge et al. 1998b). In an outdoor pool experiment, Nyström, Brönmark and Granéli (in press) compared the effect of *P. leniusculus* and *A. astacus* on benthos. All size classes of the thin-shelled *L. stagnalis* were greatly reduced, whereas the larger size classes of the hard-shelled *Bithynia tentaculata* were less preferred. As a consequence the snail species composition was altered towards larger *Bithynia* snails. In pools without crayfish, thin-shelled *Lymnaea* snails (*L. stagnalis* and *L. peregra*) dominated the snail biomass (83%), whereas the proportion in pools with *A. astacus* and *P. leniusculus* was 33% and 9%, respectively (Fig. 1a). Although both crayfish species reduced snail biomass, *P. leniusculus* had a stronger impact on snails than *A. astacus* (Fig. 1b). In a labora-

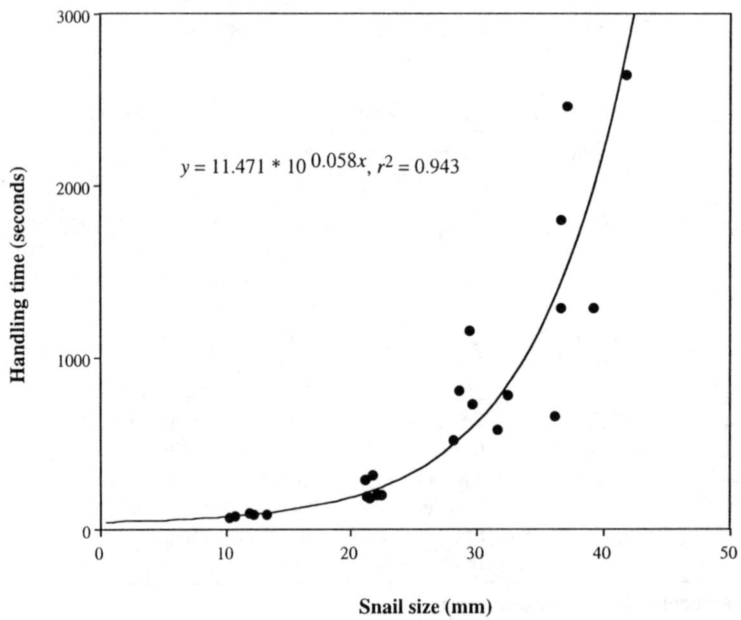

Figure 2. Handling time for *Pacifastacus leniusculus* feeding on different size classes of the pond snail *Lymnaea stagnalis*. Data from Nyström & Pérez (1998).

tory experiment, Olsen et al. (1991) compared the effects of the introduced *O. rusticus* and the natives *O. propinquus* and *O. virilis*. They found that the introduced species consumed and injured significantly more snails than the natives, and concluded that the greater declines in benthic invertebrates (and macrophytes) in lakes where *O. rusticus* had displaced its congeners could be related to the larger size and greater weight-specific consumption of *O. rusticus*.

Likely most crayfish, through predation, will reduce the number of snail species, particularly thin-shelled species. The direct impact of crayfish on actively swimming non-molluscan invertebrate species is probably less. Introduced crayfish species may have a stronger impact on snails than native species. In some areas, the reduction of snail numbers by crayfish may have an indirectly positive effect on humans. Hobbs et al. (1989) suggested that because *P. clarkii* reduce snail numbers they may also reduce vectors for human pathogens such as schistosomiasis. The genus *Lymnaea* contains many species that are vectors of parasites pathogenic to man, cattle, fishes and birds (Malek & Cheng 1974, Laitinen et al. 1996). For example, *Lymnaea trunculata* is the intermediate host for the trematode *Fasciola hepatica*.

5 VERTEBRATES

5.1 *Impact on amphibians*

There are approximately 62 amphibian species in Europe (Gasc et al. 1997), several of which are declining, and more than half are considered as threatened (Corbett 1989). As most of the amphibians exhibit very limited abilities for movement over long distances, they are particularly sensitive to habitat loss and change (Corbett 1989). Thus, the loss of suitable breeding habitats is likely to be one of the most important causes of the declines of amphibians (Blaustein et al. 1994). Amphibian mortality during the aquatic stage regulates amphibian populations (Smith 1983). The introduction of exotic aquatic predators also has been implicated in the decline of amphibian populations throughout the USA (Fisher & Shaffer 1996, Gamradt & Kats 1996). The commercial introduction of native and exotic crayfish species threatens some amphibian populations in Europe (Corbett 1989, Cedhagen & Nilson 1991, Fog et al. 1997), but few studies concern the effects of crayfish on amphibian populations.

5.1.1 *Consumption of eggs*

There are numerous reports of crayfish consuming amphibian eggs. Gamradt & Kats (1996) found that the introduced *P. clarkii* consumed embryos of the California newt (*Taricha torosa*). *Orconectes virilis* consumed single encapsulated eggs of the newt *Ambystoma maculatum*, but then were not able to consume eggs from intact egg masses (Ward & Sexton 1981). In the laboratory, both adult *A. astacus* and *P. leniusculus* consumed eggs of seven Swedish amphibian species, however, well-encapsulated eggs were less consumed (Axelsson et al. 1997). Crayfish consumption of amphibian eggs increases with elevated temperature (Axelsson et al. 1997, Fig. 3), and therefore the consumption of eggs of amphibians with an early spawning period in northern parts of Europe is probably negligible. Several amphibians deposit their eggs on the water surface and on aquatic macrophytes making the eggs less vulner-

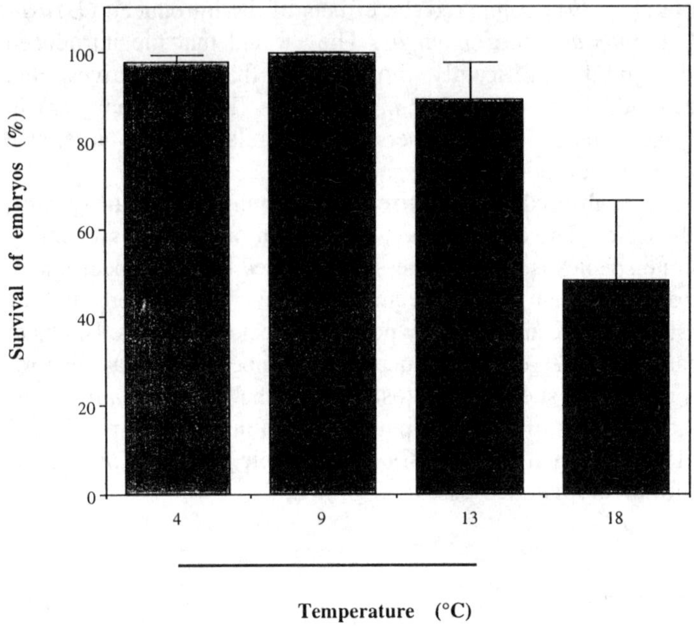

Figure 3. Percentage of surviving *Bufo bufo* embryos in aquaria with one adult *Pacifastacus leniusculus* in relation to water temperature (24 hour experiment). Initial number of eggs = 40. Percentage of embryos surviving at temperature treatments connected with a line are not statistically different (Tukey's post-hoc test, $p > 0.05$). Error bars denote 1 SE (N = 5). Data from Axelsson et al. (1997).

able to crayfish predation. The amphibian species most vulnerable to crayfish predation are those that reproduce later in the season (when crayfish are active), have eggs that are not well encapsulated, or deposit eggs on the sediment surface. Do introduced crayfish have the same impact as native crayfish species? Gamradt & Kats (1996) found that the introduced *P. clarkii* were able to consume embryos of *T. torosa*, although these embryos contain a potent neurotoxin. Then suggested that this deterrent was effective against native predators but not against introduced predators. In Europe, eggs of the common toad (*Bufo bufo*) are unpalatable to newts and predatory insects with chewing mouthparts (Henrikson 1990, Denton & Beebee 1991), but were found to be readily consumed by both *A. astacus* and *P. leniusculus* (Axelsson et al. 1997). The consumption rate of different amphibian eggs was also similar for *A. astacus* and *P. leniusculus*.

5.1.2 *Consumption of larvae and froglets*

Among the crayfish species occurring in Europe, *P. clarkii*, *P. leniusculus*, *A. astacus*, and *A. pallipes* have been reported to consume amphibian larvae under laboratory conditions (Reynolds 1978, Axelsson et al. 1997, E. Razzetti pers. comm.). In nature, however, crayfish are considered inefficient predators on swimming amphibian larvae (Holomuzki 1989, Fauth 1990, Lefcort 1996). Using experimental pools, Holomuzki (1989) studied microhabitat use by small mouthed larval salamanders (*Ambystoma texanum*) in the presence or absence of a water snake and *O. rusticus*. In crayfish pools, the larvae responded by moving from open habitats to protective *Cladophora* patches. Although crayfish preyed upon larvae, the ratio of larval predation per hour was much higher for the snake. In a natural stream in Kentucky, larvae were more common on *Cladophora* patches than on un-vegetated areas, suggesting the importance of vegetation as a predator refuge. Crayfish may reduce the biomass of *Cladophora* in streams (Hart 1992, Creed 1994) and also the biomass of macrophytes in lakes and ponds (Table 1), thus crayfish predation on amphibian larvae may increase due to grazing of macrophytes and macro-algae. Axelsson et al. (1997) studied the effects on the rare European green tree frog (*Hyla arborea*) in experimental pools and found that *P. leniusculus* reduced the habitat complexity by consuming macrophytes. Moreover, crayfish predation on the tadpoles increased when habitat complexity was reduced (Fig. 4). Figiel & Semlitsch (1991) found that habitat complexity had no significant effect on the survival of North American *H. chrysocelis* tadpoles in the presence of *P. acutus*. In a manipulative field experiment using enclosures in a pond, Nyström et al. (unpublished manuscript) assessed the impact of *P. leniusculus* on the common frog (*Rana temporaria*). They observed that tadpoles in cages containing crayfish were more active than in cages without crayfish or with rainbow trout. Crayfish impact during the tadpole stage was small compared to the trout, but crayfish reduced the number of surviving froglets, probably during metamorphosis. During metamorphosis, froglets have reduced swimming performance (Wassersug & Sperry

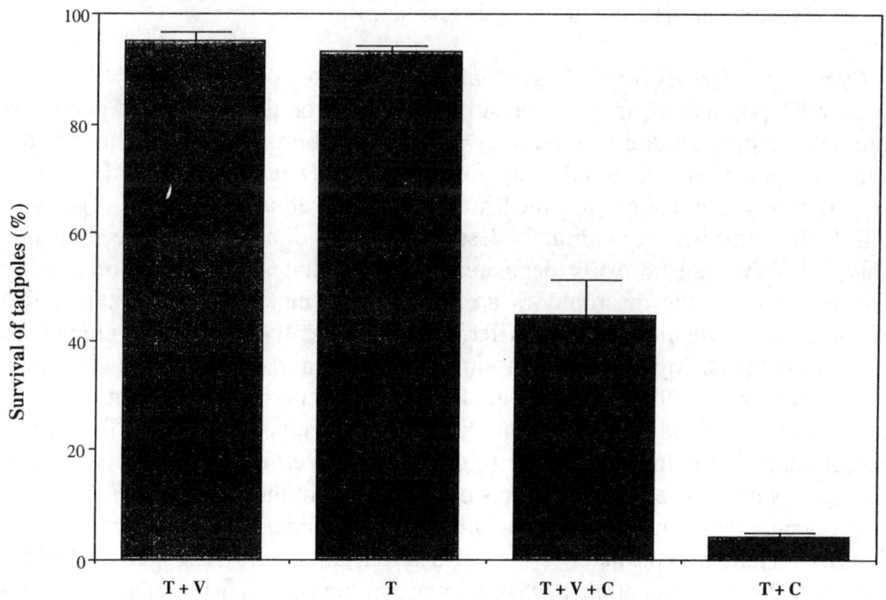

Figure 4. Percentage of *Hyla arborea* tadpoles surviving in ponds (1.3 m^2) with only tadpoles (T), tadpoles + artificial vegetation (T + V), tadpoles + artificial vegetation + two adult *Pacifastacus leniusculus* (T + V + C), and tadpoles + crayfish (T + C). Artificial vegetation consisted of nylon strings covering 50% of the pool bottom. Each pool initially stocked with 60 tadpoles (Gosner stage 25-27), and number of surviving tadpoles counted after one week. Error bars denote 1 SE (N = 5). Data from Axelsson et al. (1997).

1977, Huey 1980, Brown & Taylor 1995), which probably makes them easier for crayfish to capture. Crayfish also caused sub-lethal damage to the tadpoles, indicated by tail-loss. Tadpoles with a reduced swimming performance are likely to be easier prey for crayfish (Figiel & Semlitsch 1991).

Also the consumption of amphibian larvae by crayfish may be affected by tadpole palatability. In a laboratory experiment, Axelsson et al. (1997) observed that *A. astacus* and *P. leniusculus* preyed upon larvae of *B. bufo*, but often failed to consume them. *Bufo bufo* spawn in ponds with dense crayfish populations (e.g. *P. leniusculus* or *A. astacus*, Nyström pers. obs.). Furthermore, numerous tadpoles have been observed swimming in the ponds, indicating that *P. leniusculus* and *A. astacus* have less of an impact on *B. bufo*. Some amphibians develop granular glands during metamorphosis, which reduce predation due to their unpalatability. In aquaria, Formanowicz & Brodie (1982) showed that crayfish (*Cambarus diogenes* and *O. propinquus*) were repelled by the noxious metamorphic stages of *Rana palustris* and *Notophtalmus viridescens*, but consumed the pre-metamorphic stages of *N. viridescens* and the palatable metamorphs of *Rana clamitans*. Most studies on the effects of crayfish on amphibian larvae have been performed on larval stages, which have developed swimming ability, but as Axelsson et al. (1997) suggested, predation on less mobile hatchlings may be significant. It may be that crayfish impact on amphibian larvae is influenced by their swimming ability, the behavior of larvae, their palatability, and the presence of refuges. Therefore, crayfish predation on palatable amphibian larvae may be strong during early development and during metamorphosis, whereas predation on pre-metamorphic larvae probably is less significant.

5.1.3 *Effects of different crayfish species on amphibian populations*

Dense crayfish populations may have a strong influence on the structure of ponds, by consuming macrophytes and invertebrates. Since many European amphibians are dependent on macrophytes as a substrate for eggs (e.g. *Triturus cristatus*, Fog et al. 1997), and adults may choose to breed in habitats with abundant aquatic vegetation (Laurila 1997), crayfish may indirectly destroy the value of a pond as a breeding site. Possibly, eggs that are normally deposited on macrophytes may be more open to crayfish predation if the macrophytes are absent. The impact of different crayfish species on macrophyte biomass may differ, and consequently so may the value of the pond for amphibians. Algae may be an important food source for tadpoles (Dickman 1968, Brönmark et al. 1991, Lamberti et al. 1992), and crayfish may directly and indirectly affect this food source. Food density and availability may affect larval growth rate and size at metamorphosis (e.g. Skelly & Werner 1990). A large size at metamorphosis may have positive effects on the fitness in the terrestrial stage, by increasing juvenile performance, survivorship and by reducing time to maturation (e.g. Smith 1987). The thermal tolerance may also differ among crayfish species (Firkins & Holdich 1993, Troschel et al. 1995) that may influence crayfish predation on amphibians. The ability to capture tadpoles may also differ between crayfish species. In aquaria, adult *A. astacus* captured more tadpoles of similar sizes than *P. leniusculus* (Axelsson et al. 1997). Several amphibian species breed in temporary waters where the number of predators are few (Skelly 1996, Wellborn et al. 1996). More importantly these waters lack predatory fish. In the French upper Rhone floodplain, the highest species richness was observed in these temporary fish-less habitats (Morand

& Joly 1995). The spread of *P. clarkii* into ephemeral habitats, where no native European crayfish can survive, may pose a considerable risk for breeding amphibians in such places. The ability of crayfish to tolerate toxins also may differ between species. *Procambarus clarkii* has the power to reduce macrophytes and may directly consume amphibian eggs and tadpoles despite the presence of toxins. For example, in Italy, *P. clarkii* readily consumed *B. bufo* tadpoles despite the presence of toxins (E. Razzetti personal communication). Introduced *P. clarkii* consumed eggs of the newt *Taricha torosa*, despite the presence of a toxin. Newts disappeared from a stream invaded by *P. clarkii* (Gamradt & Kats 1996). In a wetland near Lisbon, *P. clarkii* has recently invaded, and had probably negative effects on the amphibian fauna (e.g. many urodeles suffered from tail-losses). The species of crayfish is now very abundant in the wetland and is an important prey for many nesting birds (R. Rebelo personal communication). *Procambarus clarkii* attacked adult California newts at Trancas Creek causing injuries to the tail and/or limbs. This aggressive behavior drive newts onto land at the expense of newt reproduction (Gamradt et al. 1997). The fishing methods used for catching *P. clarkii* may also have effects on native amphibians. For example, in the Guadalquivir marshes (Andalucia, Spain) the netting of *P. clarkii* kills large numbers of amphibians (Gutiérrez-Yurrita et al. 1995).

5.2 *Impact on other vertebrates*

Although predation by fish may reduce crayfish numbers (e.g. Lodge & Hill 1994), there are also examples were crayfish have had negative effects on fish populations. Guan & Wiles (1997) studied the distribution of *P. leniusculus* in a British lowland river and found a negative correlation between crayfish abundance and the abundance of two dominant benthic fish species, the bullhead (*Cottus gobio*) and the stone loach (*Noemachelius barbatulus*). By conducting laboratory experiments, they found that adult crayfish displaced both fish species from their shelters, but the crayfish also killed and consumed both fishes. In this case, like that involving tadpoles, crayfish are probably less effective in catching swimming fish (Dean 1969, Ilheu & Bernardo 1993, Xinya 1995), but they may however consume fish eggs. In aquaria, Miller et al. (1992) showed that *O. virilis* was capable of consuming rainbow trout eggs. However, when Savino & Miller (1991) quantified predation by *O. virilis* on trout eggs and sac fry, they concluded that as lake temperatures are so low during the development of trout eggs, crayfish are quite inactive and only large crayfish numbers would be expected to have any impact on trout eggs. Magnuson et al. (1975) discussed the potential impact of introduced *Orconectes* spp. on fish populations in Wisconsin lakes and suggested that crayfish may have negative effects by consuming eggs, and by reducing macro-invertebrate numbers. Rubin & Svensson (1993) suggested that *P. lenusculus* and *A. astacus* may compete for food with young trout, but by reducing macrophytes crayfish also may alter the habitats used by fish. In streams, salmonids feed on both drifting terrestrial animals and benthic prey (Dahl & Greenberg 1996). If trout consume mostly drifting terrestrial animals, then food competition between crayfish and trout may be small. In small Danish streams, mechanical weed cutting resulted in increased mortality of trout fry (Mortensen 1977). Moreover, Eklöv & Greenberg (1998) investigated the importance of in-stream cover on the density of 0+ brown trout in southern Sweden. The results indicated that cover from submerged macrophytes had a positive effect on trout densities, where other types of in-stream cover were in short supply. Rubin & Svensson (1993) studied predation by *A. astacus* on trout eggs and fry at ambient temperatures. They could not find any evidence that *A. astacus* was able to consume trout eggs placed inside gravel, by spawning trout. The temperature prior to the hatching period was very low (less than 5°C). After the emergence of fry, however, the temperature was almost 15°C and crayfish were noted chasing and consuming fry. As observed with tadpoles, crayfish also caused sub-lethal damage to the fry. In Swedish streams, the percentage of trout fry with damaged fins is normally less than 1%, but at two sites in the river of Svartån 13% and 22% of the 0+ trout had damaged fins. *Pacifastacus leniusculus* also was very abundant at these sites (A. Eklöv personal communication). Although these experiments and observations reveal that crayfish may potentially consume trout eggs and fry, the impact crayfish will have on these fish populations in nature remains to be investigated. Most studies suggest that crayfish reduce their activity in the presence of predatory fish, but crayfish may also increase the vulnerability of prey to fish by evicting them from shelters. Rahel & Stein (1988) showed that a small prey fish, the johnny darter, was forced to move by *O. rusticus*, which indirectly increased its vulnerability to small-mouth bass. The bass also increased the vulnerability of the johnny darters to crayfish by forcing the darters to seek cover under shelters occupied by crayfish.

6 CONCLUDING REMARKS

The density of cold-water crayfish species is positively correlated to shelter availability, but negatively correlated to fish numbers (Lodge & Hill 1994). The biomass of these crayfish species may, therefore, vary considerably between habitats, and thus also their impact on other trophic levels. Abundant crayfish populations will strongly affect the biodiversity of freshwater communities by reducing biomass and species richness of macrophytes and invertebrates. In the long term, crayfish grazing on macrophytes may have considerable effects on diversity and interactions within lake and pond communities and on their ontogeny. Many invertebrate species associated with macrophytes may be lost, and amphibians dependent on macrophytes for egg deposition may be negatively affected. If crayfish reduce macro-algae and macrophytes, the recruitment of commercially valuable fish stocks, such as trout, may be affected. Introduced and native crayfish species appear to have similar preferences for macrophytes. Crayfish have a strong negative impact on *Chara*, and this genus contains several vulnerable and rare species. Among the invertebrates, most snail species are reduced by crayfish predation, and vulnerable thin-shelled snails may be eliminated, whereas swimming inveretbrates such as Heteroptera, or sediment burrowers (e.g. chironomidae) may be less affected. Introduced fast-growing species, such as *P. leniusculus* may have a stronger impact on macrophyte and snail biomass than species native to Europe (e.g. *A. astacus*). Crayfish consumption rates of invertebrates (Söderbäck et al. 1988), macrophytes (Nyström & Strand 1996) and amphibian eggs (Axelsson et al. 1997) are dependent on temperature, therefore, the greater overall thermal tolerance of *Pacifastacus leniusculus* compared to *A. astacus*, and *A. leptodactylus* may contribute to the strong impact of *P. leniusculus* on other trophic levels.

Procambarus clarkii has invaded ephemeral habitats where no native crayfish species can survive, and where it has a strong effect on macrophytes and invertebrates. Many of these temporary habitats are also important for rare amphibians, but the effects of *P. clarkii* on amphibians are not well known. There are indications that amphibians unpalatable to some crayfish species may not deter *P. clarkii* from feeding on them. The invasion of *P. clarkii*, however, has positive effects on wading birds that use crayfish as food.

From a conservation perspective, ponds created for crayfish culture (i.e. shelter in excess) have little value unless harboring a native crayfish species. However, it is better to encourage the creation of new ponds rather than stocking crayfish in ponds devoid of crayfish, although being suitable for culture (e.g. lacking predatory fish). Due to human activities, the number of small ponds has declined in Europe. These fishless habitats are characterized by having high biomass of many invertebrate species, abundant submerged macrophytes, and rare bird species (e.g. Wagner 1997). The waters are also important breeding sites for many vulnerable amphibian species.

This review reveals several gaps in our knowledge about the effects of both introduced and native crayfish. Information about the impact of our native species is based on studies of *A. astacus* and *A. pallipes* only. Particularly, there is an urgent need to determine the effects of *P. clarkii* on invertebrates, macrophytes, and amphibians. Information concerning the ecological impact of *O. limosus* is also limited.

ACKNOWLEDGEMENTS

I thank Pedro Joaquln Gutiérrez-Yurrita, Rui Rebelo, and Edoardo Razzetti for valuable information. Björn Lardner, John Strand, Walter T. Momot, Ossi V. Lindqvist, and an anonymous reviewer are greatly acknowledged for their valuable comments on the manuscript. Paul Metcalfe kindly checked the language. Financial support was received from the Swedish Environmental Protection Agency and Ebba, and Sven Schwartz Foundation (grants to Wilhelm Granéli).

REFERENCES

Abrahamsson, S.A.A. 1966. Dynamics of an isolated population of the crayfish *Astacus astacus* Linné. *Oikos* 17:96-107.

Alderman, D.J. & Polglase, J.J. 1988. Pathogens, parasites and commensals. In D.M. Holdich & R.S. Lowery (eds), *Freshwater Crayfish: biology, management and exploitation*: 67-212. Croom Helm, London.

Alexander, J.E., Jr & Covich, A.P. 1991a. Predation risk and avoidance behaviour in two freshwater snails. *Biol. Bull.* 180:387-393.

Alexander, J.E. Jr & Covich, A.P. 1991b. Predator avoidance by the freshwater snail *Physella virgata* in response to the crayfish *Procambarus simulans*. *Oecologia* 87:435-442.

Anonymous, 1993. *Möjligheter att öka flodkräftbestånd i Svenska vatten*. Information från Sötvattenslaboratoriet, Drottningholm 2. 66 pp.

Axelsson, E., Nyström, P., Sidenmark, J. & Brönmark, C. 1997. Crayfish predation on amphibian eggs and larvae. *Amphib.-Reptilia* 18:217-228.

Babitt, K.J. & Jordan, F. 1996. Predation on *Bufo terrestris* tadpoles: effects of cover and predator identity. *Copeia* 1996:485-488.

Babitt, K.J. & Tanner, G.W. 1997. Effects of cover and predatory identity on predation of *Hyla squirella* tadpoles. *J. Herpetol.* 31:128-130.

Barnese, L.E., Lowe, R.L. & Hunter, R.D. 1990. Comparative grazing efficiency of pulmonate and prosobranch snails. *J. North. Amer. Benthol. Soc.* 9:35-44.

Blaustein, A.R., Wake, D.B. & Sousa, W.P. 1994. Amphibian declines: Judging stability, persistence, and susceptibility of populations to local and global extinctions. *Conserv. Biol.* 8:60-71.

Blindow, I. 1995. Kransalger. In M. Aronsson, T. Hallingbäck & J-E. Mattsson (eds), *Swedish red data book of plants 1995*:133-140. ArtDatabanken, Uppsala.

Blindow, I., Andersson, G., Hargeby, A. & Johansson, S. 1993. Long-term pattern of alternative stable states in two shallow eutrophic lakes. *Freshwater Biol.* 30:159-167.

Brown, R.M. & Taylor, D.H. 1995. Compensatory escape mode trade-offs between swimming performance and maneuvering behavior through larval ontogeny of the wood frog, *Rana sylvatica. Copeia* 1995:1-7.

Brown, P.B., Tazik, P., Hooe, M.L. & Blythe, W.G. 1990. Consumption and apparent dry matter digestability of aquatic macrophytes by male and female crayfish (*Orconectes virilis*). *Aquaculture* 89:55-64.

Brönmark, C. 1989. Interactions between epiphytes, macrophytes and freshwater snails: a review. *J. Molluscan Stud.* 55:299-311.

Brönmark, C., Rundle, S.D. & Erlandsson, A. 1991. Interactions between freshwater snails and tadpoles: competition and facilitation. *Oecologia* 87:8-18.

Butler, M.J. & Stein, R.A. 1985. An analysis of the mechanisms governing species replacements in crayfish. *Oecologia* 66:168-177.

Capelli, G.M. 1982. Displacement of northern Wisconsin crayfish by *Orconectes rusticus* (Girard). *Limnol. Oceanogr.* 27:741-745.

Carpenter, S.R. & Lodge, D.M. 1986. Effects of submersed macrophytes on ecosystem processes. *Aquat. Bot.* 26:341-379.

Carral, J.M., Celada, J.D., Gonzáles, J., Sáez-Royuela, M., Gaudioso, V.R., Fernández, R. & López-Baisson, C. 1993. Wild freshwater crayfish populations in Spain: current status and perspectives. *Freshwater Crayfish* 9:158-162.

Cedhagen, T. & Nilsson, G. (eds) 1991. *Grod-och kräldjuren i norden.* Katarina Tryck AB, Sverige, fjärde utgåvan. 141 pp.

Chambers, P.A., Hanson, J.M., Burke, J.M. & Prepas, E.E. 1990. The impact of the crayfish *Orconectes virilis* on aquatic macrophytes. *Freshwater Biol.* 24:81-91.

Chambers, P.A., Hanson, J.M. & Prepas, E.E. 1991. The effect of aquatic plant chemistry and morphology on feeding selectivity by the crayfish, *Orconectes virilis*. *Freshwater Biol.* 25:339-348.

Charlebois, B.S. & Lamberti, G.A. 1996. Invading crayfish in a Michigan stream: direct and indirect effects on periphyton and macroinvertebrates. *J. North. Amer. Benthol. Soc.* 15:551-563.

Coblenz, B.E. 1990. Exotic organisms: a dilemma for conservation biology. *Conserv. Biol.* 4:261-265.

Corbett, K. (ed.) 1989. *Conservation of European Reptiles and Amphibians*. London, Christopher Helm. 274 pp.

Covich, A.P. 1977. How do crayfish respond to plants and Mollusca as alternate food resources? *Freshwater Crayfish* 3:165-179.

Covich, A.P., Crowl, T.A., Alexander, J.E. Jr & Vaughn, C.C. 1994. Predator-avoidance responses in freshwater decapod-gastropod interactions mediated by chemical stimuli. *J. North. Amer. Benthol. Soc.* 13:283-290.

Creed, R.P. Jr 1994. Direct and indirect effects of crayfish grazing in a stream community. *Ecology* 75:2091-2103.

Cronin, G. 1998. Influence of macrophyte structure, nutrititive value, and chemistry on feeding choices of a generalist crayfish. In E.J. Jeppesen, Ma. Søndergaard, Mo. Søndergaard & K. Christoffersen (eds), *The structuring role of submerged macrophytes in lakes*: 307-317. Springer-Verlag, New York.

Crowder, L.B. & Cooper, W.E. 1982. Habitat structural complexity and the interaction between bluegills and their prey. *Ecology* 63:1802-1813.

Crowl, T.A. & Covich, A.P. 1990. Predator-induced life-history shifts in a freshwater snail. *Science* 427:949-951.

Crowl, T.A. & Schnell, G.D. 1990. Factors determining population density and size distribution of a freshwater snail in streams: effects of spatial scale. *Oikos* 59:359-367.

Dahl, J. & Greenberg, L. 1996. Impact on stream benthic prey by benthic vs drift feeding predators: a meta-analysis. *Oikos* 77:177-181.

Dean, J.L. 1969. Biology of the crayfish *Orconectes causeyi* and its use for control of aquatic weeds in trout lakes. Bureau of Sport Fisheries and Wildlife, Technical paper 24.

Denton, T. & Beebee, T.J.C. 1991. Palatability of anuran eggs and embryos. *Amphib.-Reptilia* 12:111-112.

Dickman, M. 1968. The effect of grazing by tadpoles on the structure of a periphyton community. *Ecology* 49:1188-1190.

Diehl, S. 1988. Foraging efficiency of three freshwater fish: effects of structural complexity and light. *Oikos* 53:207-214.

Eklöv, A.G. & Greenberg, L.A. 1998. Effects of artificial instream cover on the density of 0+ brown trout. *Fisheries Management and Ecology*. 5:45-53.

Fauth, J.E. 1990. Interactive effects of predators and early larval dynamics of the tree frog *Hyla chrysoscelis*. *Ecology* 71:1609-1616.

Feminella, J.W. & Resh, V.R. 1989. Submersed macrophytes and grazing crayfish: an experimental study of herbivory in a Californian freshwater marsh. *Holarctic Ecology* 12:1-8.

Figiel, C.R. Jr & Semlitsch, R.D. 1991. Effects of nonlethal injury and habitat complexity on predation in tadpole populations. *Can. J. Zool.* 69:830-834.

Firkins, I. & Holdich, D.M. 1993. Thermal studies with three species of freshwater crayfish. *Freshwater Crayfish* 9:421-248

Fisher, R.N. & Shaffer, B H. 1996. The decline of amphibians in California's great central valley. *Conserv. Biol.* 10:1387-1397.

Flint, R.W & Goldman, C.R. 1975. The effects of a benthic grazer on the primary productivity of the littoral zone of Lake Tahoe. *Limnol. Oceanogr.* 20:935-944.

Fog, K., Schmedes, A. & Rosenørn, de Lasson, D. (eds) 1997. *Nordens padder og krybdyr*. G.E.C. Gad København. 365 pp.

Formanowicz, D.R. Jr & Brodie, E.D. Jr 1982. Relative palatabilities of members of a larval amphibian community. *Copeia* 1982:91-97

Foster, J. & Slater, F.M. 1995. A global review of crayfish predation with observations on the possible loss of *Austropotamobius pallipes* in the welsh wye due to crayfish plague. *Freshwater Crayfish* 8:589-613.

France, R. 1996. Ontogenetic shift in crayfish δ^{13} C as a measure of land-water ecotonal coupling. *Oecologia* 107 ;239-242.

France, R.L., Welbourn, P.M. 1992. Influence of lake pH and macrograzers on the distribution and abundance of nuisance metaphytic algae in Ontario Canada. *Can J. Fish. Aquat. Sci.* 49:185-195.

France, R.L., Howell, E.T., Paterson, M.J. & Welbourn, P.M. 1991. Relationship between littoral grazers and metaphytic algae in five softwater lakes. *Hydrobiologia* 220:9-27.

Gamradt, S.C. & Kats, L.B. 1996. Effect of introduced crayfish and mosquito fish on California newts (*Taricha torosa*). *Conserv. Biol.* 10:1155-1162.

Gamradt, S.C., Kats, L.B. & Anzalone, C.B. 1997. Aggression by non-native crayfish deters breeding in California Newts. *Conserv. Biol.* 11:793-796.

Garvey, J.E., Stein, R.A. & Thomas, H.M. 1994. Assessing how fish predation and interspecific prey competition influence a crayfish assemblage. *Ecology* 75:532-547.

Gasc, J-P., Cabela, A., Crnobrnja-Isailovic, J., Dolmen, D., Grossenbacher, K., Haffner, P., Lescure, J., Martens, H., Martìnez Rica, J.P., Maurin, H., Oliveira, M.E., Sofianidou, T.S., Veiths, M. & Zuiderwijk, A. (eds) 1997. *Atlas of amphibians and reptiles in Europe*. Societas Europaea Herpetologica & Museum National d'Historie Naturelle (IEGB/SPN), Paris. 494 pp.

Greg, W.W. & Rose, F.L. 1985. Influences of aquatic macrophytes on invertebrate community structure, guild structure, and microdistribution. *Hydrobiologia* 128 :45-56.

Guan, R-Z. & Wiles, P.R. 1997. Ecological impact of introduced crayfish on benthic fishes in a British lowland river. *Conserv. Biol.* 11:641-647.

Gutiérrez-Yurrita, P.J. 1997. El papel ecológico del cangrejo rojo (*Procambarus clarkii*), en los ecosistemas aqu·ticos del parque nacional de Donana. Una perspectiva ecofisiológica y bioenergética. Tesis de Doctor. Universidad autónoma de Madrid, facultad de ciencias departamento de ecologìa. 213 pp.

Gutiérrez-Yurrita, P.J., Green, A., Bravo-Uteral, M.A. & Montes, C. 1995. Impacto de la pesca del cangrejo rojo (*Procambarus clarkii*) en los vertebrados del parque nacionl de donana y su Entorno. In H. Bigaré, H. de Cruz, J.A. Pascual-Trillo & T. Purcell (eds), *The conservation of wetlands in a north-south perspective: the east atlantic migration flyway*: 215-227. International conference, Sevilla, 26-29 November 1995, Spain.

Hanson, J.M., Chambers, P.A. & Prepas, E.E. 1990. Selective foraging by the crayfish Orconectes virilis and its impact on macroinvertebrates. *Freshwater Biol.* 24:69-80.

Hargeby, A., Andersson, G., Blindow, I. & Johansson, S. 1994. Trophic web structure in a shallow eutrophic lake during a dominance shift from phytoplankton to submerged macrophytes. *Hydrobiologia* 279/280:83-90.

Hart, D.D. 1992. Community organization in streams: the importance of species interactions, physical factors, and chance. *Oecologia* 9:220-228.

Hazlett, B.A., Anderson, F.E., Esman, L.A., Stafford, C. & Munro, E. 1992. Interspecific behavioral ecology of the crayfish Orconectes rusticus. *J. Freshwater Ecol.* 7:69-76

Henriksson, B.I. 1990. Predation on amphibian eggs and tadpoles by common predators in acidified lakes. *Holarct. Ecol.* 13:201-206.

Hessen D.O. & Skurdal, J. 1986. Analysis of food utilized by the crayfish *Astacus astacus* in Lake Steinsfjorden. *Freshwater Crayfish* 6:187-193.

Hessen, D.O. & Skurdal, J. 1988. Food consumption, turnover rates and assimilation in the noble crayfish (*Astacus astacus*). *Freshwater Crayfish* 7:309-317.

Hessen, D.O., Kristiansen, G. & Lid, I. 1991. Calcium uptake from food and water in the crayfish *Astacus astacus* (L., 1758), measured by radioactive ^{45}Ca (Decapoda, Astacidea). *Crustaceana* 60:77-83.

Hill, A.M., Sinars, D.M. & Lodge, D.M. 1993. Invasion of an occupied niche by the crayfish Orconectes rusticus: potential importance of growth and mortality. *Oecologia* 94:303-306.

Hobbs III, H.H., Jass, J.P. & Huner, J.V. 1989. A review of global crayfish introductions with particular emphasis on two north American species (Decapoda, Cambaridae). *Crustaceana* 56:299-316.

Hogger, J.B. 1986. A report on some of the first introductions of *Pacifastacus leniusculus* into the UK. *Freshwater Crayfish* 6:134-145.

Hogger, J.B. 1988. Ecology, population biology and behaviour. In D.M. Holdich & R.S. Lowery (eds), *Freshwater Crayfish: biology, management and exploitation*: 114-144. Croom Helm, London.

Holdich, D.M., Reeve, I.D. & Rogers, W.D. 1995. Introduction and spread of alien crayfish in British waters-implications for native crayfish populations. *Freshwater Crayfish* 8:99-112.

Holomuzki, J.R. 1989. Predation risk and macroalga use by the stream-dwelling salamander *Ambystoma texanum*. *Copeia* 1989:22-28.

Huey, R.B. 1980. Sprint velocity of tadpoles (*Bufo boreas*) through metamorphosis. *Copeia* 1980:537-540.

Huner, J.V. & Lindqvist, O.V. 1995. Physiological adaptations of freshwater crayfishes that permit successful aquacultural enterprises. *Amer. Zool.* 35:12-19.

Ilhéu, M. & Bernardo, M. 1993. Experimental evaluation of food preferences of red swamp crayfish, *Procambarus clarkii*: vegetal versus animal. *Freshwater Crayfish* 9:359-364.

Ilhéu, M. & Bernardo, M. 1995. Trophic ecology of red swamp crayfish *Procambarus clarkii* (Girard) – preferences and digestibility of plant foods. *Freshwater Crayfish* 10:132-139.

Jonsson, A. 1995. Life-history differences between crayfish *Astacus astacus* and *Pacifastacus leniusculus* in embryonic and juvenile development, laboratory experiences. *Freshwater Crayfish* 8:170-178.

Kirjavainen, J. & Westman, K. 1994. Comparative growth from length composition and mark-recapture experiments for noble crayfish (*Astacus astacus*) and signal crayfish (*Pacifastacus leniusculus*) in Finland. *Nord. J. Freshw. Res.* 69:153-161.

Laitinen, M., Siddall, R. & Valtonen, E.T. 1996. Bioelectronic monitoring of parasite-induced stress in brown trout and roach. *J. Fish Biol.* 48:228-241.

Lamberti, G.A. 1996. The role of periphyton in benthic food webs. In R.J. Stevenson, M.L. Bothwell & R.L. Lowe (eds), *Algal Ecology*: 533-572. Academic Press San Diego.

Lamberti, G.A., Gregory, S.V., Hawkins, C.P., Wildman, R.C., Ashkenas, L.R. & Denicola, D.M. 1992. Plant-herbivore interactions in streams near Mount St Helens. *Freshwater Biol.* 27:237-247.

Laurila, A. 1997. Breeding habitat selection and larval performance of two anurans in freshwater rock-pools. In A. Laurila. *Breeding habitat choice and larval ecology in two anurans* 21-40. PhD thesis, University of Helsinki, Finland.

Lefcort, H. 1996. Adaptive, chemically mediated fright response in tadpoles of the southern leopard frog, *Rana utricularia*. *Copeia* 1996:455-459.

Lirås, V., Lindberg, M., Nyström, P., Annadotter, H., Lawton, L.A. & Graf, B. 1998. Can ingested cyanobacteria be harmful to the signal crayfish (*Pacifastacus leniusculus*)? *Freshwater Biol.* 39:233-242.

Lodge, D.M. 1991. Herbivory on freshwater macrophytes. *Aquat. Bot.* 41:195-224.

Lodge, D.M. 1993. Biological invasions: lessons for ecology. *Trends Ecol. Evol.* 8:133-137.

Lodge, D.M. & Hill, A.M. 1994. Factors governing species composition, population size, and productivity of cool-water crayfishes. *Nord. J. Freshw. Res.* 69:111-136.

Lodge, D.M. & J.G. Lorman 1987. Reductions in submerged macrophyte biomass and species richness by the crayfish *Orconectes rusticus*. *Can. J. Fish. Aquat. Sci.* 44:591-597.

Lodge, D.M., G. Cronin, E. Van Donk, & Froelich, A.J. 1998a. Impact of herbivory on plant standing crop: comparisons among biomes, between vascular and non-vascular plants, and among freshwater herbivore taxa. In E.J. Jeppesen, Ma. Søndergaard, Mo. Søndergaard & K. Christoffersen (eds), *The structuring role of submerged macrophytes in lakes*: 149-174. Springer-Verlag, New York.

Lodge, D.M., Kershner, M.W., Aloi, J.E. & Covich, A.P. 1994. Effects of an omnivorous crayfish (*Orconectes rusticus*) on a freshwater littoral food web. *Ecology* 75:1265-1281.

Lodge, D.M, Brown, K.M., Klosiewski, S.P., Stein, R.A., Covich, A.P., Leathers, B.K. & Brönmark, C. 1987. Distribution of freshwater snails: spatial scale and the relative importance of physicochemical and biotic factors. *Amer. Malacol. Bull.* 5:73-84.

Lodge, D.M., Stein, R.A., Brown, K.B., Covich, A.P., Brönmark, C., Garvey, J. E. & Klosiewski, S.P. 1998b. Predicting impact of freshwater exotic species on native biodiversity: challenges in spatial scaling. *Aust. J. Ecol.* 23:53-67.

Lorman, J.G. & Magnuson, J.J. 1978. The role of crayfish in aquatic ecosystems. *Fisheries* 3:8-16.

Lowery, R.S. 1988. Growth, moulting and reproduction. In D.M. Holdich & R.S. Lowery (eds), *Freshwater Crayfish: biology, management and exploitation*: 83-113. Croom Helm, London.

Lowery, R.S. & Holdich, D.M. 1988. Pacifastacus leniusculus in north America and Europe, with details of the distribution of introduced and native crayfish species in Europe. In D.M. Holdich & R.S. Lowery (eds), *Freshwater Crayfish: biology, management and exploitation*: 283-308. Croom Helm, London.

Lund, H.M-K. 1944. A study of the food of the crayfish (*Potamobius astacus*). *Nytt Magasine For Naturvidenskab* 84:219-250.

Lövei, G.L. 1997. Global change through invasion. *Nature* 388:627-628.

Magnuson, J.J., Capelli, G.M., Lorman, J.G. & Stein, R.A. 1975. Consideration of crayfish for macrophyte control. In P.L. Brezonik & J.L. Fox (eds), *The Proceedings of a symposium on water quality management through biological control*: 66-74. Rep. No. ENV 07-75-1, University of Florida, Gainesville, University of Florida.

Malek, E.A. & Cheng, T.C. (eds) 1974. *Medical and economic malacology*. Academic Press, New York. 398 pp.

Mather, M.E. & Stein, R.A. 1993. Direct and indirect effects of fish predation on the replacement of a native crayfish by an invading congener. *Can. J. Fish. Aquat. Sci.* 50:1279-1288.

Matthews, M. & Reynolds, J.D. 1992. Ecological impact of crayfish plague in Ireland. *Hydrobiologia* 234:1-6.

Matthews, M., Reynolds, J.D. & Keatinge, M.J. 1993. Macrophyte reduction and benthic community alteration by the crayfish *Austropotamobius pallipes* (Lereboullet). *Freshwater Crayfish* 9:289-299

McCormick, P.V. 1990. Direct and indirect effects of consumers on benthic algae in isolated pools of an ephemeral stream. *Can. J. Fish. Aquat. Sci.* 47:2057-2065.

McMahon, B.R. & Hankinson, J.J. 1993. Respiratory adaptations in burrowing crayfish. *Freshwater Crayfish* 9:174-182.

Miller, J.E., Savino, J.F. & Neely, R.K. 1992. Competition for food between crayfish (*Orconectes virilis*) and the slimy sculpin (*Cottus cognatus*). *J. Freshwater Ecol.* 7:127-136.

Momot, W.T. 1988. *Orconectes* in North America and elsewhere. In D.M. Holdich & R.S. Lowery (eds), *Freshwater Crayfish: biology, management and exploitation*: 262-282. Croom Helm, London.

Momot, W.T. 1995. Redefining the role of crayfish in aquatic ecosystems. *Reviews in Fisheries Science*. 3:33-63.

Momot, W.T., Gowing, H. & Jones, P.D. 1978. The dynamics of crayfish and their role in ecosystems. *Amer. Nat.* 99:11-35.

Morand, A. & Joly, P. 1995. Habitat variability and space utilization by the amphibian communities of the French upper-Rhone floodplain. *Hydrobiologia* 300/301:249-257.

Mortensen, E. 1977. Density-dependent mortality of trout fry (*Salmo trutta* L.) and its relationship to management of small streams. *J. Fish. Biol.* 11:613-617.

Newman, R.M. 1991. Herbivory and detrivory of freshwater macrophytes by invertebrates: a review. *J. North. Amer. Benthol. Soc.* 10:89-114.

Nyström, P. & Granéli, W. 1996. The effect of food availability on survival, growth, activity and the number of mature females in crayfish populations. *Freshwater Crayfish* 11:170-180.

Nyström, P. & Pérez, J.R. 1998. Crayfish predation on the common pond snail (*Lymnaea stagnalis*): the effect of habitat complexity and snail size on foraging efficiency. *Hydrobiologia* 368:201-208.

Nyström, P. & Strand, J.A. 1996. Grazing by a native and an exotic crayfish on aquatic macrophytes. *Freshwater Biol.* 36:673-682.

Nyström, P., Brönmark, C. & Granéli, W. 1996. Patterns in benthic food webs: a role for omnivorous crayfish? *Freshwater Biol.* 36:631-646.

Nyström, P., Brönmark, C. & Granéli, W. in press. Influence of an exotic and a native crayfish species on a littoral benthic community. *Oikos*.

Olsen, T.M., Lodge, D.M., Capelli, G.M. & Houlihan, R. 1991. Mechanisms of impact of an introduced crayfish (*Orconectes rusticus*) on littoral congeners, snails and macrophytes. *Can. J. Fish. Aquat. Sci.* 481853-1861.

Pérez, J.R., Carral, J.M., Celada, J.D., Sáez-Royuela, M., Munoz, C. & Sierra, A. 1997. Current status of astaciculture production and commercial situation of crayfish in Europe. *Aquaculture Europe* 22:6-13.

Perry, W.L., Lodge, D.M. & Lamberti, G.A. 1997. Impact of crayfish predation on exotic zebra mussels and native invertebrates in a lake-outlet stream. *Can. J. Fish. Aquat. Sci.* 54:120-125.

Qvenild, T., Bohl, E., Fürst, M. & Skurdal, J. 1989. Broodstock, stocking and ecological impact. In J. Skurdal, K. Westman & P.I. Bergan (eds), *Crayfish culture in Europe*: 29-38. Report from the workshop on crayfish culture, 16-19 Nov. 1987, Trondheim, Norway.

Rahel, F.J. & Stein, R.A. 1988. Complex predator-prey interactions and predator intimidation among crayfish, piscivorous fish, and small benthic fish. *Oecologia* 75:94-98.

Reynolds, J.D. 1978. Crayfish ecology in Ireland. *Freshwater Crayfish* 4:215-220.

Rickett, J.D. 1974. Trophic relationships involving crayfish of the genus *Orconectes* in experimental ponds. *The Progressive Fish-Culturist* 36:207-211.

Rubin, J.F. & Svensson, M. 1993. Predation by the noble crayfish, *Astacus astacus* (L.), on emerging fry of sea trout, *Salmo trutta* (L.). *Nordic J. Freshw. Res.* 68:100-104.

Saffran, K.A. & Barton, D.R. 1993. Trophic ecology of *Orconectes propinquus* (Girard) in Georgian Bay (Ontario, Canada). *Freshwater Crayfish* 9:350-354.

Savino, J.F. & Miller, J.E. 1991. Crayfish (*Orconectes virilis*) feeding on young lake trout (*Salvelinus namaycush*); effect of rock size. *J. Freshwater Ecol.* 6:161-169.

Skelly, D.K. 1996. Pond drying, predators, and the distribution of *Psuedacris* tadpoles. *Copeia* 1996:599-605.

Skelly, D.K. & Werner, E.E. 1990. Behavioral and life-historical responses of larval American toads to an odonate predator. *Ecology* 71:2313-2322.

Smith, D.C. 1983. Factors controlling tadpole populations of the chorus frog (*Pseudacris triseriata*) on Isle Royale, Michigan. *Ecology* 64:501-510.

Smith, D.C. 1987. Adult recruitment in chorus frogs:effects of size and date at metamorphosis. *Ecology* 68:344-350.

Steinman, A.D. 1996. Effects of grazers on freshwater benthic algae. In R.J. Stevenson, M.L. Bothwell & R.L. Lowe (eds), *Algal Ecology*: 341-373. Academic Press, San Diego.

Söderbäck, B. 1995. Replacement of the native crayfish *Astacus astacus* by the introduced species *Pacifastacus leniusculus* in a Swedish lake: possible causes and mechanisms. *Freshwater Biol.* 33:291-304.

Söderbäck, B., Appelberg, M., Odelström, T. & Lindqvist, U. 1988. Food consumption and growth of the crayfish *Astacus astacus* L in laboratory experiments. *Freshwater Crayfish* 7:145-153.

Thompson, D.J. 1987. Regulation of damselfly populations:the effects of weed density on larval mortality due to predation. *Freshwater Biol.* 17:367-371.

Troschel, H.J. & Dehus, P. 1993. Distribution of crayfish species in the Federal republic of Germany, with special reference to *Austropotamobius pallipes*. *Freshwater Crayfish* 9:390-398.

Troschel, H.J., Schulz, U. & Berg, R. 1995. Seasonal activity of stone crayfish *Austropotamobius torrentium*. *Freshwater Crayfish* 10:196-169.

Tulonen, J., Erkamo, E. & Kirjavainen, J. 1995. Growth rate, survival and reproduction of noble crayfish (*Astacus astacus* (L.)) and signal crayfish (*Pacifastacus leniusculus* (Dana)) under similar rearing conditions. *Freshwater Crayfish* 10:623-629.

Wagner, B.M.A. 1997. Influence of fish on the breeding of the red-necked grebe *Podiceps grisegena* (Boddaert, 1783). *Hydrobiologia* 344:57-63.

Ward, D. & Sexton, J. 1981. Anti-predator role of salamander egg membranes. *Copeia* 1981:724-726.

Warner, G.F., Wood, J.C. & Orr-Ewing, R.H. 1995. Signal crayfish (*Pacifastacus leniusculus*) feeding on pond snails: optimal foraging? *Freshwater Crayfish* 8:352-359.

Wassersug, R.J. & Sperry, D.G. 1977. The relationship of locomotion to differential predation on *Pseudacris triseriata* (Anura: Hylidae). *Ecology* 58:830-839.

Weber, L.M. & Lodge, D.M. 1990. Periphytic food and predatory crayfish: relative roles in determining snail distribution. *Oecologia* 82:33-39.

Wellborn, G.A., Skelly, D.K. & Werner, E.E. 1996. Mechanisms creating community structure across a freshwater habitat gradient. *Annu. Rev. Ecol. Syst*. 27:337-363.

Westman, K., Savolainen, R. & Pursiainen, M. 1993. A comparative study on the growth and moulting of the noble crayfish, *Astacus astacus* (L.), and the signal crayfish *Pacifastacus leniusculus* (Dana) in a small forest lake in southern Finland. *Freshwater Crayfish* 9:451-465.

Westman, K., Pursiainen, M., Järvenpää, T. & Westman, P. 1989. Situation of the crayfish stocks and crayfish culture in Europe. In J. Skurdal, K. Westman & P.I. Bergan (eds), *Crayfish culture in Europe*: 175-192. Report from the workshop on crayfish culture, 16-19 Nov. 1987, Trondheim, Norway.

Wetzel, R.G. (ed.) 1983. *Limnology*, second edition. Saunders college publishing, Fort Worth. 776 pp.

Xinya, S. 1995. Effect of the crayfish *Procambarus clarkii* on the survival of the fry and fingerlings of fishes cultivated in China. *Freshwater Crayfish* 8:528-532.

Diseases of European freshwater crayfish, with particular emphasis on interspecific transmission of pathogens

GÜNTER VOGT
Department of Zoology 1 (Morphology/Ecology), University of Heidelberg, Heidelberg, Germany

ABSTRACT

An overview of the most common diseases of native and naturalized crayfish in Europe is presented. It mainly focuses on findings of the last decade covering viruses, bacteria, protozoans, fungi, DRIPs, and branchiobdellids. It further addresses the problem of interspecific transmission of crayfish diseases in the wild and presents some examples of pathogen transfer from introduced to native crayfish species and vice versa. These examples demonstrate that transmission of pathogens among crayfish species is not only a theoretical threat but has repeatedly happened in European waters. Particularly the documented introduction of different genotypes of the crayfish plague fungus, *Aphanomyces astaci*, together with their American host species *Pacifastacus leniusculus* and *Procambarus clarkii* into Europe and the proven transmission of at least one of these genotypes to native crayfish underlines the risk of introduction of alien crayfish species for our indigenous fauna.

1 INTRODUCTION

In nature, diseases play an important role as regulators of animal populations. As long as disease agents live in balanced relationships with their hosts, they have stabilizing effects on ecosystems and can therefore be considered as positive. Permanent stress caused by pollution or high competition for food and space can irreversibly change this delicate equilibrium in favor of the pathogens. Such a shift normally results in decline of the affected host population and may even have adverse effects on the ecosystem level. The most severe threatening of animal populations by pathogens, however, appears to come from alien disease agents, which are accidentally introduced into a new environment together with their natural hosts. Such pathogens can completely eradicate indigenous populations since new hosts are usually less resistant to a disease than the natural ones. An excellent example for the devastating effects of alien diseases in new geographical regions is the crayfish plague, a fungal disease of North American origin, which has wiped out most of the European freshwater crayfish populations since the last decades of the 19th century (Alderman & Polglase 1988).

This article gives an overview of the most common diseases of native and naturalized crayfish in Europe, presents some documented examples of interspecific transmission of crayfish pathogens in the wild and examines the risk of introduction of alien crayfish species into Europe from a pathologist's view.

2 OVERVIEW OF DISEASES OF NATIVE AND NATURALIZED FRESHWATER CRAYFISH IN EUROPE

The diseases of European freshwater crayfish have repeatedly been reviewed in the last few decades (Unestam 1973a, b, Vey & Vago 1973, Vey 1976, 1978, Schäperclaus et al. 1990, Unestam 1981, Smith & Söderhäll 1986, Vey 1986, Söderhäll 1988, Westman et al. 1990). The most comprehensive compilation was made by Alderman & Polglase (1988). Since these reviews summarize all of the relevant literature up to the late 1980s, I have mainly concentrated on the developments of the last 10 years. Only some key references are cited from earlier periods. The illustrations shown in this paper focus on the morphology and life cycle of the pathogens rather than on their histopathological effects.

The most catastrophic epidemics of freshwater crayfish are caused by obligate pathogens. Under conditions of nutritional and environmental stress, sporadic disease outbreaks can also be induced by opportunistic pathogens such as fouling organisms and saprophytes. The most virulent disease agents belong to the fungi. Other widespread parasites are microsporidia, DRIPs, and certain branchiobdellids. Much less is reported on viruses, and bacterial diseases are mainly known from aquaculture facilities. In general, crayfish diseases may be subdivided into four categories: 1. Diseases with no or little visible impairment to the host (e.g. psorospermiasis); 2. Diseases which are lethal for the infected individuals but cause no severe damage on the population level due to the low prevalence of the pathogen (e.g. porcelain disease); 3. Diseases which eradicate infected populations completely (e.g. crayfish plague); and 4. Diseases which cause only little damage to infected specimens but result in considerable economic losses because the gross signs of the disease make the crayfish unmarketable (e.g. burn spot disease).

2.1 *Viruses*

Research on viruses in freshwater crayfish is a comparably recent field, perhaps because detection of viruses requires some sophisticated equipment like an electron microscope. Thus far, only two crustacean viruses have been described for European crayfish, a 20 nm virus from the gills of *Astacus astacus* (Pl. 1c) (Halder & Ahne 1988a, b) and a non-occluded baculovirus of 340 × 70 nm from the midgut and hepatopancreas of the same species (Pl. 1a, b) (Edgerton 1996b, Edgerton et al. 1996). The 20 nm virus was found only once in a single specimen in southern Germany. The baculovirus was sampled from five populations of noble crayfish in central Finland. Its prevalence rate varied from 53.3% to 100%, and the intensities of infection were usually low. The virus caused lesions in the midgut and hepatopancreas but, at present, it is not known whether or not it causes mortality in wild populations.

Astacus astacus can also act as a vector of Infectious Pancreatic Necrosis Virus

(IPNV) of fish. It was shown to harbor this virus in cytoplasmic granules of the haemocytes (Pl. 1d) for more than 42 weeks and to continuously excrete virus particles into the water (Ahne & Halder 1988, Halder & Ahne 1988a, b).

2.2 *Bacteria*

Pathogenic bacteria can infect the cuticle, the gills, the blood or internal organs such as the hepatopancreas (Pl. 2a). Chitinoclastic and gill infecting bacteria usually induce heavy melanization of the cuticle and the gills. Septicaemias are particularly caused by species of the genera *Pseudomonas*, *Aeromonas*, and *Vibrio* that enter the haemocoel through wounds. *Citrobacter*, *Enterobacter*, and *Pseudomonas* species were shown to be responsible for infections of the gut and hepatopancreas (Smith & Söderhäll 1986, Vey 1986, Alderman & Polglase 1988). Such enteric conditions, for instance granulomatous hepatopancreatitis of *Astacus astacus* (Pl. 2a, b), can result in lysis of larger areas of the hepatopancreatic epithelium and can induce intense formation of granulomas (Vogt & Rug 1996). All bacteraemias identified so far are apparently caused by facultative pathogens that are regular components of the aquatic microflora or even of the gut flora of the crayfish. They seem to become pathogenic only in weakened crayfish or if they enter the haemolymph or internal organs through wounds.

2.3 *Protists*

The porcelain disease, which is caused by the microsporidian *Thelohania contejeani* (Pl. 2c, d), has a wide distribution range in Europe. So far, *Thelohania contejeani* has been detected in *Astacus astacus*, *Astacus leptodactylus*, *Austropotamobius pallipes*, and *Orconectes limosus* (Cossins 1973, Alderman & Polglase 1988, Westman et al. 1990, Fischer 1992). The parasite is preferably found in the cytoplasm of the abdominal and claw skeletal musculature (Pl. 2c) where it multiplies by schizogony and sporogony. In heavily infected crayfish up to 90% of the muscle fibers can be parasitized (Oidtmann et al. 1996). The heart musculature is less intensely infected, and the parenchyma cells of other organs are not affected at all. However, spores from disrupted muscle fibers can be spread to other tissues and organs where they are then found extracellularly in the interstitium (Oidtmann et al. 1996). There is apparently no host immune response against this parasite.

The life stages of *Thelohania contejeani* (Pl. 2c, d) have been well investigated by light and electron microscopy, particularly the 3.5 × 1.8 μm spores (Cossins 1973, Vey & Vago 1973, Cossins & Bowler 1974, Fischer 1992, Westman et al. 1992, Oidtmann et al. 1996). In contrast, the route of transmission of the parasite is not yet resolved. Neither direct infection by uptake of infected muscle tissue nor indirect infection, for instance by spores which have passed through the digestive system of a translator fish, is convincingly proven (Fischer 1992, Oidtmann et al. 1996). Infection with *Thelohania contejeani* results in successive destruction of the myofilaments and finally in a milky coloration of the musculature. Clinical symptoms appear when loss of muscle function is caused by the spore masses in the muscle fibers (Oidtmann et al. 1996). After 1-2 years, infected specimens die of the disease (Skurdal et al. 1990). Due to the low prevalence of 1-2% in the wild thelohaniasis has no devastating effects on natural crayfish populations. However, in aquaculture this disease can reach a prevalence rate of roughly 30%, and hence, can cause considerable financial losses (Schäperclaus et al. 1990, Vey 1986).

2.4 *Fungi*

The most severe crayfish disease is the crayfish plague, which is lethal to all native European species (Unestam 1973a, b, Vey 1976, 1978, Unestam 1981, Alderman et al. 1987, Cerenius et al. 1988). This epizootic is caused by the oömycete fungus *Aphanomyces astaci* (Pl. 3). In the last decades of the 19th century the crayfish plague wiped out many of the crayfish populations in mainland Europe (Hofer 1904). It has entered Great Britain, Ireland and Turkey only in the 80s of this century and has struck a number of native populations of *Austropotamobius pallipes* and *Astacus leptodactylus*, respectively (Holdich 1988, Rahe & Soylu 1989, Matthews & Reynolds 1991, Alderman 1993). Regional outbreaks of the plague are observed almost every year somewhere in Europe. *Aphanomyces astaci* originates from North America and is carried as a latent infection by certain populations of the naturalized American species *Orconectes limosus*, *Pacifastacus leniusculus*, and *Procambarus clarkii* (Söderhäll 1988, Alderman et al. 1990, Diéguez-Uribeondo et al. 1995, Holdich et al. 1995, Nylund & Westman 1995a). *Pacifastacus leniusculus* was shown to harbor viable fungal hyphae within melanized spots of the cuticle. If such crayfish are subjected to stress conditions, the fungus can grow out and sporulate, and the infected specimens can even die from the plague (Cerenius et al. 1988). Procedures on isolation, culture, and diagnosis of *Aphanomyces astaci* can be found in Alderman and Polglase (1986), Cerenius et al. (1988), and Halder et al. (1989).

The biology of *Aphanomyces astaci* has chiefly been investigated by the laboratory of Torgny Unestam and Kenneth Söderhäll (Uppsala, Sweden). The fungus usually infects its host through uncalcified parts of the cuticle (Smith & Söderhäll 1986, Cerenius et al. 1988). In an early stage of infection, the vegetative mycelium that is unseptate, frequently branching, and 7-10 μm in width (Pl. 3a) is found in the intersegmental joints of the abdomen and in the cornea of the eyes. Later, it can also penetrate internal organs via the connective tissue (Alderman & Polglase 1988, Oidtmann et al. 1996). Within approximately two weeks, the infection causes torpidity, impaired mobility, paralysis and eventually death (Smith & Söderhäll 1986). Since these symptoms can already occur before the hyphae have penetrated internal organs, they are thought to be induced by a fungal toxin (Smith & Söderhäll 1986, Oidtmann et al. 1996). In the final phase of the disease, the fungus sporulates (Pl. 3b, c) and releases bi-flagellate zoospores (Pl. 3d) which are viable for a few days only. These spores are attracted to crayfish by low molecular weight components of crayfish exudates (Cerenius & Söderhäll 1984). They attach to the cuticle of a new host, encyst, and produce a germ tube that grows out to form a mycelium (Alderman & Polglase 1986, Cerenius et al. 1988). If the zoospores do not find an appropriate host substrate they can re-encyst at least three times, which lengthens the potential time span of infectivity (Cerenius et al. 1988).

Despite the bulk of information collected on the life cycle, infectivity, and pathogenicity of the crayfish plague fungus there are still some crucial unanswered questions, particularly on the spectrum of possible modes of transmission in aquatic ecosystems and on the reasons for the persistence of the disease in rivers and lakes for decades (Nylund et al. 1993, Fürst 1995). It is generally believed that crayfish plague is mainly transmitted by man through spore-containing fishing gear or by deliberate release of infected American crayfish. However, Häll & Unestam (1980), Alderman et al. (1987), and Ahne & Halder (1988) have presented evidence that fish may also act as a transmitter. Häll & Unestam (1980) and later also Ahne & Halder (1988) observed in the laboratory that zoospores of *Aphanomyces astaci* are able to form a mycelium on the slimy surface of scales from salmonid and cyprinid fishes (Pl. 3e).

A further widespread but less harmful fungal disease of native and introduced species is the burn spot disease which according to Mann and Pieplow (1938) is caused by *Ramularia astaci* in *Astacus astacus*, *Cephalosporium leptodactyli* in *Astacus leptodactylus*, and *Didymaria cambari* in *Orconectes limosus*. Alderman and Polglase (1988) have pointed out that *Ramularia astaci* is very similar to *Fusarium tabacinum*. The burn spot disease is characterized by black spots with reddish rims on the exoskeleton. Apparently, this disease does not adversely affect tissues and organs other than the cuticle. However, successive dissolution and final breach of the cuticle in the centre of a burn spot facilitate access for secondary opportunistic disease agents into the crayfish's body. Hofmann (1980) reported that burn spots disappear after a few moultings if the crayfish are regularly fed with rotten leaves of poplar, beech or alder.

Our investigation of burn spots in *Astacus astacus* revealed a rather complicated cuticular histopathology. In the black surface layer of a burn spot that is supposedly derived from the epicuticle, we have found a fungal mycelium with 3 µm wide hyphae and plenty of curious, thick-walled, unicellular organisms of approximately 10 µm in diameter (Pl. 4b). The width of the hyphae fits well to data given for *Ramularia astaci* (Mann & Pieplow 1938). Beneath the relatively thin black surface layer there was a thick white layer composed of softened cuticle which included unmelanized mycelia of the 3 µm fungus, other fungi with thicker hyphae, and bacteria. In the innermost black layer of a burn spot the fungi were heavily melanized. Small melanized spots without hyphae were also frequent in the underlying epidermis and connective tissue. These spots may include bacteria or fungal spores.

Fungal infections are also responsible for rosty spots on the exoskeleton of crayfish. In the center of a rosty spot of the carapace of *Pacifastacus leniusculus* and in the underlying epidermis I have found heavily melanized mycelia (Pl. 4a). However, rosty spots can also be the result of bacterial infections.

Sporadic mortality can also be caused by opportunistic fungal parasites (Alderman & Polglase 1988), for instance of the genera *Saprolegnia*, *Trichosporon*, and *Fusarium*. *Saprolegnia parasitica*, a pathogen of fish, was shown to be able to kill *Astacus astacus*, *Pacifastacus leniusculus*, and *Procambarus clarkii*. In healthy, non-injured crayfish, around 20% of the infected specimens died. If the epicuticle of the experimental crayfish was abraded before challenge with zoospores, mortality increased three times. The above listed species can carry this fungus in melanized spots and act as a potential vector (Diéguez-Uribeondo et al. 1994, Diéguez-Uribeondo 1995). *Trichosporon beigelii* (Pl. 4c), a worldwide saprophyte in soil and water and an opportunistic parasite of animals and man, was found in Swedish *Astacus astacus* within the cuticle. The hyphae had a diameter of 3-4 µm. The virulence of *Trichosporon beigelii* for crayfish is not yet evaluated, but stress seems to significantly enhance the infection rate (Söderhäll et al. 1993).

Species of the genus *Fusarium,* which infect the cuticle and underlying tissue of the exoskeleton and gills, usually induce heavy melanization (Alderman & Polglase 1988). As a result, the cuticle is often locked to the tissue beneath by voluminous granulomas. These solid interconnections can be fatal during the next molting, because the exuvia can not be shed. A typical member of this genus is *Fusarium oxysporum*, which infects the gills of crayfish and induces heavy melanization (Maestracci & Vey 1988). Penetration of the fungus depends on integumentary wounds. In *Astacus leptodactylus* and *Austropotamobius pallipes,* development of the hyphae provoked a localized necrosis and physiological disturbances such as decrease of osmotic pressure and concentration of Na^+ and Cl^- ions in the haemolymph. *Fusarium oxysporum* can cause high mortalities particularly in wounded crayfish (Maestracci & Vey 1988).

Fusarium solani (Pl. 4d), a saprophyte or parasite of marine and freshwater crustaceans, insects or plants, was identified as causative agent of the brown abdomen disease of *Astacus leptodactylus* (Chinain & Vey 1988a, b, c). This fungus with 4-6 µm wide hyphae was shown to cause considerable mortality in cultured *Astacus leptodactylus* and *Pacifastacus leniusculus*. In laboratory experiments, the success of infection depended on haemolymph exudates on the surface of wounds. The hyphae were present only in the exoskeleton and the tissue immediately beneath. Despite the fact that the infections remained localized, many of the experimentally infected crayfish died. Injections of filtrates of *Fusarium solani* into *Astacus leptodactylus* and an insect resulted in mortality in both species, suggesting that *Fusarium solani* produces a lethal toxin (Chinain & Vey 1988c). Mortality can also be attributed to abnormalities at exuviation as described above for the genus *Fusarium* (Chinain & Vey 1988c). Interestingly, *Fusarium solani* strains isolated from French *Astacus leptodactylus* and *Pacifastacus leniusculus* showed a high degree of similarity in enzyme electrophoretic studies, but differed greatly from isolates of *Penaeus japonicus*, *Melolontha melolontha* and green peas (Chinain & Vey 1988a).

2.5 *DRIPs*

The most enigmatic of all crayfish parasites is *Psorospermium haeckeli* (Pl. 5a), which was detected already in 1857 by Ernst Haeckel. This organism is widespread throughout Europe (Grabda 1934, Cerenius & Söderhäll 1993, Diéguez-Uribeondo et al. 1993, Henttonen et al. 1994a, 1995, Nylund & Westman 1995b, Taugbøl & Skurdal 1995, Henttonen 1996, Vogt et al. 1996). It has been found in all native and naturalized species from Spain to Finland (Nylund & Westman 1995b, Henttonen 1996, Vogt et al. 1996). In general, older crayfish seem to be more heavily infected than younger ones. The prevalence of the parasite in natural and cultured populations was most intensively studied in Finland (Henttonen 1996), Norway (Taugbøl & Skurdal 1995) and Lithuania (Bucinskiene 1996). Paula Henttonen reported that, in Finland, *Psorospermium haeckeli* exists in *Astacus astacus* in all kinds of waters, including small brooks, rivers, ponds and lakes. From the 29 populations investigated, 22 were found to be infected, with prevalence rates ranging from a few percent to 100%. Of the 2299 crayfish examined, 52% carried the parasite (Henttonen 1996). In Norway, 78.6% of the populations examined were found to be infected with prevalence rates of 1.8-93% (Taugbøl & Skurdal 1995).

Psorospermium haeckeli occurs as a thick-walled resting sporocyst (Pl. 5a) that is particularly abundant in the collagenous wall of the thoracic arteries and the subepidermal connective tissue beneath the carapace (Rug & Vogt 1994, Vogt & Rug 1995, Henttonen 1996). So far, in Europe two different morphotypes have been detected which measure approximately 100 × 60 μm and 150 × 50 μm. These morphotypes are thought to belong to different species (Rug & Vogt 1995, Henttonen 1996, Vogt et al. 1996). By phylogenetic analysis of the ssu-rDNA of the sporocysts, *Psorospermium haeckeli* has only recently been assigned to the DRIPs, a novel clade of the Animalia that, in the tree of organisms, diverges close to the animal-fungal dichotomy (Ragan et al. 1996).

The life cycle of *Psorospermium haeckeli* includes a histozoic phase within the crayfish, which is characterized by the development of the sporocyst from a small amoeboid cell (Grabda 1934, Rug & Vogt 1995, Henttonen 1996, Henttonen et al. 1997), and a post-mortem phase of free living and encysted amoebas (Pl. 5b-e) (Vogt & Rug 1999). These amoebas are believed to infect crayfish either via the digestive tract or by penetration through soft parts of the cuticle. Since *Psorospermium haeckeli* induces no external signs of a disease, its pathogenicity is the subject of controversial discussion. Cerenius et al. (1991) and Thörnqvist & Söderhäll (1993) have presented clear experimental evidence that the parasite can start immune defense responses. In our samples, the sporocysts of *Psorospermium haeckeli* were mostly not melanized or encapsulated by haemocytes. However, in specimens harboring other pathogens in addition to *Psorospermium haeckeli*, for instance fungi on the gills, the percentage of melanization of the sporocysts could reach roughly 80% (Vogt & Rug 1995). Under aquaculture conditions (Söderhäll 1988) and during molting (Vey 1986) *Psorospermium haeckeli* can even cause mortalities.

2.6 *Platyhelminthes, Aschelminthes and Annelida*

Branchiobdellids (Pl. 6a-c) are by far the most frequent worms of European crayfish (Gelder 1996, Moon et al. 1996). Earlier, they were mostly regarded as a single species, *Branchiobdella astaci*. Meanwhile, several species of this leech-like annelid of 1-12 mm in length have been described (Gelder et al. 1994, Nesemann 1994). Quite often, two or more species are found sympatrically on the same host individual. In the Kammlach river in Bavaria, for instance, we have found *Branchiobdella parasita*, *B. balcanica*, and *B. pentodonta* on the same specimens. The complete developmental cycle of the branchiobdellids takes place on the crayfish and includes cocoons attached to the cuticle and the gill filaments (Pl. 6b, c), juveniles, and adults (Pl. 6a). The latter are preferably found on the cuticle, the mouthparts, and in the gill chamber, depending on species.

Most of the Branchiobdellida are considered to be commensals (Alderman & Polglase 1988). Our own investigations support this view, since the three species *Branchiobdella parasita*, *B. balcanica*, and *B. pentodonta,* which were sampled from the exoskeleton and mouthparts of *Astacus astacus* and *Austropotamobius torrentium* from different localities, had all diatoms, rotifers, chironomid larvae, and detritus in their digestive tract. However, Grabda & Wierzbicka (1969) demonstrated by means of histology and radioactive labeling that at least *B. hexodonta* feeds on crayfish tissue and sucks blood from the gills. Vey (1986) and ourselves have repeatedly found plenty of wounds on the gill filaments of heavily infested crayfish corroborating the parasitic behavior of those gill-dwelling branchiobdellids.

Metacercaria of digenean trematodes, cysticercoids of cestodes, cysts of acanthocephalans, and nematodes are occasionally found in European crayfish species but appear to exert no or only little damage on the host's health (Vey 1986, Alderman & Polglase 1988).

2.7 *Nutritional and environmental diseases*

Nutritional and environmental diseases of European crayfish have not yet been investigated in depth. However, it is well known that inadequate nutrition and poor water quality can principally weaken crayfish and thereby enhance their susceptibility to infectious diseases.

3 INTERSPECIFIC TRANSMISSION OF PATHOGENS

Regarding transfer of pathogens from one species to another or from one geographical region to another, there are several well-documented examples for cultured shrimp (Lightner 1990, 1996). Much less data are available for crayfish, particularly on transmission of pathogens in the wild. Transmission of pathogens in nature can be studied either by examination of sympatric crayfish populations or by molecular analysis of pathogens sampled from disease outbreaks. This chapter presents examples for the transmission of pathogens between native European species, from native European to naturalized American species, and from American to native species.

3.1 *Transmission between native European species*

Examination of a naturally mixed population of *Astacus astacus* and *Austropotamobius torrentium* in the Kammlach river (Bavaria, Germany) revealed infection of both species by the elongate morphotype of *Psorospermium haeckeli* (Pl. 7a, b) (Vogt et al. 1996). The distribution pattern of the parasite in the host tissues was similar in both species. However, differences were noted with respect to the immune response: in the noble crayfish there was no sign of immune defense whereas in the stone crayfish approximately 10% of the sporocysts were encapsulated by haemocytes. Since the elongate morphotype of *Psorospermium haeckeli* is typical of *Astacus astacus* of that region and since we have not found *Psorospermium haeckeli* in stone crayfish sampled from other locations, it seems probable that the parasite has been transmitted to *Austropotamobius torrentium* from *Astacus astacus*. In this context, it may also be of interest that both species of crayfish harbored the same branchiobdellids, *Branchiobdella parasita*, *B. pentodonta*, and *B. balcanica*.

3.2 *Transmission from European to American species*

Transmission of branchiobdellids from *Austropotamobius torrentium* to introduced American *Orconectes limosus* has been observed in a mixed population in the Steinbach creek (Hesse, Germany). The main water course of the Steinbach is inhabited by

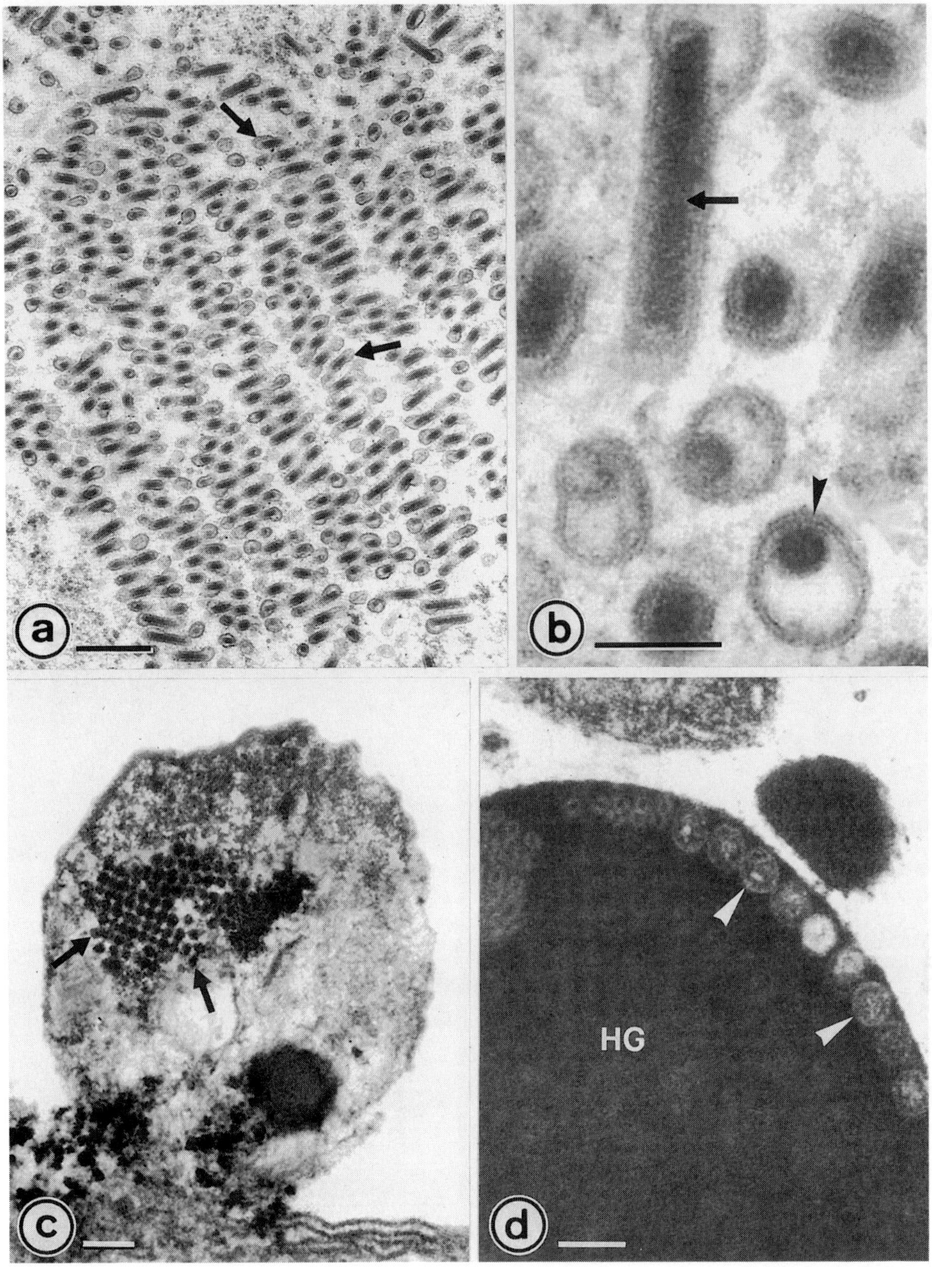

Plate 1. Viruses from *Astacus astacus*. a. Cluster of bacilliform viruses (arrows) in hepatopancreas cell. Bar: 500 nm. b. Longitudinal and cross sections of bacilliform viruses displaying nucleocapsid (arrow) and envelope (arrowhead). Bar: 100 nm. c. 20 nm viruses (arrows) in connective tissue of gill. Bar: 100 nm. d. IPNV-fish viruses (arrowheads) in granule of a crayfish haemocyte (HG). Bar: 100 nm. a and b from Edgerton et al. (1996); c and d from Halder & Ahne (1988a).

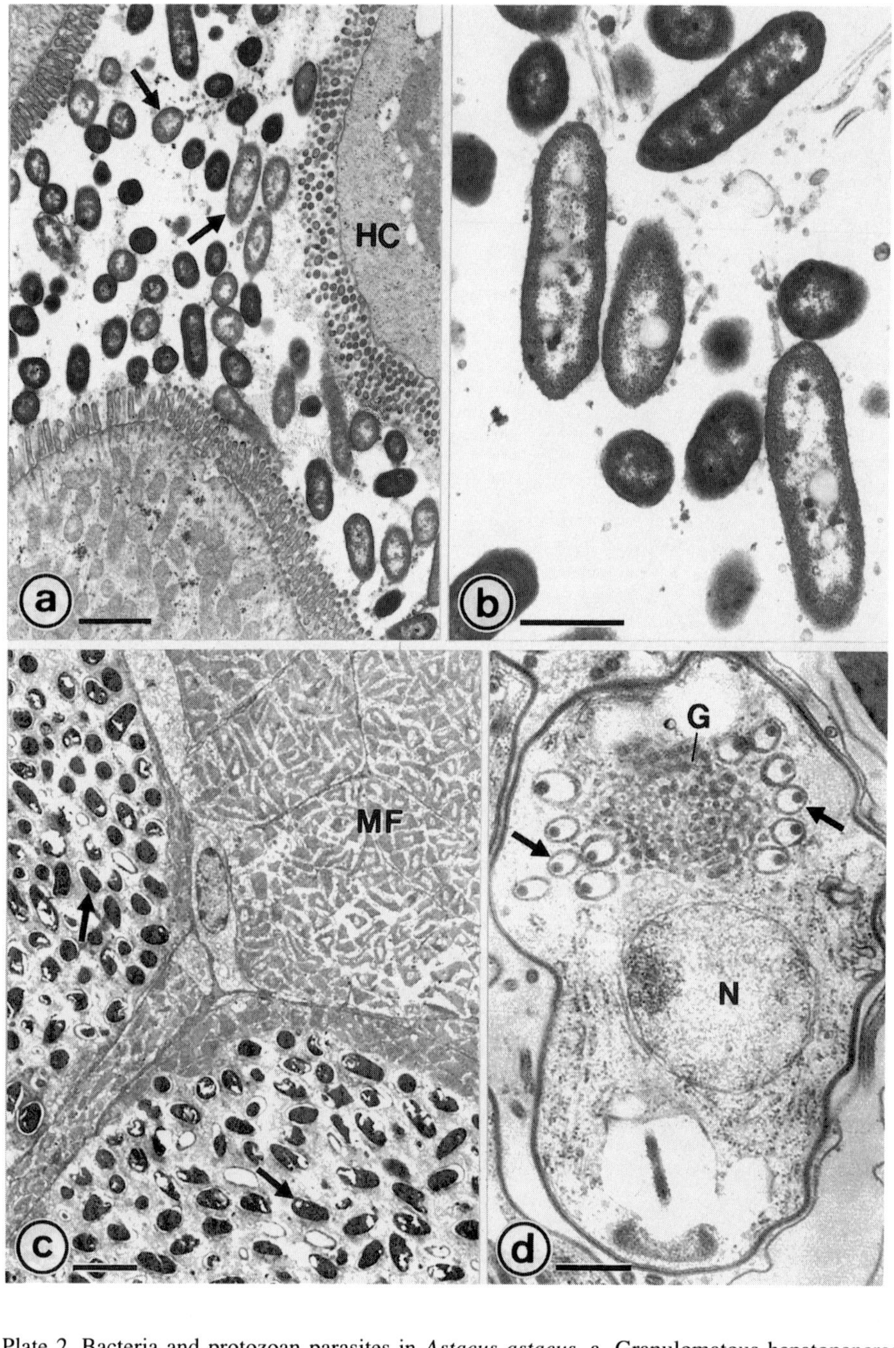

Plate 2. Bacteria and protozoan parasites in *Astacus astacus*. a. Granulomatous hepatopancreatitis bacteria (arrows) in tubular lumen of hepatopancreas. HC: hepatopancreatic cell. Bar: 2 µm. b. Higher magnification of GH-bacteria. Bar: 1 µm. c. Electron micrograph of intact (MF) and with *Thelohania contejeani* (arrows) infected muscle fibres. Bar: 50 µm. d. Ultrastructure of sporoblast of *Thelohania contejeani* with nucleus (N), Golgi apparatus (G), and young polar filament (arrows). Bar: 500 nm. c from Cossins (1973); d from Vey & Vago (1973).

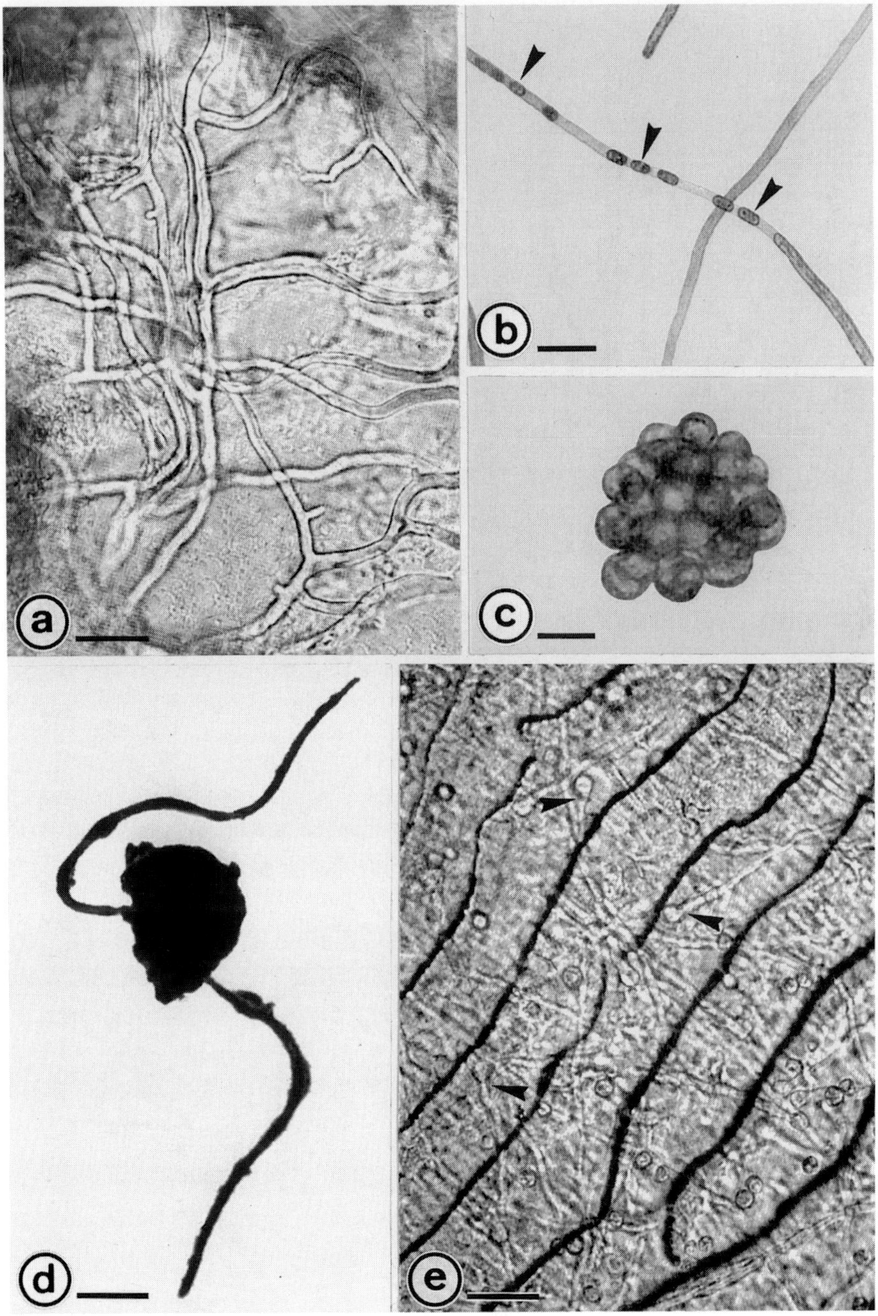

Plate 3. The crayfish plague fungus *Aphanomyces astaci*. a. Mycelium in the cuticle of *Astacus astacus*. Bar: 50 µm. b. Formation of spores (arrowheads) in a hypha. Bar: 50 µm. c. Typical spore cluster at the end of a hypha. Bar: 10 µm. d. Biflagellate zoospore. Bar: 5 µm. e. Outgrowth on fish scale of hyphae from spores (arrowheads) 10 h after inoculation. Bar: 50 µm. a and d from Westman et al. (1992), d originally provided by T. Unestam (Uppsala); b from Halder et al. 1989; c from Diéguez-Uribeondo et al. (1994); and e from Ahne & Halder (1988).

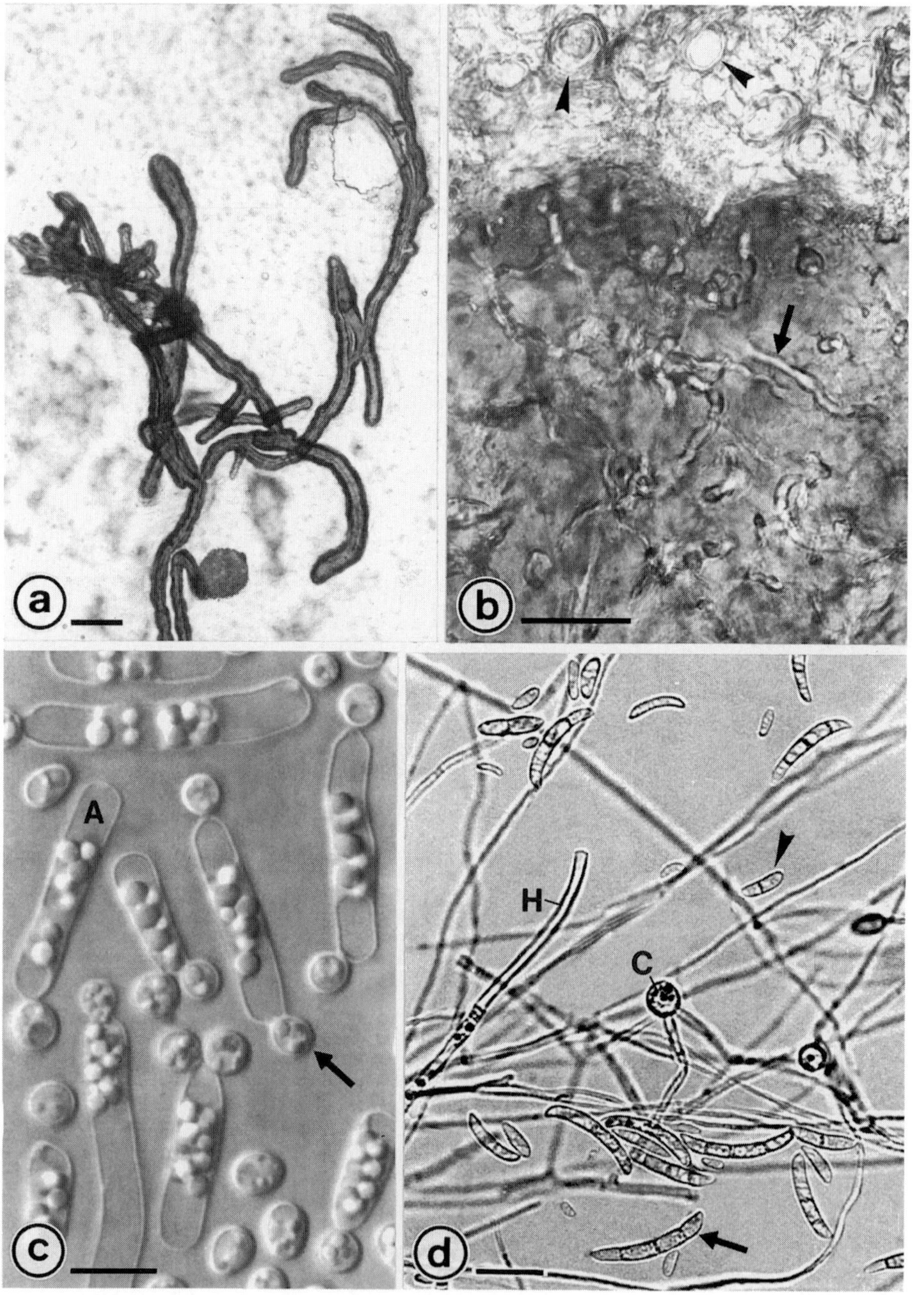

Plate 4. Further fungi from European crayfish. a. Melanized fungus in centre of rosty spot of carapace of *Pacifastacus leniusculus*. Bar: 50 µm. b. Fungal mycelium (arrow) and thick-walled unicellular objects (arrowheads) in surface layer of a burn spot of *Astacus astacus*. Bar: 20 µm. c. Arthroconidia (A) and blastoconidia (arrow) of cultured *Trichosporon beigelii* isolated from *Astacus astacus*. Bar: 10 µm. d. Cultured *Fusarium solani* isolated from brown abdominal patches of *Astacus leptodactylus* with hyphae (H), chlamydospores (C), microconidia (arrowhead), and macroconidia (arrow). Bar: 20 µm. *c from Söderhäll et al. (1993); d from Chinain & Vey (1988a).*

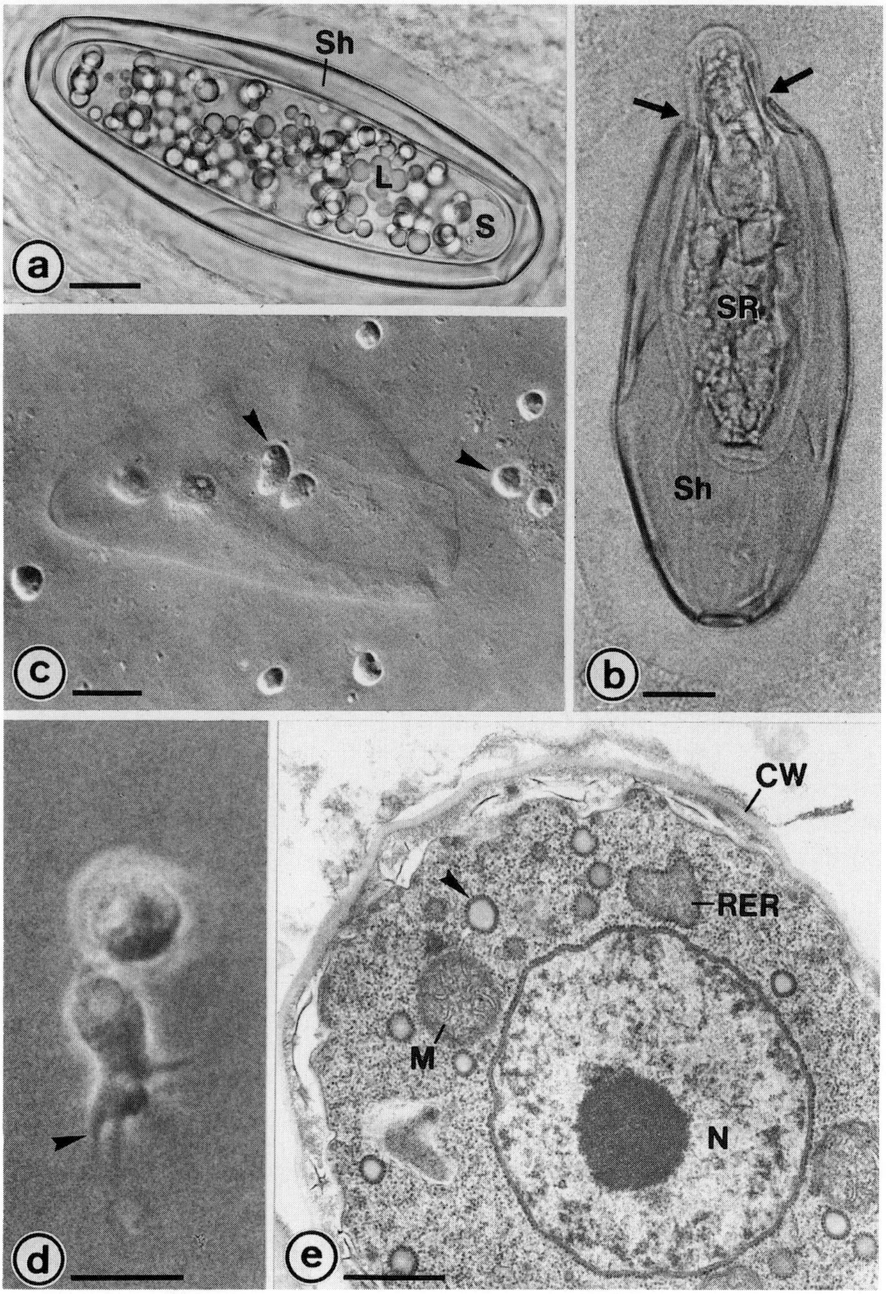

Plate 5. *Psorospermium haeckeli* from *Astacus astacus*. a. Sporocyst in subepidermal connective tissue with thick shell (Sh), lipid globules (L) and spores (S). Bar: 20 µm. b. Hatching (arrows) of spore receptacle (SR) from shell in freshwater. Bar: 20 µm. c. Release of spores (arrowheads) from ruptured spore receptacle. Bar: 20 µm. d. Free motile amoeboids with pseudopodia (arrowhead). Bar: 10 µm. e. Ultrastructure of free encystating amoeboid with nucleus (N), cyst wall (CW), mitochondrium (M), rough endoplasmic reticulum (RER), and granule-studded vesicles (arrowhead). Bar: 1 µm. b-e from Vogt & Rug (1999).

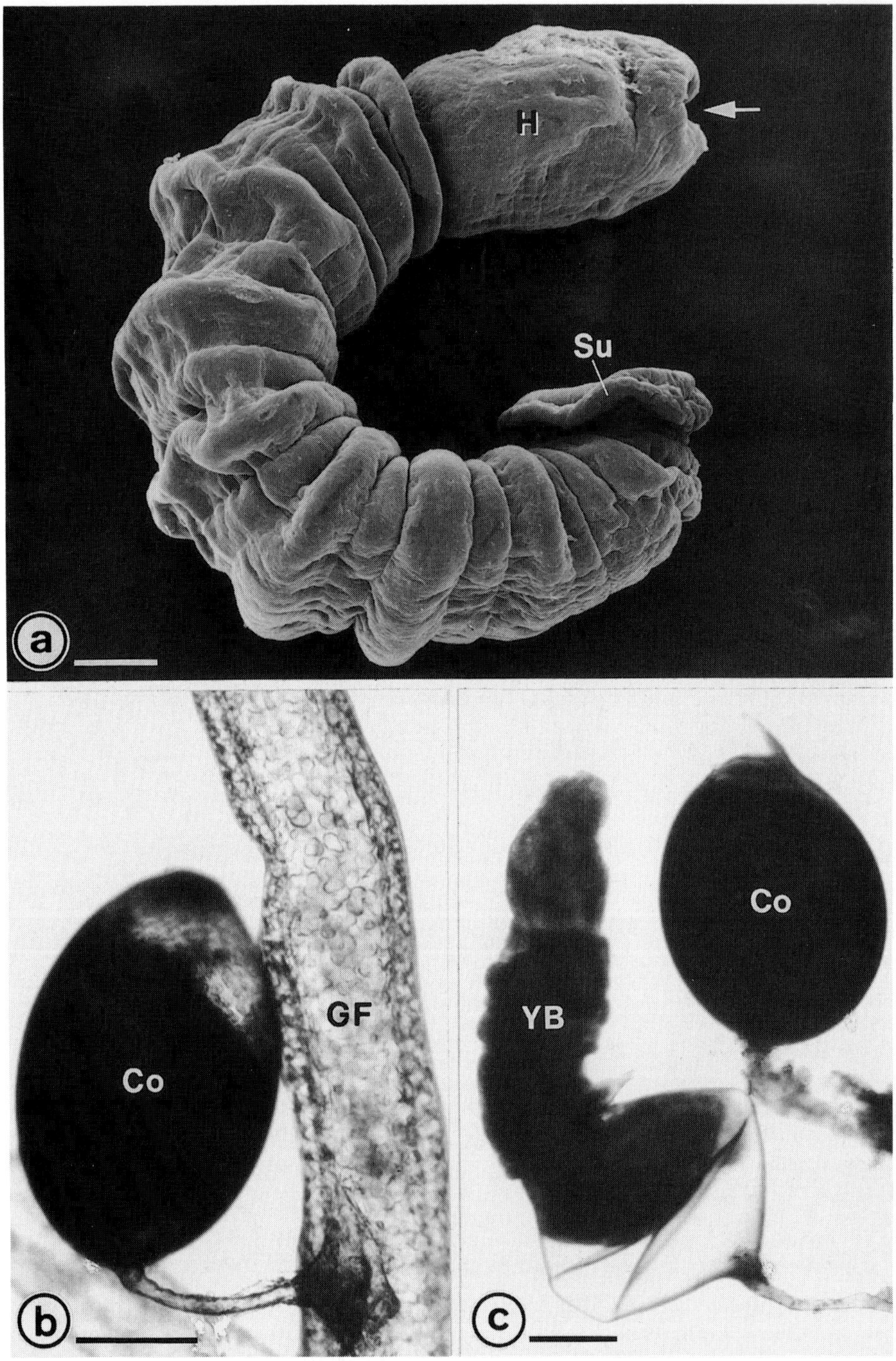

Plate 6. Branchiobdellids from *Astacus astacus*. a. Scanning electron micrograph of adult branchiobdellid displaying head (H) with mouth (arrow), body segments, and posterior sucker (Su). Bar: 100 µm. b. Cocoon (Co) attached to gill filament (GF). Bar: 100 µm. c. Hatching of young branchiobdellid (YB) from cocoon. Bar: 100 µm.

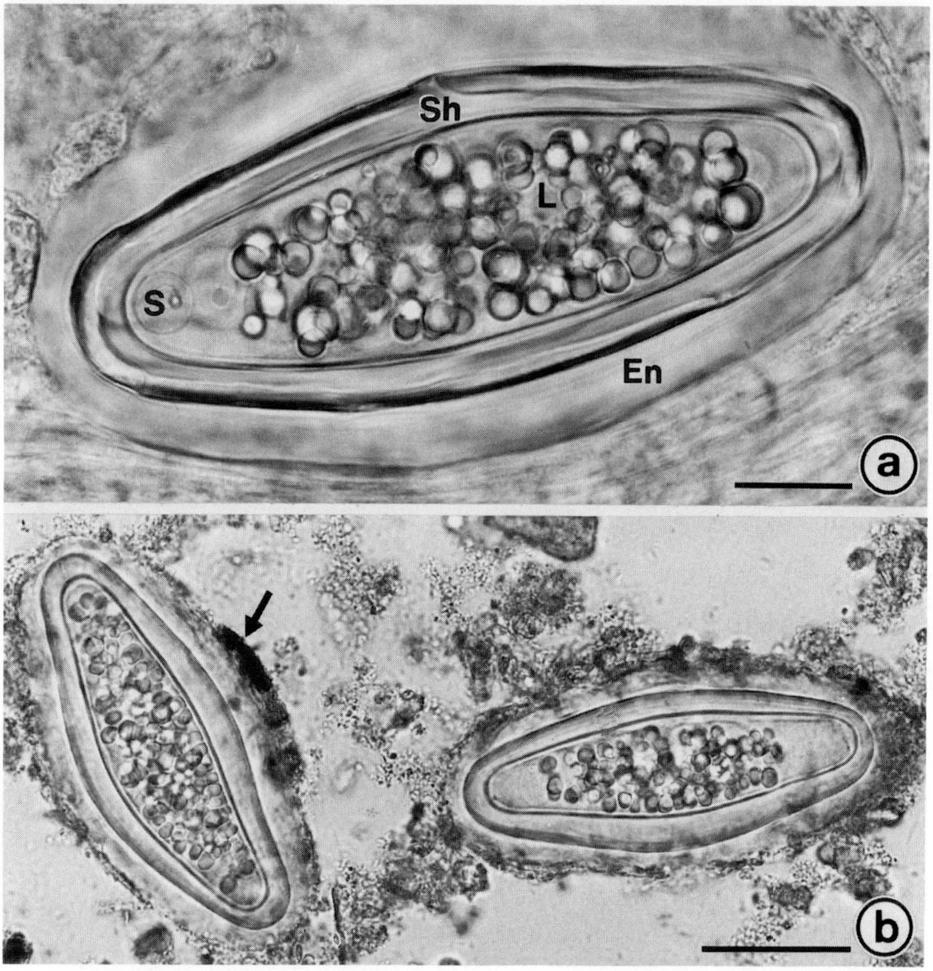

Plate 7. *Psorospermium haeckeli* from sympatric populations of *Astacus astacus* and *Austropotamobius torrentium* in the Kammlach river (Bavaria, Germany). a. Sporocyst in subepidermal connective tissue of noble crayfish. En: envelope of host connective tissue; L: lipid globules; S: spore; Sh: shell. Bar: 20 µm. b. Sporocysts in stone crayfish. Arrow indicates melanized area. Bar: 50 µm. The sporocysts found in both species belong to the elongate morphotype of *Psorospermium haeckeli* which, in that region, is typical of *Astacus astacus*. Since *Austropotamobius torrentium* is normally free of *Psorospermium haeckeli*, the parasite has apparently been transmitted from the noble to the stone crayfish. a and b from Vogt et al.(1996).

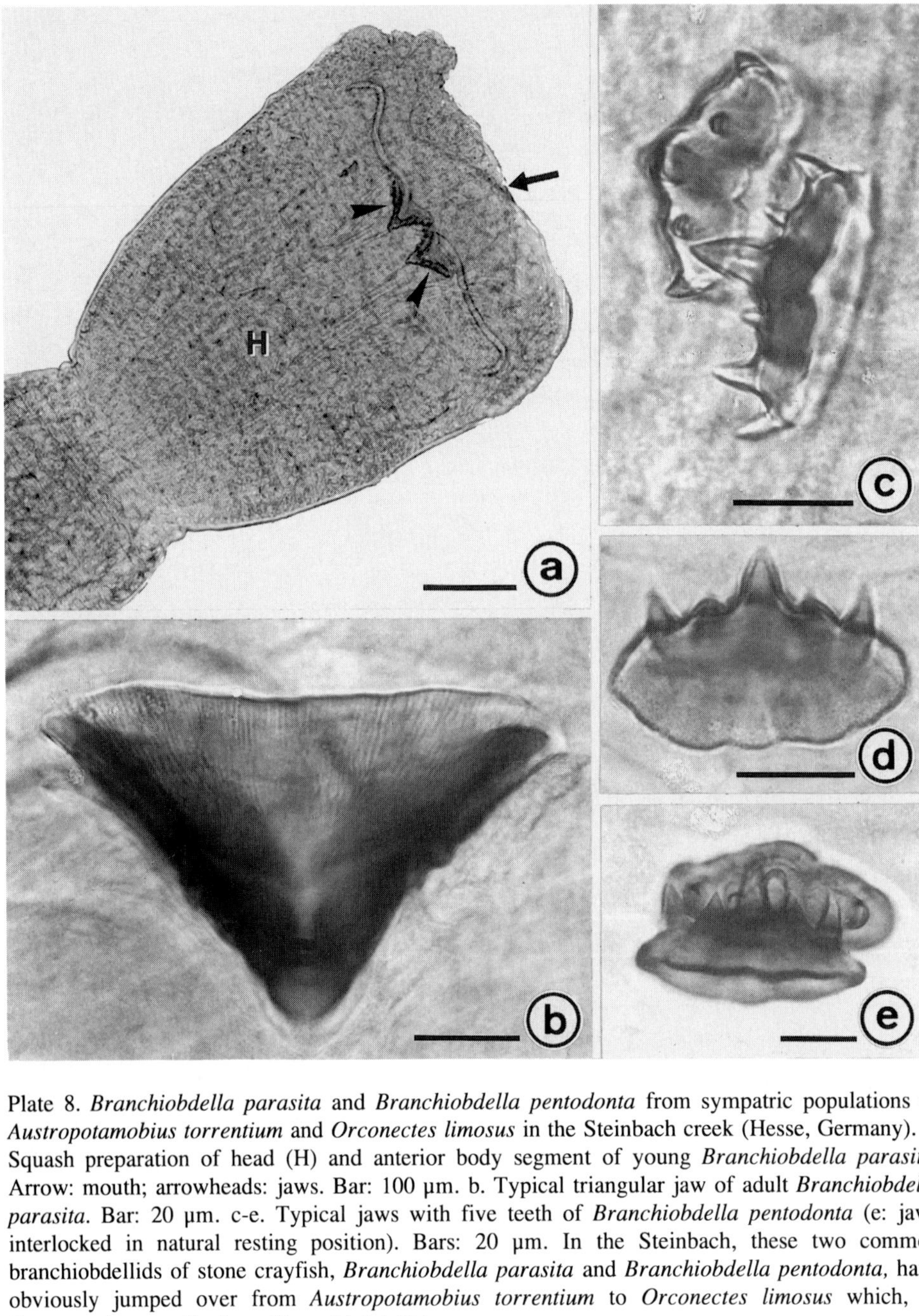

Plate 8. *Branchiobdella parasita* and *Branchiobdella pentodonta* from sympatric populations of *Austropotamobius torrentium* and *Orconectes limosus* in the Steinbach creek (Hesse, Germany). a. Squash preparation of head (H) and anterior body segment of young *Branchiobdella parasita*. Arrow: mouth; arrowheads: jaws. Bar: 100 µm. b. Typical triangular jaw of adult *Branchiobdella parasita*. Bar: 20 µm. c-e. Typical jaws with five teeth of *Branchiobdella pentodonta* (e: jaws interlocked in natural resting position). Bars: 20 µm. In the Steinbach, these two common branchiobdellids of stone crayfish, *Branchiobdella parasita* and *Branchiobdella pentodonta,* have obviously jumped over from *Austropotamobius torrentium* to *Orconectes limosus* which, in Germany, is normally free of branchiobdellids.

a thriving population of *Austropotamobius torrentium,* but a tributary of approximately 200 m in length is populated by both *Austropotamobius torrentium* and *Orconectes limosus* in a proportion of ca 1:1. The spiny-cheek crayfish originates from a small pond located at the upper end of this tributary where they were introduced some years ago.

Examination for branchiobdellids of *Austropotamobius torrentium* from both water courses of the Steinbach revealed plenty of small *Branchiobdella pentodonta* (Pl. 8c-e) on the mouthparts and fewer but much larger *B. parasita* (Pl. 8a, b) on the carapace. Both species can easily be determined under the microscope by the morphology of their jaws (compare Pl. 8b with 8c) (Nesemann 1994). Interestingly, the *Orconectes limosus* investigated here carried the same branchiobdellid species as the stone crayfish, and these branchiobdellids even displayed the same pattern of spatial segregation on the crayfish's body. In contrast, spiny-cheek crayfish sampled from the Rhine river, the Neckar river, and several gravel pits were totally free of branchiobdellids. These findings strongly suggest that both *Branchiobdella pentodonta* and *B. parasita,* which are also very common on stone crayfish of other locations, have been transmitted from native *Austropotamobius torrentium* to naturalized American *Orconectes limosus*.

3.3 *Transmission from American to European species*

For a long time it was hypothesized that the crayfish plague fungus *Aphanomyces astaci* which first appeared in Europe in the 1860s is of North American origin (Unestam 1973a, b, Schäperclaus et al. 1990, Ackefors this volume). This assumption was mainly deduced from the high resistance against the plague of the American species *Orconectes limosus* and *Pacifastacus leniusculus,* which had been introduced into Europe for aquaculture and for restocking of crayfish depleted rivers and lakes (Holdich 1988, Hobbs et al. 1989, Holdich 1993, Ackefors & Lindqvist 1994, Henttonen & Huner this volume). By means of molecular genetics it was recently shown that different genotypes of the crayfish plague fungus have repeatedly been introduced into Europe together with their natural American host species (Huang et al. 1994, Diéguez-Uribeondo et al. 1995, 1997, Oidtmann et al. 1997). Aside from the original crayfish plague fungus that appeared in Europe in the 1860s, a genetically different genotype has been imported together with signal crayfish in the late 1960s when large scale stocking of *Pacifastacus leniusculus* began in Sweden (Fig. 1) (Huang et al. 1994). A third genotype was brought into Sweden together with signal crayfish via Pitt Lake, Canada, in the late 1970s (Fig. 1) (Huang et al. 1994), and a fourth, warm-adapted genotype was introduced into Spain together with the red swamp crayfish *Procambarus clarkii*, first introduced in 1973 (Diéguez-Uribeondo et al. 1995, 1997).

The crayfish plague genotype imported into Sweden together with the signal crayfish *Pacifastacus leniusculus* from Lake Tahoe, USA, has already been transmitted to native *Astacus astacus* in Sweden in the early 1970s. This was demonstrated by analysis of randomly amplified polymorphic DNA from 15 isolates of *Aphanomyces astaci* (Fig. 1) (Huang et al. 1994). Meanwhile, the same genotype has been identified as the causative agent of recent disease outbreaks in *Astacus astacus* in southern Germany (Oidtmann et al. 1997) and in *Austropotamobius pallipes* in Herefordshire,

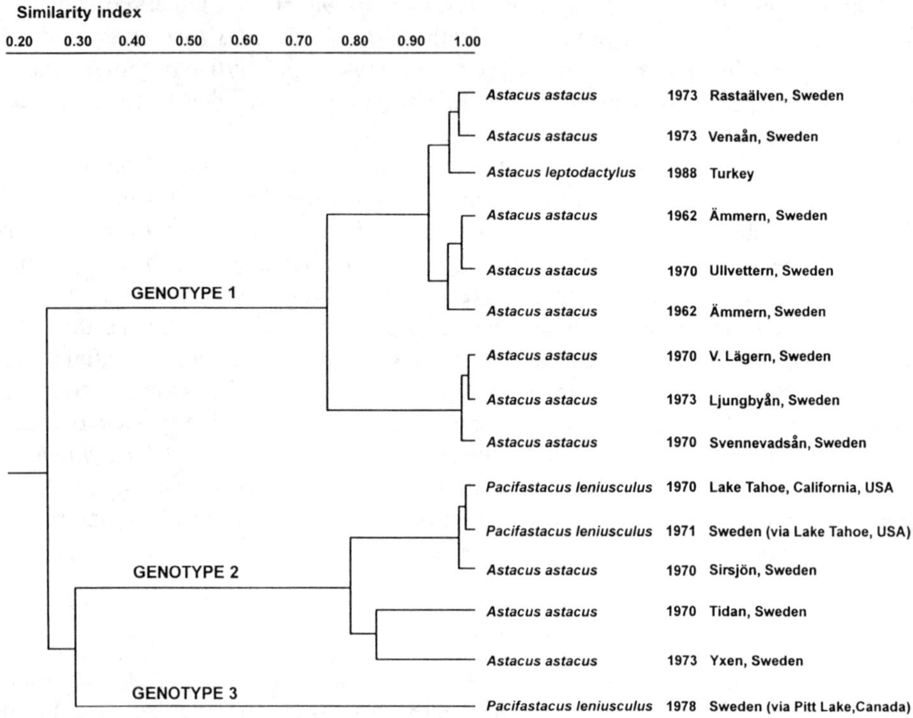

Figure 1. Genetic diversity of 15 isolates of *Aphanomyces astaci* from *Astacus astacus* and *Pacifastacus leniusculus*. The dendrogram shows that there are now three different genotypes of the crayfish plague fungus in Sweden. The most frequent genotype (genotype 1) is obviously the old type introduced with the first plague outbreak. The two other genotypes can clearly be traced back to imports of *Pacifastacus leniusculus* via Lake Tahoe (USA) (genotype 2) and Pitt Lake (Canada) (genotype 3). The Lake Tahoe genotype 2 has already been transmitted from *Pacifastacus leniusculus* to at least three populations of *Astacus astacus*. From Huang et al. (1994), redrawn.

England (Lilley et al. 1997). These findings demonstrate that at least one genotype of *Aphanomyces astaci,* which has been brought into Europe with deliberate introductions of an American crayfish species, has been transmitted to indigenous species. Moreover, it has already been disseminated in Europe. The other two genotypes of the crayfish plague fungus, isolated from *Pacifastacus leniusculus* imported via Pitt Lake and from *Procambarus clarkii,* have not yet been found in native European species.

4 DISCUSSION

Despite the many reports published in the last four decades on diseases of freshwater crayfish, our understanding of crayfish pathology is still very incomplete. The pathogens responsible for the most severe disease outbreaks are thought to have been largely identified. However, we have only very little information on virus diseases, and bacterial diseases are only known from culture facilities, not from the wild. And even for several of the better studied crayfish pathogens we neither know the complete life cycle nor the complete spectrum of infection and transmission mechanisms in nature. Since in cultured and wild shrimp the most disastrous diseases are caused by viruses and rickettsiae (Lightner 1996, Vogt & Štrus 1998), one may assume that a certain percentage of the mass mortality of crayfish with unconfirmed aetiology, which have been attributed to the crayfish plague in the past, may have been caused by viral and rickettsial agents. A further gap of knowledge concerns the geographical distribution of the diseases that last have been compiled by Alderman & Polglase (1988) and Westman et al. (1990). The tables presented in these papers clearly indicate that relatively good databases are only available for countries that have at least one crayfish pathology expert group, and that the geographical range of a disease is only well recorded, if its typical symptoms are detectable with the naked eye. The distribution pattern of pathogens such as viruses and bacteria, which require more sophisticated technical equipment for their detection, is only fragmentarily documented.

In aquaculture, diseases can be largely avoided by using specimens from disease-free hatcheries, quarantine prior to stocking (e.g. crayfish plague), careful examination of the stocking material (e.g. burn spot disease), repeated removal of diseased animals (e.g. thelohaniasis), or good feeding and water management (e.g. opportunistic pathogens). Already infected individuals may be treated with antibiotics, fungicides and other chemicals that are commercially available for aquaculture purposes (Alderman & Polglase 1984, Westman et al. 1992, Alderman et al. 1994). In future, new techniques for the diagnosis and control of crayfish diseases are also expected from the field of crustacean immunology which has made great progress in the last decade (Söderhäll & Cerenius 1992, Chisholm & Smith 1995, Söderhäll et al. 1996). In the wild, disease prevention and control is much more difficult and appears impossible with respect to crayfish plague. Establishment of genetic reservoirs in remote river sections and ponds with pathogen-free water obtained from springs and deep wells is one of the few possible measures to conserve the most endangered species. In Germany, most of the remaining natural habitats of indigenous species are located in headwaters of river catchments, which are separated from the main water course by a series of smaller dams. In these particular streams, the establishment of fish ladders should be avoided in order to prevent natural upstream spreading of crayfish diseases, particularly the plague.

It has often been suspected that crayfish pathogens are not species specific and may easily be transferred from one crayfish species to another. Respective laboratory experiments strongly corroborate this hypothesis. For instance, crayfish plague was successfully transmitted from *Pacifastacus leniusculus* and *Orconectes limosus* to *Austropotamobius pallipes* and *Astacus astacus* (Alderman 1993, Keller pers. comm.), Branchiobdellidae from *Austropotamobius torrentium* to *Pacifastacus leniusculus* (author's results), *Psorospermium haeckeli* from *Pacifastacus leniusculus* to *Astacus astacus* and vice versa (Gydemo 1996), and temnocephalids from imported Australian *Cherax destructor* to *Austropotamobius pallipes*, *Orconectes limosus* and *Procambarus clarkii* (Quaglio et al. 1997). Demonstration of transmission of pathogens in the wild is more difficult. This topic can be studied by examination of sympatric natural populations or by genetic investigation of pathogens isolated from disease outbreaks. The results obtained so far indicate transmission of *Psorospermium haeckeli* from native *Astacus astacus* to native *Austropotamobius torrentium* (Vogt et al. 1996), of *Branchiobdella pentodonta* and *B. parasita* from native *Austropotamobius torrentium* to American *Orconectes limosus* (author's results), of *Aphanomyces astaci* from American *Pacifastacus leniusculus* to native *Astacus astacus* (Huang et al. 1994) and *Austropotamobius pallipes* (Lilley et al. 1997), and of an unidentified ostracode from *Pacifastacus leniusculus* to *Astacus astacus* (Henttonen et al. 1994b). The branchiobdellids *Xironogiton instabilius* and *Cambarincola mesochoreus* have been introduced into Europe together with *Pacifastacus leniusculus* (Westman et al. 1992) and *Procambarus clarkii* (Gelder et al. 1994), respectively, but have not yet been found in native species.

These examples from the laboratory and the field clearly show that most of the common crayfish pathogens are not species specific. Therefore, the risk of transmission of an alien crayfish disease to our indigenous species is very high once it has entered European waters. The pathogens in question are not only new strains of crayfish plague but also new viruses, bacteria, further species of *Thelohania*, branchiobdellids and temnocephalids. Such potential pathogens that could be brought into Europe with live imports of crayfish are: *Thelohania cambari* and other *Thelohania* species found in American Cambaridae (Sprague & Couch 1971), baculoviruses detected in California in native *Pacifastacus leniusculus* and cultured *Cherax quadricarinatus* imported from Australia (Hedrick et al. 1995), *Cherax quadricarinatus* bacilliform virus observed in Australia in native *Cherax quadricarinatus* (Anderson & Prior 1992, Edgerton & Owens 1997) and also in North America in imported *Cherax quadricarinatus* (Groff et al. 1993), systemic parvo-like virus (Edgerton et al. 1997) and a new bacilliform virus found in *Cherax destructor* in Australia (Edgerton 1996a), *Cherax* Giardiavirus-like virus detected in *Cherax quadricarinatus* in Australia (Edgerton & Owens 1997), and temnocephalids in *Cherax destructor* brought to Italy (Quaglio et al. 1997). *Cherax*-species are already present in several European countries in research laboratories and farms, and in Spain there is even a wild population established in one location (Gutiérrez-Yurrita 1997). These Australian crayfish may carry one or the other of the above listed viruses and may therefore be a new potential threat for our native species.

5 CONCLUSIONS AND RECOMMENDATIONS

1. The greatest gaps of knowledge in crayfish pathology concern viral and rickettsial diseases and the geographical distribution of crayfish pathogens. Much more information is also needed on the life cycle and mode of transmission of some common disease agents such as *Thelohania contejeani*. These gaps should be closed by intensified research on those topics, systematic field surveys, and a central disease register as suggested by Nylund & Westman (1995c).

2. In future, research on crayfish immunology and genetics is expected to produce new measures of disease diagnosis and control (Söderhäll et al. 1996). A goal of prime importance should be breeding of plague resistant crayfish.

3. Transplantation of diseases by introduction of crayfish into a new geographical region is a very high risk for the indigenous populations, as exemplified by crayfish plague. In order to suspend this dangerous 'jumbo jet vector of disease' (Lightner

1990), transplantations of any crayfish outside of its native range should be avoided (see Holdich this volume).

ACKNOWLEDGEMENTS

I wish to express my thanks to Winfried Ahne (Munich, Germany), Andrew Cossins (Liverpool, England), Brett Edgerton (Townsville, Australia), Viljo Nylund (Helsinki, Finland), Kenneth Söderhäll (Uppsala, Sweden), Alain Vey (St-Christol-les-Ales, France), Academic Press (San Diego, USA), Bayerische Landesanstalt für Wasserforschung (Wielenbach, Germany), Elsevier Science (Amsterdam, Netherlands), Inter Research (Amelinghausen, Germany), Kiviksgårdens Förlag (Kivik, Sweden), and Terra Verlag (Konstanz, Germany) for permission to reproduce published material as indicated in the Plate and Figure legends. I am particularly grateful to Gisela Adam (Heidelberg) for reproduction of the photographs, and to David Holdich (Nottingham, England), Kenneth Söderhäll (Uppsala, Sweden), and Paula Henttonen (Kuopio, Finland) for valuable comments on the manuscript.

REFERENCES

Ackefors, H. & Lindqvist, O. 1994. Cultivation of freshwater crayfishes in Europe. In J.V. Huner (ed.), *Freshwater crayfish aquaculture in North America, Europe and Australia. Families Astacidae, Cambaridae, and Parastacidae*: pp. 157-216. Binghamton: Food Products Press.

Ahne, W. & Halder, M. 1988. *Virologische und mykologische Untersuchungen an Krebsbeständen verschiedener Herkunft.* München: Bayerische Landesanstalt für Wasserforschung, 147 pp.

Alderman, D.J. 1993. Crayfish plague in Britain, the first twelve years. *Freshwater Crayfish* 9:266-272.

Alderman, D.J. & Polglase, J.L. 1984. A comparative investigation of the effects of fungicides on *Saprolegnia parasitica* and *Aphanomyces astaci*. *Trans. Br. Mycol. Soc.* 83:313-318.

Alderman, D.J. & Polglase, J.L. 1986. *Aphanomyces astaci*, isolation and culture. *J. Fish Dis.* 9:367-379.

Alderman, D.J. & Polglase, J.L. 1988. Pathogens, parasites and commensals. In D.M. Holdich & R.S. Lowery (eds), *Freshwater crayfish: biology, management and exploitation*: pp. 167-212. London & Sydney: Croom Helm/Portland, Oregon:Timber Press.

Alderman, D.J., Holdich, D.M. & Reeve, I.D. 1990. Signal crayfish as vector in crayfish plague in Britain. *Aquaculture* 86:3-6.

Alderman, D.J., Polglase, J.L. & Frayling, M. 1987. *Aphanomyces astaci* pathogenicity under laboratory and field conditions. *J. Fish Dis.* 10:385-393.

Alderman, D.J., Rosenthal, H., Smith, P., Stewart, J. & Weston, D. 1994. Chemicals used in mariculture. *ICES Cooperative Research Report* 202:1-97.

Anderson, I.G. & Prior, H.C. 1992. Baculovirus infections in the mud crab, *Scylla serrata*, and a freshwater crayfish, *Cherax quadricarinatus*, from Australia. *J. Invertebr. Pathol.* 60:265-273.

Bucinskiene, R. 1996. Occurrence and relative abundance of *Psorospermium haeckeli* in Lithuanian crayfish. *Freshwater Crayfish* 11:494-499.

Cerenius, L. & Söderhäll, K. 1984. Chemotaxis in *Aphanomyces astaci*, an arthropod-parasitic fungus. *J. Invertebr. Pathol.* 43:278-281.

Cerenius, L. & Söderhäll, K. 1993. The distribution of *Psorospermium haeckeli* in Sweden: a preliminary survey. *Freshwater Crayfish* 9:280-285.

Cerenius, L., Söderhäll, K., Persson, M. & Ajaxon, R. 1988. The crayfish plague fungus *Aphanomyces astaci* – diagnosis, isolation, and pathobiology. *Freshwater Crayfish* 7:131-144.

Cerenius, L., Henttonen, P., Lindqvist, O.V. & Söderhäll, K. 1991. The crayfish pathogen *Psorospermium haeckeli* activates the prophenoloxidase activating system of freshwater crayfish in vitro. *Aquaculture* 99:225-233.

Chinain, M. & Vey, A. 1988a. Comparative studies on strains of *Fusarium solani* parasitic for crayfish or isolated from other hosts. *Freshwater Crayfish* 7:179-186.

Chinain, M. & Vey, A. 1988b. Experimental study of *Fusarium solani*: infections in *Astacus leptodactylus* and *Pacifastacus leniusculus* (Crustacea, Decapoda). *Dis. Aquat. Org.* 5:215-223.

Chinain, M. & Vey, A. 1988c. Infection caused by *Fusarium solani* in crayfish *Astacus leptodactylus*. *Freshwater Crayfish* 7:195-202.

Chisholm, J.R.S. & Smith, V.J. 1995. Comparison of antibacterial activity in the hemocytes of different crustacean species. *Comp. Biochem. Physiol.* 110A:39-45.

Cossins, A.R. 1973. *Thelohania contejeani* Henneguy, microsporidian parasite of *Austropotamobius pallipes* Lereboullet – a histological and ultrastructural study. *Freshwater Crayfish* 1:151-164.

Cossins, A.R. & Bowler, K. 1974. A histological and ultrastructural study of *Thelohania contejeani* Henneguy, 1892 (Nosematidae), microsporidian parasite of the crayfish *Austropotamobius pallipes* Lereboullet. *Parasitology* 68:81-91.

Diéguez-Uribeondo, J. 1995. Adaptation to parasitism of some animal pathogenic Saprolegniaceae. *Acta Univ. Upsaliensis* 122:1-43.

Diéguez-Uribeondo, J., Cerenius, L. & Söderhäll, K. 1994. *Saprolegnia parasitica* and its virulence on three different species of freshwater crayfish. *Aquaculture* 120:219-228.

Diéguez-Uribeondo, J., Huang, T., Cerenius, L. & Söderhäll, K. 1995. Physiological adaptation of an *Aphanomyces astaci* strain isolated from the freshwater crayfish *Procambarus clarkii*. *Mycol. Res.* 99:574-578.

Diéguez-Uribeondo, J., Huang, T.-S., Cerenius, L. & Söderhäll, K. 1997. The swamp crayfish *Procambarus clarkii* as carrier of an *Aphanomyces astaci* strain physiologically adapted to warm temperatures. Abstract Volume of the International Workshop 'The Introduction of Alien Species of Crayfish in Europe: how to make the best of a bad situation?, pg. 46. Florence, September 24-27, 1997.

Diéguez-Uribeondo, J., Pinedo-Ruiz, J., Cerenius, L. & Söderhäll, K. 1993. Presence of *Psorospermium haeckeli* (Hilgendorf) in a *Pacifastacus leniusculus* (Dana) population of Spain. *Freshwater Crayfish* 9:286-288.

Edgerton, B. 1996a. A new bacilliform virus in Australian *Cherax destructor* (Decapoda, Parastacidae) with notes on *Cherax quadricarinatus* bacilliform virus (equals *Cherax* baculovirus). *Dis. Aquat. Org.* 27:43-52.

Edgerton, B. 1996b. *Viruses of freshwater crayfish.* Doctoral Thesis. Townsville: Department of Biomedical and Tropical Veterinary Sciences, James Cook University of North Queensland, 152 pp.

Edgerton, B. & Owens, L. 1997. Age at first infection of *Cherax quadricarinatus* by *Cherax quadricarinatus* bacilliform virus and *Cherax* Giardiavirus-like virus, and production of putative virus-free crayfish. *Aquaculture* 152:1-12.

Edgerton, B., Webb, R. & Wingfield, M. 1997. A systemic parvo-like virus in the freshwater crayfish *Cherax destructor*. *Dis. Aquat. Org.* 29:73-78.

Edgerton, B., Paasonen, P., Henttonen, P. & Owens, L. 1996. Description of a bacilliform virus from the freshwater crayfish, *Astacus astacus*. *J. Invertebr. Pathol.* 68:187-190.

Fischer, H. 1992. *Untersuchungen zur Porzellankrankheit des Edelkrebses (Astacus astacus L.) verursacht durch Thélohania contejeani H.* Dissertation. München: Tierärztliche Fakultät, Universität München, 107 pp.

Fürst, M. 1995. On the recovery of *Astacus astacus* L. populations after an epizootic of the crayfish plague (*Aphanomyces astaci* Shikora). *Freshwater Crayfish* 8:565-576.

Gelder, S.R. 1996. A review of the taxonomic nomenclature and a checklist of the species of the Branchiobdellae (Annelida, Clitellata). *Proc. Biol. Soc. Washington* 109:653-663.

Gelder, S.R., Delmastro, G.B. & Ferraguti, M. 1994. A report on Branchiobdellidans (Annelida, Clitellata) and a taxonomic key to the species in Northern Italy, including the first record of

Cambarincola mesochoreus on the introduced American red swamp crayfish. *Boll. Zool.* 61:179-184.

Grabda, E. 1934. Recherches sur un parasite de l'écrevisse (*Potamobius fluviatilis* L.) connu sous le nom de *Psorospermium haeckeli* Hlgd. *Mém. Acad. Pol. Sci. Lett., Cracovie (Sér. B)* 6:123-142.

Grabda, E. & Wierzbicka, J. 1969. The problem of parasitism of the genus *Branchiobdella* Odier, 1823. *Pol. Arch. Hydrobiol.* 16:93-104.

Groff, J.M., McDowell, T., Friedman, C.S. & Hedrick, R.P. 1993. Detection of a nonoccluded baculovirus in the freshwater crayfish *Cherax quadricarinatus* in North America. *J. Aquat. Animal Health* 5:275-279.

Gutiérrez-Yurrita, P.J. 1997. The status of crayfish populations in Spain and Portugal. Abstract Volume of the International Workshop 'The Introduction of Alien Species of Crayfish in Europe: how to make the best of a bad situation?, pp. 24-27. Florence, September 24-27, 1997.

Gydemo, R. 1996. Signal crayfish, *Pacifastacus leniusculus*, as a vector for *Psorospermium haeckeli* to noble crayfish, *Astacus astacus*. *Aquaculture* 148:1-9.

Halder, M. & Ahne, W. 1988a. *Astacus astacus* L. identified as IPNV-vector. *Freshwater Crayfish* 7:303-308.

Halder, M. & Ahne, W. 1988b. Freshwater crayfish *Astacus astacus* – a vector for infectious pancreatic necrosis virus (IPNV). *Dis. Aquat. Org.* 4:205-209.

Halder, M., Ahne, W. & Thomsen, I. 1989. Die Krebspest: Diagnose und Isolierung des Erregers *Aphanomyces astaci*. *Tierärztliche Umschau* 44:79-82.

Häll, L. & Unestam, T. 1980. The effects of fungicides on survival of the crayfish plague fungus *Aphanomyces astaci* oomycetes growing on fish scales. *Mycopathol.* 72:131-134.

Hedrick, R.P., McDowell, T.S. & Friedman, C.S. 1995. Baculoviruses found in two species of crayfish from California. Abstract Volume of 'World Aquaculture '95', pg. 135. San Diego, USA, February 1-4, 1995.

Henttonen, P. 1996. The parasite *Psorospermium* in freshwater crayfish. *Kuopio Univ. Publ. C, Nat. Environm. Sci.* 48:1-78.

Henttonen, P., Huner, J.V. & Lindqvist, O. 1994a. Occurrence of *Psorospermium* sp. in several North American crayfish species, with comparative notes on *Psorospermium haeckeli* in the European crayfish, *Astacus astacus*. *Aquaculture* 120:209-218.

Henttonen, P., Huner, J.V. & Lindqvist, O.V. 1994b. Possible movement of epibionts, parasites and diseases with crayfish introductions. Abstract Volume of 'World Aquaculture '94', pg. 244. New Orleans, USA, January 14-18, 1994.

Henttonen, P., Huner, J.V. & Lindqvist, O.V. 1995. Observations on *Psorospermium haeckeli* in noble crayfish *Astacus astacus* (L.) (Decapoda, Astacidae), populations in Central Finland. *Freshwater Crayfish* 10:339-351.

Henttonen, P., Huner, J.V., Rata, P. & Lindqvist, O.V. 1997. A comparison of the known life forms of *Psorospermium* spp. in freshwater crayfish (Arthropoda, Decapoda) with emphasis on *Astacus astacus* L. (Astacidae) and *Procambarus clarkii* (Girard) (Cambaridae). *Aquaculture* 149:15-30.

Hobbs III, H.H., Jass, J.P. & Huner, J.V. 1989. A review of global crayfish introductions with particular emphasis on two North American species (Decapoda, Cambaridae). *Crustaceana* 56:299-316.

Hofer, B. 1904. *Handbuch der Fischkrankheiten*. München: Verlag der Allgemeinen Fischerei-Zeitung, 359 pp.

Hofmann, J. 1980. *Die Flußkrebse*, 2. Aufl. Hamburg: Parey, 110 pp.

Holdich, D.M. 1988. The dangers of introducing alien animals with particular reference to crayfish. *Freshwater Crayfish* 7:XV-XXX.

Holdich, D.M. 1993. A review of astaciculture: freshwater crayfish farming. *Aquat. Living Resour.* 6:307-317.

Holdich, D.M., Reeve, I.D. & Rogers, W.D. 1995. Introduction and spread of alien crayfish in British waters – implications for native crayfish populations. *Freshwater Crayfish* 8:99-112.

Huang, T.S., Cerenius, L. & Söderhäll, K. 1994. Analysis of genetic diversity in the crayfish plague fungus, *Aphanomyces astaci*, by random amplification of polymorphic DNA. *Aquaculture*

126:1-10.

Lightner, D.V. 1990. Viroses section: introductory remarks. In F.O. Perkins & T.C. Cheng (eds), *Pathology in marine science*: pp. 3-6. San Diego: Academic Press.

Lightner, D.V. 1996. *A handbook of shrimp pathology and diagnostic procedures for diseases of cultured penaeid shrimp*. Baton Rouge: World Aquaculture Society.

Lilley, J.H., Cerenius, L. & Söderhäll, K. 1997. RAPD evidence for the origin of crayfish plague outbreaks in Britain. *Aquaculture* 157:181-185.

Maestracci, V. & Vey, A. 1988. Fungal infection of gills in crayfish: histological, cytological and physiopathological aspects of the disease. *Freshwater Crayfish* 7:187-194.

Mann, H. & Pieplow, U. 1938. Die Brandfleckenkrankheit bei Krebsen und ihre Erreger. *Zeitschrift für Fischerei* 36:225-240.

Matthews, M.A. & Reynolds, J.D. 1991. Ecological impact of crayfish plague in Ireland. *Hydrobiologia* 204:1-6.

Moon, S.Y., Kim, C.B., Gelder, S.R. & Kim, W. 1996. Phylogenetic positions of the aberrant Branchiobdellidans and Aphanoneurans within the Annelida as derived from 18S ribosomal RNA gene sequences. *Hydrobiologia* 324:229-236.

Nesemann, H. 1994. Die Krebsegel im Gebiet der Oberen Donau (Österreich, Deutschland) mit Bestimmungsschlüssel zu den europäischen Arten (Clitellata, Branchiobdellida). *Lauterbornia* 19:79-93.

Nylund, V. & Westman, K. 1995a. Frequency of visible symptoms of the crayfish plague fungus (*Aphanomyces astaci*) on the signal crayfish (*Pacifastacus leniusculus*) in natural populations in Finland in 1979-1988. *Freshwater Crayfish* 8:577-588.

Nylund, V. & Westman, K. 1995b. On the chemical structure and taxonomic position of the crayfish parasite *Psorospermium haeckeli* Hilgendorf. *Freshwater Crayfish* 10:352-362.

Nylund, V. & Westman, K. 1995c. The crayfish mortality register as an aid in the control of crayfish diseases in Finland. *Freshwater Crayfish* 10:363-373.

Nylund, V., Kirjavainen, J., Tulonen, J. & Westman, K. 1993. The spread of crayfish plague (*Aphanomyces astaci*) and its effects on the noble crayfish (*Astacus astacus*) population in the Lake Ormajärvi waterway in Finland in 1988-1991. *Freshwater Crayfish* 9:273-279.

Oidtmann, B., Cerenius, L., Schmid, I., Hoffmann, R. & Söderhäll, K. 1997. Classification of two German isolates of the crayfish plague fungus, *Aphanomyces astaci*, by random amplification of polymorphic DNA. Abstract Volume of the International Workshop 'The Introduction of Alien Species of Crayfish in Europe: how to make the best of a bad situation?, pg. 40. Florence, September 24-27, 1997.

Oidtmann, B., El-Matbouli, M., Fischer, H., Hoffmann, R., Klärding, K., Schmid, I. & Schmidt, R. 1996. Light microscopy of *Astacus astacus* L. under normal and selected pathological conditions, with special emphasis to porcelain disease and crayfish plague. *Freshwater Crayfish* 11:465-480.

Quaglio, F., Nobile, L. & Mazzoni, D. 1997. Temnocephalans (Platyhelminthes) ectosymbionts in Australian freshwater crayfish (*Cherax destructor* Clark) introduced to Northern Italy. Abstract Volume of the International Workshop 'The Introduction of Alien Species of Crayfish in Europe: how to make the best of a bad situation?, pp. 41-42. Florence, September 24-27, 1997.

Ragan, M.A., Goggin, C.L., Cawthorn, R.J., Cerenius, L., Jamieson, A.V.C., Plourde, S.M., Rand, T.G., Söderhäll, K. & Gutell, R.R. 1996. A novel clade of protistan parasites near the animal-fungal divergence. *Proc. Natl. Acad. Sci. USA* 93:11907-11912.

Rahe, R. & Soylu, E. 1989. Identification of the pathogenic fungus causing destruction to Turkish crayfish stocks (*Astacus leptodactylus*). *J. Invertebr. Pathol.* 54:10-15.

Rug, M. & Vogt, G. 1994. Developmental stages of the crayfish parasite *Psorospermium haeckeli* in thoracic arteries of *Astacus astacus*. *J. Invertebr. Pathol.* 64:153-155.

Rug, M. & Vogt, G. 1995. Histology and histochemistry of developing and mature spores of two morphotypes of *'Psorospermium haeckeli'*. *Freshwater Crayfish* 10:374-384.

Schäperclaus, W., Kulow, H. & Schreckenbach, K. 1990. *Fischkrankheiten*, 5. Aufl. Berlin: Akademie-Verlag, 1123 pp.

Skurdal, J., Qvenild, T., Taugbøl, T. & Fjeld, E. 1990. A 6-year study of *Thelohania contejeani* parasitism of the noble crayfish, *Astacus astacus* L., in Lake Steinsfjorden, S.E. Norway. *J. Fish Dis.* 13:411-415.

Smith, V. & Söderhäll, K. 1986. Crayfish pathology: an overview. *Freshwater Crayfish* 6:199-211.

Söderhäll, K. 1988. Fungal parasites and other diseases on freshwater crayfish. In: J. Kovanen & R. Lappalainen (eds), *Raputalous 2000*: pp. 23-46. Jyväskylä (Finland): Keski-Suomen kalastuspiirin kalastustoimisto. Tiedotus No. 5.

Söderhäll, K. & Cerenius, L. 1992. Crustacean immunity. *Ann. Rev. Fish Dis.* 2:3-23.

Söderhäll, K., Cerenius, L. & Johansson, M.W. 1996. The prophenoloxidase activating system in invertebrates. In K. Söderhäll, S. Iwanaga & G.R. Vasta (eds), *New directions in invertebrate immunology*: pp. 229-253. Fair Haven: SOS Publications.

Söderhäll, K., Rantamäki, J. & Constantinescu, O. 1993. Isolation of *Trichosporon beigelii* from the freshwater crayfish *Astacus astacus*. *Aquaculture* 116:25-31.

Sprague, V. & Couch, J. 1971. An annotated list of protozoan parasites, hyperparasites and commensals of decapod crustacea. *J. Protozool.* 18:526-537.

Taugbøl, T. & Skurdal, J. 1995. Occurrence and relative abundance of *Psorospermium haeckeli* in Norwegian noble crayfish *Astacus astacus* populations. *Freshwater Crayfish* 10:388-395.

Thörnqvist, P.-O. & Söderhäll, K. 1993. *Psorospermium haeckeli* and its interaction with the crayfish defence system. *Aquaculture* 117:205-213.

Unestam, T. 1973a. Fungal diseases of Crustacea. *Rev. Med. Mycol.* 8:1-20.

Unestam, T. 1973b. Significance of diseases on freshwater crayfish. *Freshwater Crayfish* 1:135-150.

Unestam, T. 1981. Fungal diseases of freshwater and terrestrial crustacea. In E.W. Davidson (ed.), *Pathogenesis of invertebrate microbial diseases*: pp. 485-507. Totowa: Allanheld, Osmun.

Vey, A. 1976. Mycoses des Crustacés. *Oceanis* 3:209-218.

Vey, A. 1978. Infections fongiques chez l'écrevisse *Astacus leptodactylus* Esch. *Freshwater Crayfish* 4:403-410.

Vey, A. 1986. Disease problems during aquaculture of freshwater Crustacea. *Freshwater Crayfish* 6:212-222.

Vey, A. & Vago, C. 1973. Protozoan and fungal diseases of *Austropotamobius pallipes* Lereboullet in France. *Freshwater Crayfish* 1:165-179.

Vogt, G. & Rug, M. 1995. Microscopic anatomy and histochemistry of the crayfish parasite *Psorospermium haeckeli*. *Dis. Aquat. Org.* 21:79-90.

Vogt, G. & Rug, M. 1996. Granulomatous hepatopancreatitis: immune response of the crayfish *Astacus astacus* to a bacterial infection. *Freshwater Crayfish* 11:451-464.

Vogt, G. & Rug, M. (1999). Life stages and life cycle of *Psorospermium haeckeli*, a species of the novel DRIPs clade from the animal-fungal dichotomy. *J. Exp. Zool.* 283:31-42.

Vogt, G. & Štrus, J. (1999). Diseases of the shrimp *Palaemon elegans* (Crustacea, Decapoda) in the Bay of Piran, Adriatic Sea. *J. Nat. Hist.* 32:1795-1806.

Vogt, G., Keller, M. & Brandis, D. 1996. Occurrence of *Psorospermium haeckeli* in the stone crayfish *Austropotamobius torrentium* from a population naturally mixed with the noble crayfish *Astacus astacus*. *Dis. Aquat. Org.* 25:233-238.

Westman, K., Ackefors, H. & Nylund, V. 1992. *Kräftor. Biologi, Odling, Fiske*. Kivik: Kiviksgården, 152 pp.

Westman, K., Pursiainen, M. & Westman, P. (eds) 1990. *Status of crayfish stocks, fisheries, diseases and culture in Europe*. Report of the FAO European Inland Fisheries Advisory Commission (EIFAC) Working Party on Crayfish. *Kalatutkimuksia-Fiskundersökningar (Helsinki)* 3:1-206.

Case studies of alien crayfish in Europe

The situation in Italy

FRANCESCA GHERARDI
Dipartimento di Biologia Animale e Genetica 'Leo Pardi', Università di Firenze & Centro di Studio per la Faunistica ed Ecologia Tropicali del C.N.R., Firenze, Italy
GILBERTO N. BALDACCINI & PAOLO ERCOLINI
A.R.P.A.T., Servizio Locale della Versilia, Località Pollino, Pietrasanta (LU), Italy
SILVIA BARBARESI
Dipartimento di Biologia Animale e Genetica 'Leo Pardi', Università di Firenze, Firenze, Italy
GIORGIO DE LUISE
Via XXIV Maggio, Reana del Rojale (UD), Italy
DAVID MAZZONI
Dipartimento di Protezione e Valorizzazione Agroalimentare, Università di Bologna, Bologna, Italy
MARIO MORI
Istituto di Anatomia Comparata, Università di Genova, Genova, Italy

ABSTRACT

In Italy, the accidental or intentional transplantation of crayfish species as part of the inland aquatic community is widespread and has produced many significant changes in the structure of the ecological community and the function of the ecosystem. The two Nearctic species, *Procambarus clarkii* and *Orconectes limosus*, have recently appeared as breeding populations in many northern and central Italian freshwater habitats, while there are scattered reports in the wild of *Astacus leptodactylus* and *Pacifastacus leniusculus*. Several aquaculture farms are cultivating exotic species, including the Australian *Cherax destructor* and *C. quadricarinatus*, without adequate precautions to prevent them from escaping. There are numerous risks for the native species, especially for the most widely distributed *Austropotamobius pallipes* (*A. torrentium* and *Astacus astacus* are reported in a few water bodies). Alien species can displace the native crayfish by direct and/or indirect competition and can transmit aphanomycosis or other diseases. Dangers for the ecosystem are also high. For example, in Lake Massaciuccoli (Tuscany), where studies are in progress, *P. clarkii* is destroying the submersed macrophytes, with heavy damage to the whole aquatic community. Moreover, its burrowing habit may result in structural deterioration of river banks and agricultural fields, as seen in the suburbs of Florence. Thus, homogenization of the Italian freshwater aquatic biota is underway and steps should be taken to slow this process. On October 23, 1997, the Italian government ratified the European Union directive 92/43/EEC Habitats. This new regulation is a great advance for crayfish conservation in Italy. New Regional Laws are to be drawn up according to the European directives and the native crayfish (now called by their specific names), although not designated as priority species, are recognized as species of interest for the EU and are subject to management actions. The new regulation declares that the introduction of alien species can be authorized only if a proper study has been conducted and only if the study has been favorably evaluated by the National Institute for Wild Fauna (INFS) or by another competent technical-scientific organization, assuring that neither natural habitats nor wild flora and fauna will be damaged. These Articles of the law focus on the primary actions capable of slowing the homogenization of Italian freshwater systems (or, from a more idealistic point of view, to stop it completely), i.e. that government authorities should: a) Promote the increase in scientific knowledge of the biology of both alien and native crayfish species, and b) Be able and ready to translate the acquired knowledge into practice.

1 INTRODUCTION

Italy holds the dubious record of being the first country in Europe affected by the crayfish plague of the 1860s (Ninni 1865, Ackefors this volume), possibly due to infected crayfish being released in ballast water from a North American ship (Unestam 1973, Schäperclaus 1992). From Italy, the disease spread across most of Europe, causing widespread depletion of native crayfish stocks. Identification of the fungus, *Aphanomyces astaci*, as the etiologic agent of the plague was established with certainty only in the 1930s (Schäperclaus 1935). After the disappearance of a crayfish population from a habitat, fungus spores can survive for 1-2 weeks at 14°C, up to two months at 2°C (Mancini 1986). Moreover, spores can be carried on damp fishing equipment and mud (Alderman & Polglase 1988); in Italy the main vector is the fisherman or crayfish trader himself, but other potential vectors are animals, from birds to fish (Alderman et al. 1987, Heitz et al. 1997, Vogt this volume). This explains why in Italy the plague has infected some *Austropotamobius pallipes* populations through the introduced North American *Procambarus clarkii* (although North American crayfish have low susceptibility to the plague, they are nevertheless potential carriers of the infectious agent, Unestam 1969), even though the two species do not share the same habitat at present.

In a discussion of the future of *A. pallipes* in Ireland, Matthews & Reynolds (1995) claimed that the species has been dramatically reduced in both numbers and distribution due to habitat destruction, pollution and an outbreak of crayfish plague (Holdich, Rogers & Reynolds this volume). A population decline of this native European species, without however a substantial change in its distribution (Mancini 1986, Westman et al. 1990, Holdich 1996), has also been recorded in Italy over the last 20-30 years, during which inland waters have suffered various kinds of modifications. Industrialization has given rise to water pollution, with acidification (noted rather early in central Europe, Nolte 1933), eutrophication and toxic discharges from agriculture and industry (Chaisemartin 1976, De Luise 1991, Arrignon 1996). Physical habitat modifications (dam construction, drainage, dredging, canalization and embanking) have caused large-scale changes in native fauna, but apparently have had less dramatic effects, thus obtaining less publicity (Lindqvist 1988). Over-fishing and poaching (Arrignon 1996) and meteorological events like drought and floods (Mori et al. 1996) have also contributed to the decrease of crayfish numbers. At present, native species are under the threat from alien crayfish, deliberately or accidentally introduced (Holdich 1988, 1995, this volume). On the one hand, they can displace the native crayfish by direct and/or indirect competition, while on the other hand they favor the spread of the lethal plague and other diseases.

2 THE PRESENT DISTRIBUTION OF NATIVE SPECIES

Figure 1 shows the present distribution of the largest freshwater decapods in Italy. It is apparent that the two species dominating Italian waterbodies are the crayfish, *Austropotamobius pallipes*, and the river crab, *Potamon fluviatile*.

The distribution of the former species ranges from Calabria to Piedmont (Froglia 1978, Laurent 1988), with the exception of part of Calabria and Puglia. In northern Italy, it is common in the rivers between Venice and Trieste draining into the Adriatic Sea (De Luise 1988) and in Liguria (Mori et al. 1988, 1989, Salvidio et al. 1993). Despite its wide distribution, the ecology of *A. pallipes* in Italy has been studied only by Pascutto et al. (1991), Mazzoni & Minelli (1996), Mori et al. (1996) and Gherardi et al. (1997a).

The river crab, *P. fluviatile*, ranges from southern Italy, including Sicily, to the Ligurian-Tuscan-Emilian Apennines (Froglia 1978). In Liguria, it is only present in theprovince of La Spezia (Capra 1953), while in northern Italy there are scattered rec-

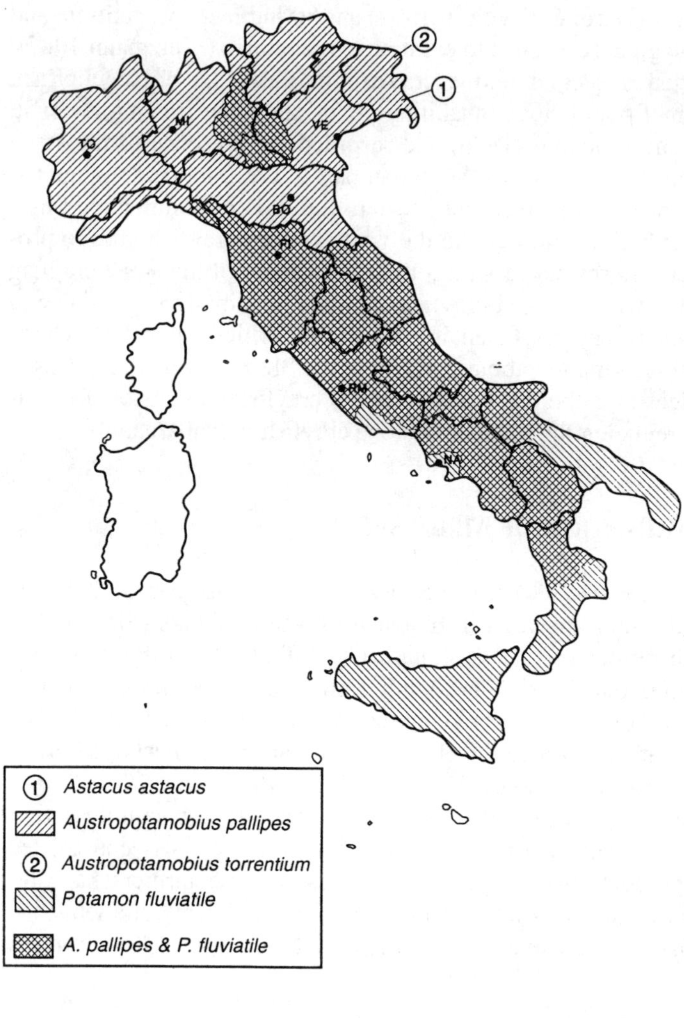

Figure 1. Distribution of the native largest freshwater decapods in Italy.

ords of its presence in Romagna (Capra 1953, Tedaldi et al. 1997), Lombardia (Bettoni 1884, Garbini 1894) and Veneto (De Betta 1863).

The colonization by these two dominant decapods of the Italian water bodies is a story of competition between them (Barbaresi & Gherardi 1997). According to Pretzmann's (1987) paleogeographical reconstruction, both Astacidae and river crabs originated in the southeastern areas of what is now Europe, the former in the early Jurassic period and the latter in the Upper Cretaceous. The two crustaceans converged into Italy along two different migration routes, the astacids in the early Oligocene and the potamids in the Upper Miocene. During the last glaciation, river crabs survived in small refuges in southern Europe (Calabria and Sicily included), while Astacidae, coming from northern and eastern Europe, occupied those Italian areas left free by the river crabs. After the end of the Pleistocene, the potamids migrated northwards to the Ligurian-Tuscan-Emilian Apennines. In the area where the distributions still overlapped, *A. pallipes* was displaced to the higher streams in the Apennines. Pretzmann (1987) and Barbaresi & Gherardi (1997) agree on the hypothesis that the clear exclusion of crayfish by river crabs in the areas of geographical overlap was caused by the dominance of *Potamon* over *Austropotamobius* in direct (Dardi & Gherardi 1994) and/or indirect competition, and not by a different physiological resistance to extreme temperatures (Pretzmann 1987).

Froglia (1978) reported a second native crayfish in Italy, the noble crayfish, *Astacus astacus,* with small populations inhabiting ponds and slow-flowing rivers in eastern Venezia-Giulia, and Machino (1996) and recorded the presence of the stone crayfish, *Austropotamobius torrentium*, in the Italian part of the Danube drainage basin in the province of Udine (Slizza River, northeastern Italy). For unknown reasons – he reports – this area was believed to contain the white-clawed crayfish and, to protect this species, the river was restocked with it to the detriment of the stone crayfish (generally smaller than the white-clawed crayfish and thus potentially subordinate to it). In addition, the countermeasures taken by Italian authorities against poachers, consisting in blocking all information about crayfish in the Slizza River basin, caused serious negative side-effects, i.e. the legal destruction of crayfish habitats by civil engineering works, further reducing the number of stone crayfish in that area.

3 THE PRESENT DISTRIBUTION OF ALIEN SPECIES

Breeding populations of the two Nearctic species, *Procambarus clarkii* and *Orconectes limosus*, have recently appeared in Italian freshwater habitats (Fig. 2). The native distribution of the former species is climatically different from the Italian region; it ranges from northeastern Mexico and south central USA westwards to Texas, eastwards to Alabama and northwards into Tennessee and Illinois (Huner 1988). The latter species comes from an area where the climate is similar to south central Europe (the eastern coast of USA, from Maine to Virginia) (Hobbs 1989).

The presence of these species in the wild is always related to aquaculture activities. Figure 3 shows the present distribution of crayfish farms in Italy, constructed either for commercial production or experimental research. The farmers have attempted to cultivate a wide range of species originating from different continents, i.e. the North European (native exclusively in northern Italy, Froglia 1978) *Astacus*

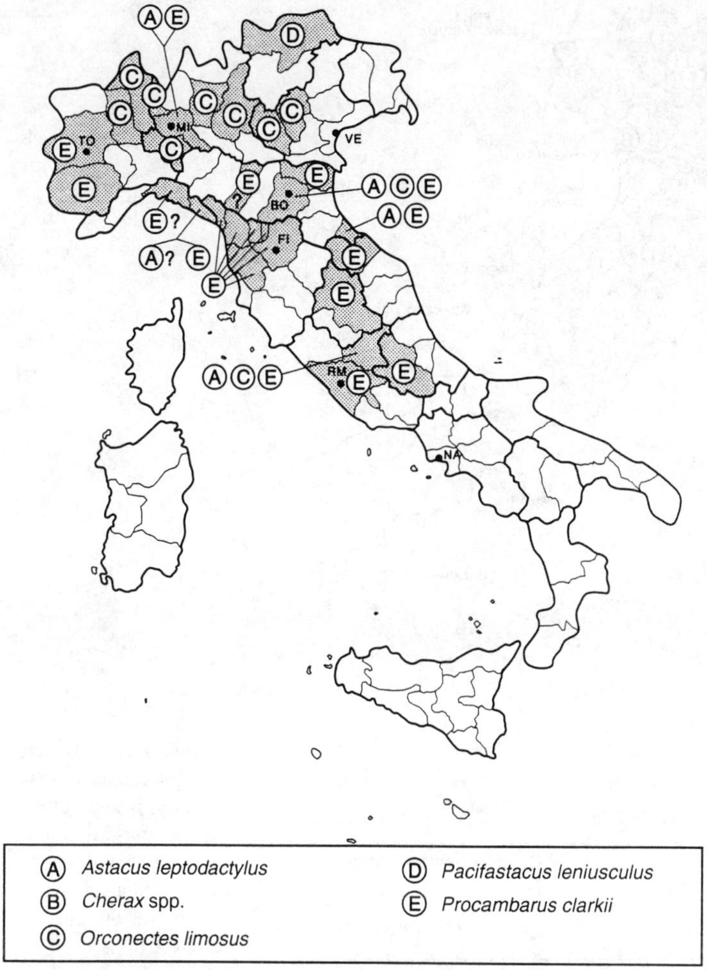

Figure 2. Distribution of alien species of crayfish in Italy. Shaded area represents the interested provinces.

astacus, the East European *A. leptodactylus*, the American *Orconectes limosus*, *Pacifastacus leniusculus* and *Procambarus clarkii*, and the Australian *Cherax* spp.

3.1 *Procambarus clarkii*

3.1.1 *Species range*

Procambarus clarkii is the most cosmopolitan crayfish, having been introduced to all continents except Australia and Antarctica (Huner 1977, Huner & Avault 1979). This species is the object of intensive culture in the southern USA (48,500 ha of culture ponds in Louisiana alone, Hobbs et al. 1989, and an annual production exceeding 60,000 tons, Huner & Lindqvist 1995) and recently in the People's Republic of China (annual production: 40,000 tons, Huner & Lindqvist 1995). The US importation of Chinese tail meat, which is less expensive than the products from the American farms, has had dramatic effects on the Louisiana crayfish industry (Avery 1996).

Figure 3. Distribution of aquaculture farms in Italy, where cultivation aims at both commerce and research. Shaded area represents the interested regions.

Following these profitable examples, the species has been imported into Italy, mostly from Spain.

Some of the life history traits of *P. clarkii*, i.e. adaptability to burrow environments, resistance to air exposure, polytrophism, rapid growth, high fecundity, and disease resistance (Huner & Lindqvist 1995, Lindqvist & Huner this volume), make its cultivation commercially feasible and its invasion of the wild extremely successful. This is especially true in a country like Italy where most farmers have failed to take adequate precautions in their cultivation methods to prevent the crayfish from escaping from farm enclosures.

At present, *P. clarkii* is well established in northern and central Italy. It is undergoing a great expansion in some sectors of the Po River drainage basin, in western Piedmont. One of these is in the Banna River, a tributary of the Po, where it appeared in 1989 after escaping from a farm where the species was experimentally reared from 1977-1985 (Delmastro 1992b). Another occurs in the Reno River drainage area in eastern Emilia-Romagna since 1995 (in a 1 km long effluent of a fishing pool near

Torre Pedrera, Rimini, in a 460 ha marsh in Campotto, Ferrara, and in the neighboring network of channels with slimy and muddy bottoms, from where it has spread northwest for 30 km into the province of Bologna (Mazzoni et al. 1996). One potential danger is its propagation towards the rice field ponds of Vercelli and Pavia, a favorite habitat of this species (Correia & Ferreira 1995).

In central Italy, it is widespread in Tuscany, especially in the 112 km^2 Massaciuccoli Lake after the establishment in 1990 of a farm at the border of the Regional Park 'Migliarino, San Rossore, Massaciuccoli'. When the farm went bankrupt in 1993, the entire cultivated stock of *P. clarkii* poured into the lake, causing a sudden habitat disequilibrium (Baldaccini 1995, Ercolini et al. in press). It seems plausible that all the populations appearing in the wild in Tuscany during the following years originated from an irresponsible translocation by man from Massaciuccoli Lake. Thus, *P. clarkii* is to be found in the streams of the Versilia coast (the provinces of Lucca, Massa and Carrara, A. Raddi and G. Salvi pers. comm.) up to the Magra River (in neighboring Liguria). It also occurs in the province of Genoa, in channels and ponds of the Bientina marshland (Lucca and Pisa provinces) (G.N. Baldaccini personal communication). It also can be found in irrigation ditches of Osmannoro and Sesto Fiorentino (the suburbs of Florence) (Gherardi et. al. 1997b), in low-flowing rivers of Pistoia and Prato provinces (S. Barbaresi pers. comm.), and in the 1800 ha Fucecchio pond (S. Barbaresi pers. comm.). Other regions with published records of breeding populations of *P. clarkii* are Umbria, the Marches (Gabucci et al. 1990) and Latium (Welcomme 1988, Delmastro & Laurent 1997), in particular the provinces of Rome (in the Tiber River basin, Giucca 1997, Gelosi et al. in press, and at Civitavecchia, G.N. Baldaccini pers. comm.) and Rieti (Gelosi et al. in press). In the latter province, a large population inhabits (since 1995) a tributary of the Velino River (Delmastro & Laurent 1997). In Italy, the species seems to be physiologically resistant to low temperatures, being found in several cold lakes of Abruzzo at high altitudes (G. De Luise pers. comm.).

3.1.2 *Pros and cons of the naturalization of P. clarkii in Italy*

The risks to aquatic ecosystems resulting from the naturalization of this alien species are numerous. Momot et al. (1978) claim that at least in oligotrophic lakes where detritus and benthic algae are a source of first-hand energy *P. clarkii* constitutes a major pathway for the conversion of primary production into animal flesh.

Indeed the meat of this crayfish has a high nutritional value and is low in calories: the abdominal muscle provides 14-25% of the yield (17.9% in Italian populations, Mazzoni et al. 1996), with a protein content of 17.13% (vs 17.89% in Italian populations, Mazzoni et al. 1996; vs 7.09% in *Austropotamobius pallipes*), a lipid content of 2.83% (vs 0.24% in Italian populations, Mazzoni et al. 1996; vs 2.45% in *A. pallipes*) and an energy content of 732 cal kg^{-1} (vs 4336 cal kg^{-1} in *A. pallipes*). It is also rich in B vitamins and minerals (potassium, calcium, sodium, magnesium, in order of content) (Goddard 1988).

However, these favorable qualities of *P. clarkii* contrast with what was observed in Massaciuccoli Lake, once densely populated by aquatic macrophytes. Indeed the recent deterioration of the community of *Nymphaetum* and the disappearance of *Utricularia australis* have been imputed to *P. clarkii*'s voracious feeding behavior. In addition, its burrowing habit may result in structural damage to river banks and agricultural fields (e.g. the case of rice field areas in Portugal, Correia & Ferreira 1995).

A positive side-effect of the introduction of *P. clarkii* is that several avian species, among which some threatened ones, have reappeared or increased in number in some waterbodies, having found in this species a new and rich food item. As in the Ebro region, Ardeidae like *Botaurus stellaris*, *Egretta garzetta*, and *Ardea purpurea* have increased in Massaciuccoli Lake, together with cormorants (*Phalacrocorax carbo*) (N.E. Baldaccini personal communication). In contrast, *P. clarkii* seems to play only a secondary role in the diet of fish (the stomach contents of a sample of 19 Centrarchidae, the large-mouth bass, *Micropterus salmoides,* revealed only one example of predation on the red swamp crayfish).

The danger that *P. clarkii* could compete with the native *A. pallipes* and displace it is low in Tuscany, since the latter species lives exclusively in streams and brooks at medium-high altitude, possibly as a result of competition with the river crab, *P. fluviatile* (Barbaresi & Gherardi 1997). It must still be ascertained which species (*P. clarkii* or *P. fluviatile*) will be dominant when their ranges overlap. In Bologna province, where the river crab is absent, the danger of competition and displacement is instead high, because *A. pallipes* reaches the limit of the lowland, being found also at altitudes lower than 100 m (Mazzoni & Minelli 1996, Mazzoni et al. 1997b). In laboratory experiments, *P. clarkii* has been found to be dominant in agonistic interactions with similar-sized *A. pallipes* (Dardi et al. 1996).

A further risk from the presence of *P. clarkii* is that the species is an intermediate host for some pathogens; in addition to crayfish plague (it may host the fungus within its body in a chronic or latent infection, Huner & Lindqvist 1995), *Procambarus* contributes to the spread of numerous parasitic helminths of vertebrates (Sogandares-Bernal 1965). Mazzoni et al. (1997a) found the protozoan *Epystilis* sp. on its gills (a pathogen only when highly diffused within the gills, since it can inhibit gas exchange) and isolated the fungus *Saprolegnia* sp. from its eggs (not a pathogen except in cases of intensive culture when females are unable to fan their eggs, Söderhäll et al. 1991). A serious infection affecting the red swamp crayfish is caused by the bacterium *Pseudomonas* sp., which produces necrosis involving the whole body (Mancini 1986). At the moment, we do not know if *P. clarkii* can transmit these parasites to the native crayfish. Ectoparasites are branchiobdellidans (Annelida, Clitellata), usually present also in the gill chambers of the native *Austropotamobius*. These are not considered true pathogens, but in large numbers they can weaken the crayfish, increasing its vulnerability to mycoses and bacterial infections (Mancini 1986). In specimens of *P. clarkii* from a wild population of northern Italy, Gelder et al. (1994) isolated the North American branchiobdellidan *Cambarincola mesochoreus*. Within a scenario of an overlap of the distribution ranges of alien and native species, it is possible that Nearctic branchiobdellidans will be found on a Holarctic host, or vice versa (Gelder et al. 1994), as has already occurred in Germany for *Branchiobdella pentodonta*, which was transmitted from the native *Austropotamobius torrentium* to the introduced *Orconectes limosus* (Vogt 1997).

3.1.3 *Population size at two sites in Tuscany*

Two preliminary studies have been undertaken to monitor the population size of *P. clarkii* at two sites in Tuscany: irrigation ditches at Osmannoro, in the suburbs of Florence, and the Massaciuccoli Lake. At Osmannoro (research coordinated by F.

Gherardi), a population living in a 380 m long ditch has been monitored year round (1996-1997). Five baited traps (80 cm long, with an aperture 28 cm in diameter, 2 mm mesh size) were located about 95 m apart on the bottom of the ditch, these positions remaining constant throughout the sampling period. Records were taken after one day from the positioning of the trap and subsequently 2-5 times per month (on average, 2.7 times). The number of captured specimens gave an estimate of the weight of the catch (the weight of an average-sized crayfish is 40 g). Figure 4 shows the trend in average density of crayfish per month (in g m^{-2}) throughout the year. It is evident that there are few active individuals during the winter (average temperature = 10°C), which they spend inside burrows excavated in the banks, while the peak density occurs in May, when both the higher temperature (21°C on average) and the availability of water favours their activity outside the burrows. The reduction in number during July and September can be attributed to the drying up of the ditches, but other biological factors may induce them to stay inside the burrows, such as ecdysis (moulting specimens occur from May to December) and the incubation of eggs by the females (occurring between August and October).

At each catch, the animals were marked with numbered plastic tags, which permit the estimation of the whole (active) population size in that area. Petersen's method was followed, with the formula:

$$P = m\,(r + u)\,/\,r$$

where m = specimens captured and marked one time, r = specimens captured two times and u = unmarked specimens, which provides the most likely value of the (active) population size (Fig. 5). After the Willks test, the two distributions did not significantly differ ($G = 6.238$, $df = 8$, ns), confirming a good agreement between observed and expected densities throughout the year.

In Massaciuccoli Lake (research coordinated by G.N. Baldaccini), six unbaited

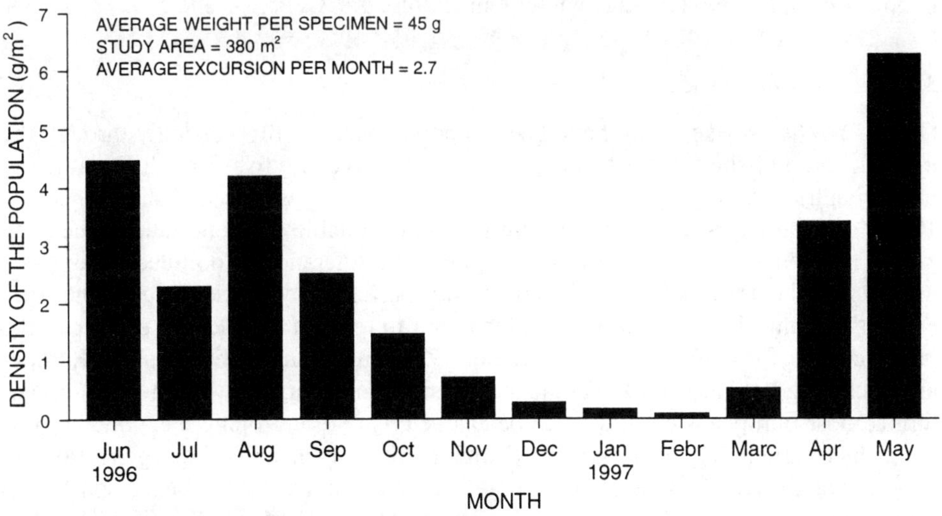

Figure 4. Density of one population of *Procambarus clarkii*, monitored one year round (1996-1997) at Osmannoro, suburbs of Firenze (Tuscany).

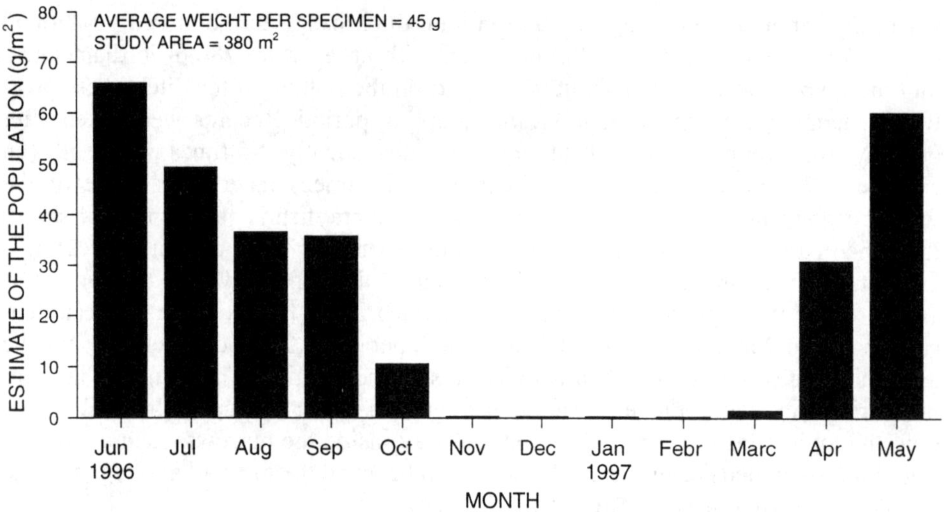

Figure 5. Estimate of the active population of *Procambarus clarkii* in the same habitat and period as Figure 4. The values were obtained applying the Petersen's method.

traps were used (each 1 m long, with an aperture 50 cm in diameter, 6-8 mm mesh size), positioned at a depth of 70-150 cm. Records of the weight of the captured specimens were taken after three days from the positioning of the traps along a 2750-m-long transect, during May-July and October-December in 1995 and February-March and May-September in 1996. These data were compared with the weight of *Procambarus* captured by professional fishermen in the same area in 1997 (Fig. 6). At least in 1995 and in 1997, an increase was evident from May to June, followed by a decrease in July and August, in good agreement with the data recorded at Osmannoro. Likewise, the number of catchable crayfish decreased during the winter months in both 1995 and 1996 (average winter temperature = 9°C, Pensabene 1992).

3.2 *The other alien species*

Orconectes limosus seems to have been imported accidentally into Italy through the introduction by fishermen of fish stocks containing live crayfish from different European countries, where this American species has been well established since the 1890s. *Orconectes* is highly productive and quite suitable for pond culture, because of fast reproduction, resistance to the plague and a tolerance of polluted water (Momot 1988). The drawback is that it is much less preferred by gourmets than other species on account of its size (most specimens fail to reach a total length of 10 cm, the preferred size for commercial use, Momot 1988) and its narrow chelipeds (Europeans, including Italians, love to eat the meat inside the chelipeds, Momot 1988). In addition, most Europeans consider it to be characteristic of eutrophic, i.e. polluted, waters, which further degrades its image (Momot 1988). Its meat has been generally less appreciated than that of other alien crayfish (Welcomme 1988), being considered 'mediocre' (Laurent & Forest 1979) or even 'tasteless' (Mancini 1986). Thus, although Momot (1988) claims that *O. limosus* will play an important future role in the

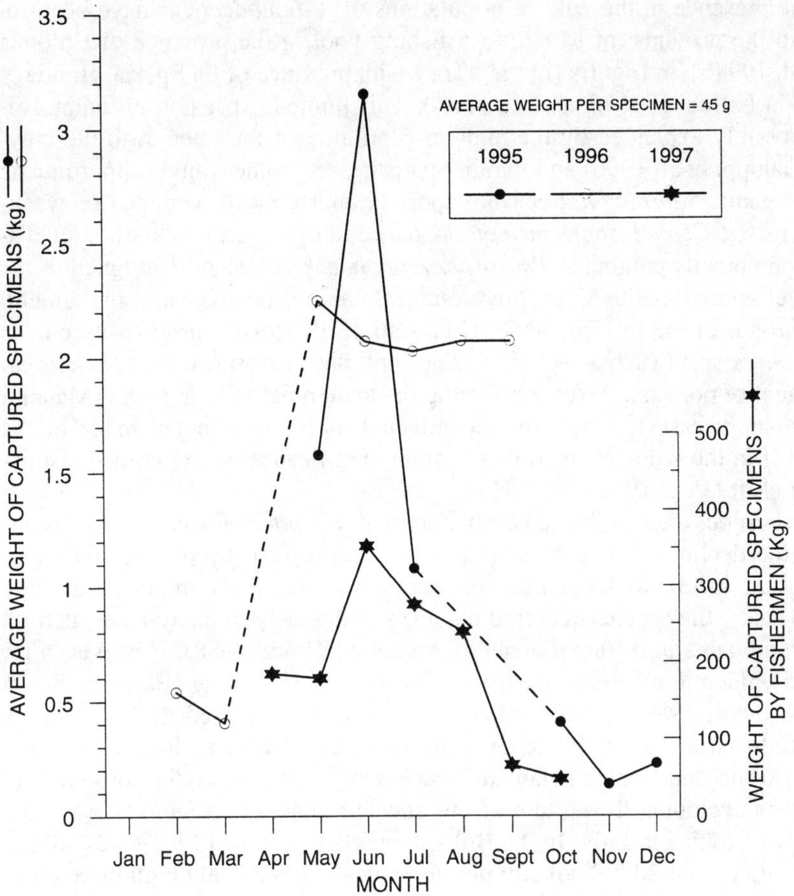

Figure 6. Average weight of *Procambarus clarkii* monitored in 1995 and 1996, compared with the weight of catches of the same species by professional fishermen in 1997, in Lake Massaciuccoli.

gastronomic aspects of European astacology, at present the commercial culture of this species remains doubtful, at least in Italy (only one farm, Fig. 3). It would also seem to be risky, since this crayfish is a vector of several parasitic diseases, including aphanomycosis, burn spot disease (caused by the fungus *Dydimaria cambari*, Vogt 1997) and the still poorly known *Psorospermium haeckeli* (Mancini 1986, Vogt this volume).

The first Italian record of this species in the wild was made in the Iseo Lake in 1991 (Delmastro 1992a) and later in the Po River basin (close to Pavia in 1992, Groppali 1993, and Verona in 1994, Confortini & Natali 1995) and in the Reno River basin (in a 1 ha fishing pool near S. Giovanni in Persiceto and in the Santerno River, province of Bologna, M. Trentini pers. comm.). Recently, it has been reported in the Salto Lake (Rieti, Latium) (Quattrocchi et al. 1996), which is also inhabited by *Astacus leptodactylus* (Gelosi et al. in press), while the native *Austropotamobius pallipes* lives in some tributaries of the same lake. It will be interesting to monitor the situation in this area over the years to see if and how the different species interact.

In Italy, the presence in the wild of populations of *A. leptodactylus* have been recorded only in the province of Milan, in a fishing pool in the province of Bologna (Mazzoni et al. 1997b), in Liguria (in the Vara basin, province of La Spezia, Braida et al. 1997) and in Latium (Gelosi et al. in press). This limited expansion in Italian waterbodies is possibly explained by the high susceptibility of this species to the crayfish plague (Galuppi et al. 1996) and to other parasites, e.g. other fungi (opportunistic species of the genus *Saprolegnia*, causing sporadic mortality of wounded or weakened specimens, or *Cephalosporium leptodactyli*, etiologic agent of burn spot disease), the taxonomically enigmatic *Psorospermium haeckeli* (recently assigned to the DRIPs, a novel animal clade, Vogt, this volume), unionid bivalves and the annelid *Branchiobdella* sp. in the gill chambers (Mazzoni et al. 1997a), protozoans on the gills (*Lagenophrys* sp.) (Mazzoni et al. 1997a), and the microsporidium *Thelohania contejeani*, agent of porcelain disease affecting the abdominal muscle fibers (Mancini 1986). The latter disease is lethal for the infected individuals, but damage at the population level in the wild seems to be of minor significance; its route of transmission is not yet clear (Vogt 1997).

In western Europe, interest in the commerce of *A. leptodactylus* has markedly increased with the decline of *A. astacus* populations, destroyed by plague and water pollution (Köskal 1988). In Italy, live specimens were regularly imported directly from Turkey, where the species occurred naturally and widely in many lakes, natural ponds, rivers and dams in different areas of Anatolia (Köskal 1988). However, after 1984 when an epidemic of plague destroyed the catches in Turkey (Baran & Soylu 1989, Rahe & Soylu 1989, Ackefors this volume), Italy decreased the importation from that country from about 90 to 30 tons (Köskal 1988) and looked to other sources, like Bosnia and France (Mancini 1986). Only a few farms in northern and central Italy are developing the culture of this species, although *A. leptodactylus* is a rather expensive crayfish in Italy. In the Bologna market in 1994-1996, it was sold at 36,000 Italian lira (about 20 US dollars) per kg, which is a relatively high price compared to 32,000 lira (about 19 US dollars) per kg in 1997 for *Astacus astacus* imported from Albania and to 25,000 lira (about 15 US dollars) per kg in 1996-1997 for *Procambarus clarkii* of local production.

In Italian freshwater systems, the occurrence of the signal crayfish, *Pacifastacus leniusculus,* native to northwestern North America (Lowery & Holdich 1988), was recorded in 1994 by Machino (1997) in a small creek near Brunico (province of Bolzano, in the Adige River basin). If the presence of this species is confirmed, the native *A. pallipes* will no doubt be displaced from that area. This has already occurred in Great Britain (Lowery & Holdich 1988, Holdich & Domaniewski 1995, Holdich et al. 1995b) and a similar process has taken place in Sweden with *A. astacus* (Söderbäck 1995). In addition, the spread of this alien species will lead to all the catastrophic damage it has caused in northern Europe, since it is a permanent vector of aphanomycosis and it destroys weed cover and fish spawning sites, with the consequent decline of local sport fishing (Lowery & Holdich 1988, Holdich this volume). It is noteworthy that, in Sweden and France, *P. leniusculus* is used to keep water bodies free of vegetation (Lowery & Holdich 1988). Fortunately, as far as we know, this species is cultivated only in two farms, one in the province of Alessandria (D. Mazzoni personal communication) and one in Sila (Calabria, southern Italy) (Bianchini et al. 1991).

The Australian *Cherax* spp. have never been recorded in the wild, although some authors (De Luise & Sabbadini 1997) predict their growing presence in our water bodies. Since 1996, there are experimental cultivations of the yabby, *Cherax destructor*, in Bologna (Quaglio et al. 1997) and Udine (F. Quaglio pers. comm.), where its potential as an aquaculture species in Italy is being assessed. There are also several centers for the commercial aquaculture of the yabby and the redclaw crayfish, *C. quadricarinatus*, in Umbria (province of Perugia) (E. Gelosi pers. comm.), Abruzzo, Campania, Emilia-Romagna, Tuscany and Veneto. These species have invasive and destructive habits plus a number of described diseases like the protozoan microsporid *Thelohania* (Mancini 1986) and the temnocephalidean platyhelminths (Quaglio et al. 1997). Hence they may cause problems if introduced into new areas (Sokol 1988), since they are likely to become pests (E. Wallis pers. comm.).

4 ITALIAN REGULATIONS

4.1 *Concerning the conservation of native species*

Astacus astacus, Austropotamobius pallipes, and *A. torrentium* are listed as protected species in Annex III of the 'Convention on the conservation of European wildlife and natural habitats' (Bern, 1982). The aim of the Bern Convention was to 'conserve wild flora and fauna and their natural habitats', giving 'particular emphasis to endangered and vulnerable species'. The 'Council directive 92/43/ECC of 21 May 1992 on the conservation of natural habitats and of wild flora and fauna' had a specific objective to establish a network of protected areas to ensure effective implementation of the Bern Convention. Annex II designates *A. pallipes* as 'a species whose conservation requires designation of special areas', while Annex V defines *A. astacus*, *A. pallipes*, and *A. torrentium* as 'species whose taking in the wild and exploitation may be subject to management measures'. The consideration of these crayfish as endangered species by the IUCN (The World Conservation Union) (Groombridge 1993) represents a milestone in invertebrate conservation (New 1995). This even though the report is only a technical guidance document (Glowka & de Klemm 1996), indicating existing or potential losses and thereby seeking to focus attention on the taxa that need urgent conservation measures (Fitter & Fitter 1987).

Before October 1997, with a few exceptions, these European and International suggestions were not taken into account by Italian legislation. This was so even though crayfish (without any indication of species) have been considered, since 1914, as a crustacean decapod that should be protected and whose fishing should be regulated (Royal Decree, RD, November 22, 1914, No. 1486).

More recently, *A. pallipes* and the other native species could be fished in most Italian regions and provinces (administrative institutions whose duties include the control of inland waters) and were treated as fish with respect to management and legislation. Thus, while in Emilia-Romagna, Latium, Liguria, Lombardia, and Piedmont, the European directives were soon incorporated into Regional Laws and the catching of crayfish was prohibited throughout the year, in Tuscany (to cite one opposite example), the Regional Law of April 24, 1984 (No. 25) declared that: 1. Only licensed fishermen have the right to catch crayfish (still a general term) in some lakes and rivers; 2. The legal catching season is from July 1 to March 31; 3. Baited traps are prohibited; 4. No more than 20 crayfish can be caught by one fisherman per day; 5. Nocturnal catches are prohibited; and 6. The minimum legal size is 70 mm total length (the smallest in all Europe). Criticism of these regulations was directed at both the catching season (when most biological events occur, e.g. reproduction and molting) and the minimum legal size (the frequency of both ovigerous females and functionally mature males reaches a maximum exactly in correspondence to the minimum legal size) (Gherardi et al. 1997a).

On October 23, 1997, the 'Gazzetta Ufficiale della Repubblica Italiana' (Supplement No. 248) published the 'Decree of the President of the Republic' (D.P.R.) of September 8, 1997 (No. 357) which 'disciplines – it declares – the procedures of adoption of the measurements examined by the directive 92/43/EEC Habitats' (Article 1.1), providing seven Annexes (A-G). In conformity with the EU directive, crayfish are not designated as priority species for whose conservation the Union has particular responsibility, thus requiring rigorous protection (fauna and flora included in Annex D). Nevertheless, this new regulation is a great advance for crayfish conservation in Italy. New Regional Laws are to be compiled according to the European directives and the native crayfish (now called by their specific names) are recognized as species of interest for the EU and subject to management actions.

4.2 *Concerning the importation of alien species*

Among global multi-lateral treaties, the Earth Summit in Rio de Janeiro (1992) offers the best opportunities for development of a comprehensive global approach to the intentional and accidental introduction of alien species and their eradication or control (Glowka & de Klemm 1996). As far as possible and as appropriate, it requires its signatories 'to prevent the introduction of, control and eradicate those alien species which threaten ecosystems, habitats or species' (Article 8h) (Glowka & de Klemm 1996).

Moreover, the 'Council directive 92/43/ECC of 21 May 1992 on the conservation of natural habitats and of wild flora and fauna' declares in Article 22b that Member States shall 'ensure that the deliberate introduction into the wild of any species which is not native to their territory, is regulated so as not to prejudice natural habitats within their natural range or the wild native fauna and flora and, if they consider it necessary, prohibit such introduction'.

In Italy, before October 23, 1997, the basis for the regulation of the importation of alien species was the 'Convention on International Trade in Endangered Species of Wild Fauna and Flora' (CITES) (Washington, March 3, 1973), ratified in Italy by Law No. 874 of December 19, 1975, which provided a list of endangered species whose trade was prohibited entirely, unless special export and import permits were granted under highly exceptional circumstances (Appendix I, only some molluscs among invertebrates), and species whose trade could have been a factor of increasing vulnerability and for which the extent of trade should have been monitored through permit systems (Appendix II, some molluscs and insects among invertebrates).

No crayfish were included in any CITES Appendix. Thus, in Italy live specimens could easily have been imported from other countries. They must be accompanied by a health certificate stating that they were not affected by aphanomycosis in the cases of *Astacus* spp., *Austropotamobius pallipes*, and *Procambarus clarkii* (D.P.R. No. 555, December 30, 1992, ratifying the directive of sanitary policy for aquaculture products 91/67/EEC, January 28, 1991).

Only Regional Laws referred to the introduction of alien species. To cite one example, in Tuscany the Regional Law of May 26, 1993 (No. 34) declared (Article 1) that 'the utilization of species extraneous to the regional aquatic flora and fauna is allowed after a favorable opinion of the Regional Consortium of Hydrobiology and Fishing (CRIP) [...] which can prescribe the adoption of possible measures to avoid uncontrolled diffusion of the species'. Since 1996, this role has been exercised by the Tuscan Regional Agency for Environmental Protection (ARPAT).

With the recent Italian government ratification of the directive 92/43/ECC, Article 12.3 of D.P.R. No. 357 declares that the introduction of alien species can be authorized only if a proper study has been conducted and only if the study has been favorably evaluated by either the National Institute for Wild Fauna (INFS), or by another competent technical-scientific organization, assuring that neither natural habitats nor wild flora and fauna will be damaged. This again requires that Regional Laws must change accordingly, to designate a competent technical-scientific organization whose responsibility will become vital for the conservation of Italian freshwater ecosystems.

5 DISCUSSION

In Italy, the accidental and intentional transplantation of crayfish species as part of the inland aquatic community is widespread and has produced many significant changes in the structure of the ecological community and the function of the ecosystem.

Homogenization (Lodge & Hill 1994) of the Italian freshwater aquatic biota is underway and steps should be taken to slow this process.

The first step is to weigh the costs and benefits of the introduction of alien species. However, there is a poor statistical basis for risk analysis. Organisms placed in novel environments sometimes act in unexpected ways or have unexpected interactions and effects (A. Cohen personal communication). For example, different stages of one species can occupy different niches (Clout et al. 1996) and some species change their biology and become tolerant of different climatic regimes. Thus, they inhabit regions considerably colder than their native range (A. Cohen personal communication), viz., the case of *Procambarus clarkii* well acclimatized in several cold lakes of Abruzzo. Moreover, the ecological and economic cost of some exotic species is unpredictable and usually high. To cite one example, *P. clarkii* has become a pest in Massaciuccoli Lake, one consequence being the deterioration of submersed macrophytes with an ecological loss within the whole lake system. It has also destroyed the only European cultivation of lotus lilies near the lake, with a monetary loss of 400 million lira, about 230,000 US dollars. Thus, we concur in recommending that government agencies should assume that any species that is a candidate for introduction is guilty until proven innocent (Ruesink et al. 1995).

This issue contrasts with the recent importation into Italy for aquaculture purposes and commercialization of potential pests like *Cherax* spp. and *Pacifastacus leniusculus*. The former species behave in the wild as agents of destruction of water bodies, with a fast diffusion rate thanks to their capacity for air-breathing (they can walk outside water for hundreds of meters) (E. Wallis personal communication). The latter species has displaced the native *Austropotamobius pallipes* and *Astacus astacus* in many water bodies of Great Britain and Sweden, respectively, through plague transmission, direct competition and reproductive interference in the form of interspecific matings (Holdich et al. 1995a, Söderbäck 1996). The other alien crayfish present in Italy are dangerous because they are vectors of parasites other than the only one considered by Italian regulation, *Aphanomyces astaci*. At the moment, the effects on the ecological community, humans included, of some of these parasites are poorly known (Vogt this volume).

As a consequence, some obvious questions are raised. Is it commercially advantageous to cultivate alien crayfish in Italy? Does the extent of Italian crayfish consumption justify the high number of farms and/or is exportation to other European countries a profitable business? Unfortunately, the answers to these questions are lacking since the statistics furnished by Eurostat are biased for Italy (Laurent 1997) by the linguistic confusion between prawns, shrimps and lobsters on the one hand and crayfish on the other (in Italy, all macrurans are called 'gamberi' or 'gamberetti'). However, there is no doubt that it is difficult to assess the value placed on the conservation of native biodiversity and natural ecosystems in relation to the improvement of human economics and well-being (Daehler & Gordon 1997).

Therefore, provided that aquaculture is a source of profit in Italy, the second step in retarding the loss of biodiversity in Italian freshwater ecosystems is to impose severe rules for the location and architecture of cultivation farms to prevent the escape of specimens and the establishment of breeding populations in the wild. Hence, following the plans for the conservation of native crayfish in operation in Great Britain (Palmer 1994, Wynne et al. 1995, Holdich, Rogers & Reynolds this volume), the following measures should be required: 1. Control of the quality of effluent from crayfish farms (Rogers & Holdich 1995); 2. Effective containment measures (any farm should be confined to enclosed waterbodies surrounded by appropriate perimeter fencing, Holdich et al. 1995b); and 3. The prohibition of keeping alien crayfish in certain specified areas (no-go areas) except under license (Rogers & Holdich 1995). These seem to be the minimum number of rules, aimed at anticipating and repairing the damage caused by inappropriate introductions. 'Such rules should be based' – de Klemm (1995) states – 'on principles that are now universally recognized, especially the principles of prevention and precaution, and the polluter-pays principle'.

From this point of view, the Italian government should have an exact and complete picture of the distribution of aquaculture centres operating in Italy. However, so far the only measure taken by the Ministry of Health (Circular No. 23 of November 5, 1994) was to invite the Regions to take a census of all the structures where aquaculture is carried out, including the deposits of fish and crayfish and sport fishing pools, using the Veterinary Operative Units of the Local Sanitary Agencies.

The third step in slowing the homogenization of Italian freshwater habitats is to stimulate either restocking or reintroduction of the native crayfish, now relegated to being limited populations in small pockets of the Italian territory, and to reduce to the minimum their withdrawal from where they are still present. It would be wise to make the measures expressed in Article 10 (2) of D.P.R. No. 357 more restrictive, given that the removal from the wild and the exploitation of specimens of our native crayfish is not compatible with the maintenance of a favorable conservation status (Article 10 (1)).

It is astonishing that no Italian farm is cultivating *A. pallipes* for either commercial use or reintroduction purposes. Indeed, *A. pallipes* does not have a high economic value in the European market, as do *A. astacus* and *A. leptodactylus* (Ackefors this volume), and is subject to only a restricted consumption. Nevertheless, at the court of Lorenzo 'il Magnifico' de' Medici (15th century) it was an exquisite dish, worthy of the most important banquets (P. Brunetti pers. comm.), while in the 18th century Princess Violante Beatrix of Bavaria (the wife of Ferdinando de' Medici) built one of the first aquaculture farms, where *A. pallipes* (called the 'black crayfish' in the documents of those times) was cultivated, a few kilometers from Florence (M. Casprini personal communication).

However, apart from their economic value, the restoration of native crayfish is needed for the sake of environmental policy, since they represent good indicator species for the quality of the aquatic ecosystems, and makes good sense from the standpoint of conservation policy (Lindqvist 1988).

The fourth step would be to increase education and effective communication to a large audience as a means of increasing the knowledge of crayfish and the resources to foster their conservation (New 1995). Environmental education could slow down the current rates of introduction and translocation of alien species, but, as Hobbs et al. (1989) observed, it would be naive to believe that this strategy alone could stop the process. At times, the promotion of animals as 'pets', which certainly has a considerable educational value (New 1995), has led to negative effects on the habitat. For example, aquaria have been responsible for 65% of the 46 exotic fish species established as breeding populations in open waters of the USA (Courtenay & Stauffer 1990). We do not have statistics at hand, but the habit of rearing *P. clarkii* in aquaria (it is often sold as a pet at several fairs at the price of 10,000-12,000 lira, about 6-7 US dollars, per specimen) might explain part of its wide diffusion in the wild. Family members often cannot bear to kill their aquarium pets when faced with moving house or a decline in interest or the pet's increased size, and they often release them in the nearest water body (also see Pullin 1994).

The fifth step is related back to the first step. In order to prove the 'innocence' of some alien crayfish and to provide a list of 'acceptable' species (Lodge et al. in press), Clout et al. (1996) have focused on the need for scientific and technical work. This would include: 1. Elaborate an international database of the distributions of invasive species, plus their characteristics and environmental impact, on the basis of both macro- and micro-scale surveys (Lindqvist 1988); the database would assist with risk assessment, preventive action, the choice of methods of eradication or control, and the education of key quarantine and conservation personnel; 2. Improve knowledge of the ecology of the invasion process, including the lag effect; 3. Improve knowledge of the risks of the transport and establishment of new species; 4. Improve the ability to predict which species are likely to become invasive in particular situations and habitats; 5. Provide better knowledge of disease vectors, sources and pathways caused by human interventions; and 6. Study effective, target-specific, humane and socially-acceptable methods of eradication or control. The transfer of information among and within countries could be a good baseline for prevention of the repetition of similar ecological errors.

In conclusion, there is no doubt that this last step is of primary importance to a slowing of the homogenization of Italian freshwater systems (or, from a more idealistic point of view, to stopping it completely). In other words, government authorities should: a) promote the increase in scientific knowledge of the biology of both alien and native crayfish species, and b) be able and ready to translate the acquired knowledge into practice.

ACKNOWLEDGMENTS

Funds were provided by MURST ('fondi ex-60%') and by Regione Toscana to the first author. We are indebted for their invaluable help to: Dr Antonio Barsanti (Azienda USL 12, Viareggio, Lucca), Dr Massimiliano Mattioli (ARPAT, Lucca), Dr Andrea Raddi and Mr Gabriele Salvi (Dipartimento di Biologia Animale e Genetica, Università di Firenze).

REFERENCES

Alderman, D.J. & Polglase, J. 1988. Pathogens, parasites and commensals. In D.M. Holdich & R.S. Lowery (eds), *Freshwater crayfish: biology, management and exploitation*: pp. 167-212. London: Croom Helm.

Alderman, D.J., Polglase, J. & Frayling, M. 1987. *Aphanomyces astaci* pathogenicity under laboratory and field conditions. *J. Fish Dis.* 10:385-396.

Arrignon, J. 1996. *Il gambero d'acqua dolce e il suo allevamento.* Bologna: Edagricole, xv + 240 pp.

Avery, J.L. 1996. Status of the Lousiana crawfish industry. Abstract Volume, XI Symposium of the International Association of Astacology, 50.

Baldaccini, G.N. 1995. Considerazioni su alcuni macroinvertebrati dell'area umida del Massaciuccoli (Toscana). In P.E. Tomei & E. Guazzi (eds), Vol. IV *Il bacino del Massaciuccoli*: pp. 91-103. Pisa: Pacini.

Baran, I. & Soylu, E. 1989. Crayfish plague in Turkey. *J. Fish Dis.* 12:193-197.

Barbaresi, S. & Gherardi, F. 1997. Italian freshwater decapods: exclusion between the crayfish *Austropotamobius pallipes* (Faxon) and the river crab *Potamon fluviatile* (Herbst). *Bull. Fr. Pêche Piscic.* 347:731-747.

Bettoni, E. 1884. *Prodromi della faunistica bresciana*. Brescia: Apollonio.

Bianchini, M.L., Costa, C. & Lombardi, F. 1991. Progetto per lo sviluppo della produzione ittica in Sila. *Riv. Idrobiol.* 29 (1990):123-130.

Braida, L., Mori, M. & Salvidio, S. 1997. Alien crayfishes in Liguria (N.W. Italy). Abstract Volume of the International Workshop 'The introduction of alien species of crayfish in Europe. How to make the best of a bad situation?', pp. 23-24. Florence, September 24-27, 1997.

Capra, F. 1953. Il granchio di fiume, *Potamon edule* (Latr.) in Liguria. *Doriana* 1:1-7.

Chaisemartin, C. 1976. Effects des facteurs des nuisances et de pollutions sur le métabolisme des Astacidae et de leur reproduction. *La Pisc. Franç.* 48:60-65.

Clout, M., Lowe, S. & the IUCN/SSC Invasive Species Specialist Group 1996. Draft IUCN guidelines for the prevention of biodiversity loss due to biological invasion.

Confortini, I. & Natali, M. 1995. Presenza del gambero americano *Orconectes limosus* (Rafinesque 1817), in alcuni corsi d'acqua della pianura veronese (Crustacea, Decapoda, Cambaridae). *Ann. Mus. Civ. Rovereto, Sez. Arch., St., Sc. Nat.* 10:399-404.

Correia, A.M. & Ferreira, O. 1995. Burrowing behaviour of the introduced red swamp crayfish *Procambarus clarkii* (Decapoda: Cambaridae) in Portugal. *Crustacean Biol.* 15:248-257.

Courtenay, W.R. Jr & Stauffer Jr, C.R. 1990. The introduced fish problem and the aquarium trade. *J. World Aquacult. Soc.* 21:145-154.

Daehler, C.C. & Gordon, D.R. 1997. To introduce or not to introduce: trade-offs of non-indigenous organisms. *Tree* 12:424-425.

Dardi, P. & Gherardi, F. 1994. Competition and predation between the river crab *Potamon fluviatile* and the crayfish *Austropotamobius pallipes*. *Boll. Zool. Suppl.* 61:41.

Dardi, P., Villanelli, F., Raddi, A. & Gherardi, F. 1996. Competizione tra il gambero autoctono *Austropotamobius pallipes* e il gambero alloctono *Procambarus clarkii*. Atti del XVII Simposio della Società Italiana di Etologia, pp. 31-32.

De Betta, E. 1863. Materiali per una fauna veronese. *Mem. Acc. Agr. Comm. Verona* 42:1-15.

Delmastro, G.B. 1992a. Il gambero americano *Orconectes limosus* (Rafinesque), un nuovo decapode neartico nelle acque dolci del nord Italia (Crustacea Decapoda Cambaridae). *Natura Bresciana* 27 (1990-91):171-174.

Delmastro, G.B. 1992b. Sull'acclimatazione del gambero della Lousiana *Procambarus clarkii* (Girard, 1852) nelle acque dolci italiane. *Pianura* 4:5-10.

Delmastro, G.B. & Laurent, P.J. 1997. Récentes avancées des écrevisses américaines en Italie. *L'Astaciculteur de France* 50:2-6.

De Luise, G. 1988. *Indagine preliminare sulla distribuzione del gambero di acqua dolce della specie Austropotamobius pallipes italicus (Faxon) nel Friuli Venezia-Giulia. Sulle possibilità di allevamento e ripopolamento.* Reana del Rojale: Chiandetti.

De Luise, G. 1991. *Diffusione, allevamento e ripopolamento in Friuli del gambero d'acqua dolce.* Reana del Rojale: Chiandetti.

De Luise, G. & Sabbadini, A. 1997. Conservation and enhancement of native crayfish of Italy. Abstract Volume of the International Workshop 'The introduction of alien species of crayfish in Europe. How to make the best of a bad situation?', p. 45.

Ercolini, P., Baldaccini, G.N. & Mattioli, M. in press. *Procambarus clarkii* (Girard) (Crustacea, Decapoda) nella zona umida di Massaciuccoli: una specie esotica infestante o una risorsa da sfruttare? Atti del Convegno 'I biologi e l'ambiente oltre il duemila'.

Fitter, R. & Fitter, M. 1987. *The road to extinction.* Gland & Cambridge: IUCN.

Froglia, C. 1978. Decapodi (Crustacea Decapoda). In S. Ruffo (ed.) *Guide per il riconoscimento delle specie animali delle acque interne italiane*, vol. 4, AQ/1/9. Roma: Consiglio Nazionale delle Ricerche.

Gabucci, L., Para, R. & Poselli, M. 1990. *Pesci e Crostacei d'acqua dolce della Provincia di Pesaro-Urbino.* Villa Verrucchio: TPL La Pieve.

Galuppi, R., Mazzoni, D., Quaglio, F. & Tampieri, M.P. 1996. Episodio di mortalità in gamberi turchi (*Astacus leptodactylus*) di importazione. Atti del Convegno Nazionale 'Il contributo dei progetti di ricerca allo sviluppo dell'acquacoltura nazionale', pp. 47-53.

Garbini, A. 1894. Primi materiali per una monografia limnologica del Lago di Garda. *Boll. Soc. Ent. Ital.* 26:1-94.

Gelder, S.R., Delmastro, G.B. & Ferraguti, M. 1994. A report on branchiobdellidans (Annelida: Clitellata) and a taxonomic key to the species in northern Italy, including the first record of *Cambarincola mesochoreus* on the introduced American red swamp crayfish. *Boll. Zool.* 61:179-183.

Gelosi, E., Colombari, P.T., Formichi, M., Leoni, A. & Mancini, A. in press. *Contributo alla conoscenza della fauna delle nostre acque interne*, 3rd Edition. ARSIAL: Regione Lazio.

Gherardi, F., Villanelli, F. & Dardi, P. 1997a. Behavioral ecology of the white-clawed crayfish *Austropotamobius pallipes* in a Tuscan stream: preliminary results. *Freshwater Crayfish* 11:182-193.

Gherardi, F., Barbaresi, S., Raddi, A. & Salvi, G. 1997b. Distribuzione, diffusione, impatto sugli ecosistemi dulcacquicoli toscani del gambero alloctono, *Procambarus clarkii,* Technical Report No.1, Regione Toscana.

Giucca, F, 1997. Nuove segnalazioni relative all'ittiofauna del tratto urbano del fiume Tevere. I Conv. Naz. Fauna Urbana, Roma 12 Aprile 1997, Riassunti, p. 48.

Glowka, L. & Klemm, C. de 1996. International instruments, processes, organizations and non-indigenous species introductions: is a protocol to the convention on biological diversity necessary? In O.T. Sandlund, P.J. Schei & Å. Viken (eds) *Proceedings of the Norway/UN Conference on Alien Species*: pp. 211-219. Trondheim.

Goddard, J.S. 1988. Food and feeding. In D.M. Holdich & R.S. Lowery (eds), *Freshwater crayfish. Biology, management and exploitation*: pp.145-166. London: Croom Helm.

Groombridge, B. (ed.) 1993. *1994 IUCN red list of threatened animals*. Gland, Switzerland, and Cambridge, U.K.: IUCN.

Groppali, R. 1993. Sulla presenza del gambero americano *Orconectes limosus* (Rafinesque) in acque della provincia pavese (Crustacea Decapoda Cambaridae). *Riv. Piem. St. Nat.* 14:93-96.

Heitz, E., Oidtmann, B. & Hoffmann, R.W. 1997. Crayfish plague: the vector function of birds and fish for the spread of freshwater crayfish plague, a fungal infection caused by *Aphanomyces astaci*. Abstract Volume of the International Workshop 'The introduction of alien species of crayfish in Europe. How to make the best of a bad situation?', p. 32.

Hobbs, H.H. 1989. An illustrated checklist of the American crayfishes (Decapoda: Astacidae, Cambaridae and Parastacidae). *Smithsonian Contrib. Zool.* 480:iv + 1-236.

Hobbs, H.H. III, Jass, J.J. &. Huner, J.V 1989. A review of global crayfish introductions with particular emphasis on two North American species (Decapoda, Cambaridae). *Crustaceana* 56:299-316.

Holdich, D.M. 1988. The dangers of introducing alien animals with particular reference to crayfish. *Freshwater Crayfish* 7:xv-xxx.

Holdich, D.M. 1995. Problems caused by the introduction of freshwater organisms. *Direktoratet für naturforvaltning* 242:14-21.

Holdich, D.M. 1996. *Austropotamobius pallipes*. In P.J. van Helsdingen, L. Willemse & M.C.D. Speight (eds) *Background information on invertebrates of the Habitats Directive and the Bern Convention. Part I. Crustacea, Coleoptera and Lepidoptera. Nature and Environment* 79:1-8.

Holdich, D.M. & Domaniewski, J.C. 1995. Studies on a mixed population of crayfish *Austropotamobius pallipes* and *Pacifastacus leniusculus* in England. *Freshwater Crayfish* 10:37-45.

Holdich, D.M., Reader, J.P., Rogers, D.W. & Harlioglu, M. 1995a. Interactions between three species of crayfish (*Austropotamobius pallipes*, *Astacus leptodactylus* and *Pacifastacus leniusculus*). *Freshwater Crayfish* 10:46-56.

Holdich, D.M, Reeve, I.D. & Rogers, D.W. 1995b. Introduction and spread of alien crayfish in British waters – Implications for native crayfish populations. *Freshwater Crayfish* 8:99-112.

Huner, J.V. 1977. Introductions of the Louisiana red swamp crayfish, *Procambarus clarkii* (Girard): an update. *Freshwater Crayfish* 3:193-202.

Huner, J.V. 1988. *Procambarus* in North America and elsewhere. In D.M. Holdich & R.S. Lowery (eds), *Freshwater crayfish. Biology, management and exploitation*: pp. 239-261. London: Croom Helm.

Huner, J.V. & Avault Jr, J.W. 1979. Introductions of *Procambarus* spp. *Freshwater Crayfish* 4:191-194.

Huner, J.V. & Lindqvist, O.V. 1995. Physiological adaptations of freshwater crayfish that permit successful aquacultural enterprises. *Am. Zool.* 35:12-19.

Klemm, C. de 1995. Introductions of non-native organisms into the natural environment. Report to the Group of experts on legal aspects of introduction and re-introduction of wildlife species - (T-PVS (95) 17), Council of Europe, Strasbourg.

Köskal, G. 1988. *Astacus leptodactylus* in Europe. In D.M. Holdich & R.S. Lowery (eds), *Freshwater crayfish. Biology, management and exploitation*: pp.365-400. London: Croom Helm.

Laurent P.J. 1988. *Austropotamobius pallipes* and *A. torrentium*, with observations on their interactions with other species in Europe. In D.M. Holdich & R.S. Lowery (eds), *Freshwater crayfish. Biology, management and exploitation*: pp. 341-364. London: Croom Helm.

Laurent, P.J. 1997. Commerce extérieur des écrevisses en France en 1996 et coup d'oeil à cette activité en Europe. *L'Astaciculteur de France* 53:8-11.

Laurent, P.J. & Forest, J. 1979 Donnée sur les écrevisses qu'on peut recontrer en France. *Pisc. Franç.* 56:25-40.

Lindqvist, O.V. 1988. Restoration of native European crayfish stocks. *Freshwater Crayfish* 7:6-12.

Lodge, D.M. & Hill, A.M. 1994. Factors governing species composition, population size, and productivity of coolwater crayfishes. *Nordic J. Freshw. Res.* 69:111-136.

Lodge, D.M., Stein, R.A., Brown, K.M., Covich, A.P., Brönmark, C., Garvey, J.E. & Klosiewski, S.P. in press. Predicting impact of freshwater exotic species on native biodiversity: challenges in spatial scaling. *Aust. J. Ecol.* 23.

Lowery, R.S. & Holdich, D.M. 1988. *Pacifastacus leniusculus* in North America and Europe, with details of the distribution of introduced and native crayfish species in Europe. In D.M. Holdich & R.S. Lowery (eds), *Freshwater crayfish. Biology, management and exploitation*: pp. 283-308. London: Croom Helm.

Machino, Y. 1996. L'écrevisse de torrent *Austropotamobius torrentium* (Schrank, 1803) est au bord de l'extinction en Italie. *L'Astaciculteur de France* 49:9-12.

Machino, Y. 1997. Présence de l'écrevisse de Californie (*Pacifastacus leniusculus*) en Italie. *L'Astaciculteur de France* 52:2-5.

Mancini, A. 1986. *Astacicoltura. Allevamento e pesca dei gamberi d'acqua dolce*. Bologna: Edagricole.

Matthews, M.A. & Reynolds, J.D. 1995. A population study of the white-clawed crayfish *Austropotamobius pallipes* (Lereboullet) in an Irish reservoir. *Biology and Environment: Proc. R. Ir. Acad.* 95B:99-109.

Mazzoni, D. & Minelli, G. 1996. Presenza e analisi di popolazioni del gambero d'acqua dolce (*Austropotamobius pallipes italicus* Faxon) in provincia di Bologna. *Risorsa Fauna* 4:1-15.

Mazzoni, D., Minelli, G., Quaglio, F. & Rizzoli, M. 1996. Sulla presenza del gambero della Lousiana *Procambarus clarkii* (Girard, 1852) nelle acque interne dell'Emilia-Romagna. Atti del Convegno Nazionale 'Il contributo dei progetti di ricerca allo sviluppo dell'acquacoltura nazionale', pp. 75-82.

Mazzoni, D., Nobile, L., Quaglio, F. & Restani, R. 1997a. Diffusione dei gamberi di acqua dolce in Emilia-Romagna: problemi ecopatologici legati all'introduzione di specie alloctone. *Boll. Soc. It. Pat. Ittica* 21:45-51.

Mazzoni, D., Quaglio, F., Nobile, L. & Rizzoli, M. 1997b. Exotic species of freshwater crayfish in Emilia-Romagna (Northern Italy). Abstract Volume of the International Workshop 'The introduction of alien species of crayfish in Europe. How to make the best of a bad situation?', p. 38.

Momot, W.T. 1988. *Orconectes* in North America and elsewhere. In D.M. Holdich & R.S. Lowery (eds), *Freshwater crayfish. Biology, management and exploitation*: pp. 262-282. London: Croom Helm.

Momot, W.T., Gowing, H. & Jones, P.D. 1978. The dynamics of crayfish and their role in the ecosystems. *Amer. Midl. Nat.* 99:10-35.

Mori, M., Mandich, A. & Isola, G. 1988. Relazioni taglia-peso e fecondità di tre popolazioni di *Austropotamobius pallipes* (Lereboullet) (Crustacea, Decapoda, Astacidae) della provincia di Genova. *Boll. Mus. Stor. Nat. Lunigiana* 6-7:163-167.

Mori, M., Mandich, A. & Isola, G. 1989. Morfometria e maturità sessuale di due popolazioni di *Austropotamobius pallipes falcisianus* (Ninni) (Crustacea, Astacidae) nella Provincia di Genova. *Boll. Mus. Ist. Biol. Univ. Genova* 54-55:49-58.

Mori, M., Salvidio, S., Isola, G. & Cresta, P. 1996. Struttura demografica di quattro popolazioni del gambero d'acqua dolce, *Austropotamobius pallipes* (Lereboullet), della Liguria (Italia). *Ann. Mus. Civ. St. Nat. 'Giacomo Doria'* 91:341-354.

New, T.R. 1995. *Introduction to invertebrate conservation biology*. Oxford, New York, Melbourne: Oxford University Press.

Ninni, A.P. 1865. Sulla mortalità dei gamberi (*Astacus fluviatilis*, L.) nel Veneto e più particolarmente nella provincia trevigiana. *Atti Ist. Veneto Sci., Let. Arti* 10:5-12.

Nolte, W. 1933. Über die Möglichkeit, saure Gewässer mit Krebsen zu bewirtschaften. *Fish. Zeit.* 36:49.

Palmer, M. 1994. Action plan for the conservation of the native crayfish *Austropotamobius pallipes* in the United Kingdom. JNCC Report No. 193. Joint Nature Conservation Committee: Peterborough.

Pascutto, T., Maffeo, B. & Vanzi, R. 1991. Censimento preliminare di *Austropotamobius pallipes* Le. in alcune vallate delle Prealpi Biellesi. *Riv. Piem. St. Nat.* 12:216-219.

Pensabene, G. 1992. Relazione preliminare sulle caratteristiche chimico-fisiche delle acque del Lago di Massaciuccoli. In M. Cenni (ed.) *Eutrofizzazione del lago di Massaciuccoli, Atti del Convegno 'Problemi di eutrofizzazione e prospettive di risanamento del Lago di Massaciuccoli'*: pp. 63-85. Pisa: Editrice Universitaria Litografia Felici.

Pretzmann, G. 1987. A contribution to a historic analysis of Mediterranean freshwater decapods chorology. *Inv. Pesq.* (Supl.1) 51:17-25.

Pullin, R.S.V. 1994. Exotic species and genetically modified organisms in aquaculture and enhanced fisheries: ICLARM's position. *Naga, ICLARM Q.* 17:19-24.

Quaglio, F., Nobile, L. & Mazzoni, D. 1997. Temnocephalans (platyhelminthes) ectosymbionts in Australian freshwater crayfish (*Cherax destructor* Clark) introduced to northern Italy. Abstract Volume of the International Workshop 'The introduction of alien species of crayfish in Europe. How to make the best of a bad situation?', pp. 41-42.

Quattrocchi, L., Gibertini, G., Mancini, A. & Bonifazi, A. 1996. *Orconectes limosus* (Crustacea Decapoda Cambaridae), un gambero esotico d'acqua dolce, nelle acque del Centro Italia. Contributi scientifici del LVII Congresso Nazionale dell'Unione Zoologica Italiana, 36.

Rahe, R. & Soylu, E. 1989. Identification of the pathogenic fungus causing destruction to Turkish crayfish stocks (*Astacus leptodactylus*). *J. Invertebr. Pathol.* 54:10-15.

Rogers, W.D. & Holdich, D.M. 1995. Conservation and management of crayfish in Britain. *Freshwater Crayfish* 10:92-97.

Ruesink, J.L., Parker, I.M., Groom, M.J. & Karieva, P.M. 1995. Reducing the risks of non indigenous species introductions: guilty until proven innocent. *Bioscience* 45:465-477.

Salvidio, S., Cresta, P. & Doria, G. 1993. Preliminary survey on the distribution of the freshwater crayfish *Austropotamobius pallipes* in Liguria, N.W. Italy. *Crustaceana* 65:218-221.

Schäperclaus, W. 1935. Die Ursache der pestartigen Krebsterben. *Zeitschr. Fischerei* 33:343-366.

Schäperclaus, W. 1992. Aphanomycosis of crayfish. In W. Schäperclaus, H. Kulow & K. Schreckenbach (eds) *Fish diseases* 2:616-621.

Söderbäck, B. 1995. Replacement of the native crayfish *Astacus astacus* by the introduced species *Pacifastacus leniusculus* in a Swedish lake: possible causes and mechanisms. *Freshwater Biology* 33:291-304.

Söderbäck, B. 1996. Reproductive interference between *Astacus astacus* and *Pacifastacus leniusculus*. Abstract Volume, XI Symposium of the International Association of Astacology, p. 26.

Söderhäll, K., Dick, M.W., Clark, G., Fürst, M. & Constantinescu, O. 1991. Isolation of *Saprolegnia parasitica* from the crayfish *Astacus leptodactylus*. *Aquaculture* 92:121-125.

Sogandares-Bernal, F. 1965. Parasites from Louisiana crayfish. *Tulane Stud. Zool.* 12:79-85.

Sokol, A. 1988. The Australian yabby. In D.M. Holdich & R.S. Lowery (eds), *Freshwater crayfish. Biology, management and exploitation*: pp. 401-425. London: Croom Helm.

Tedaldi, G., Ciani, C. & Fabbri, R.A. 1997. Indagine preliminare sulla distribuzione del granchio di fiume *Potamon fluviatile fluviatile* (Herbst, 1785) in Romagna. *Quad. Studi Nat. Romagna* 7:17-22.

Unestam, T. 1969. Resistance to the crayfish plague in some American, Japanese and European crayfishes. *Rpt. Inst. Freshw. Res. Drottningholm* 49:202-209.

Unestam, T. 1973. Significance of diseases in freshwater crayfish. *Freshwater crayfish* 2:136-150.

Vogt, G. 1997. Diseases of native and naturalized freshwater crayfish in Europe. Abstract Volume of the International Workshop 'The introduction of alien species of crayfish in Europe. How to make the best of a bad situation?', pp.17-20.

Welcomme, R.L. 1988. International introductions of inland aquatic species. FAO Fisheries Technical Paper 294.

Westman, K., Pursiainen, M. & Westman, P. 1990. Status of the crayfish stocks in Europe. Report of the FAO European inland Fisheries Advisory Commission (EIFAC) Working Party on Crayfish. Kalatutkimuksia-Fiskundersokningar 3. Helsinki.

Wynne, G., Avery, M., Campbell, L., Gubbay, S., Hawkswell, S., Juniper, T., King, Newberry, M.P., Smart, J., Steel, C., Stones, T., Stubbs, A., Taylor, J., Tydeman, C. & Wynde, R. 1995. *Biodiversity challenge: an agenda for conservation in the UK* (2nd edition). Sandy: RSPB.

The situation in Belgium, France, and Luxembourg

JACQUES, C.V. ARRIGNON
Association Française de Limnologie, Compiègne, France
PIERRE GÉRARD
Station Recherche Forest, Gembloux, Belgium
ADY KRIER
Service de la Chasse et de la Pêche, Luxembourg
PIERRE J. LAURENT
Association des Astaciculteurs de France, Avonnex, Port à Marin, Thonon les Bains, France

ABSTRACT

Since the beginning of the 19th century, alien crayfish have been introduced into Wallonian Belgium, France, and Luxembourg. The main alien species belong to the American genera: *Orconectes*, *Pacifastacus*, and *Procambarus*. The alien species, *Astacus leptodactylus,* has also been introduced from Turkey. All are pioneer species and consequently severe competitors of the native species, *Astacus torrentium* and *Austropotamobius pallipes*. Some alien crayfish may be vectors of diseases, such as the crayfish plague fungus. *Aphanomyces astaci*. The genus *Procambarus* has a negative action on the environment due to its burrowing in dams and dykes. The national and European laws are not sufficient to regulate the populations of alien crayfish. An effort must be initiated by the scientific community to change a very bad ecological situation.

1 INTRODUCTION

The status of alien crayfish is considered in three neighboring countries having many hydrological relationships, i.e. Belgium (Wallonia), France, and Luxembourg (Fig. 1). These form part of the large quaternary plain of northwestern Europe, drained by a net of rivers and canals having closed hydrobiological characteristics.

The basins of the rivers Adour, Rhône (particularly the Camargue delta), Loire (particularly the Brière, with important marshes on the right side of the mouth), Seine, Mosella (in France and Luxembourg), and Maas (in Belgium and France) are mainly considered from an astacological point of view. Other hydrological basins are smaller but important to, e.g. polders in Belgium and in France, and coastal streams and lakes in Brittany, Normandy, Picardy, Languedoc, and Provence.

2 ALIEN CRAYFISH

Four species of alien crayfish are present in this part of Europe, *Orconectes limosus*, *Pacifastacus leniusculus*, and *Procambarus clarkii*, all introduced from America.

Figure 1. Geography of Belgium, France, and Luxembourg.

Despite its eastern origins *Astacus leptodactylus*, is also considered an alien species in Western Europe (see Holdich this volume).

2.1 Genus *Orconectes*

Max von dem Borne introduced the genus Orconectes into Germany in 1890 and subsequently manually introduced into various continental countries. Later, the crayfish became widely established in rivers, streams, and canals, even those that were polluted. It is now probably the most widespread crayfish species in Europe, although it has only been extensively studied in Poland (see review by Momot 1988).

2.1.1 *Belgium*

This crayfish has been in the main canals and streams in the whole Belgium for a long time (Gérard 1986). In Belgium, *Orconectes limosus* (Fig. 2) is present in 30 out of 120 streams and 103 out of 388 ponds having crayfish (Herman & Denoel 1997).

2.1.2 *France*

The genus (Fig. 3) was first introduced in France by C. Raveret-Wattel in Normandy, near Fécamps in November 1896 with *Orconectes limosus*, then in April 1897 with *Orconectes virilis* but without success. According to P. Buffault (1925), the first successful stocking of *Orconectes limosus* was made in the River Cher, a tributary of the River Loire in the central part of France by Lesaule, a German citizen. The stocking took place from 1911 to 1913 with 2000 individuals imported from Germany (Arrignon 1997).

The species lives now in 874 locations distributed in almost all the French area (Vigneux 1997). It is present since 1945 in the large River Saône (a tributary of the R. Rhône). Commercial fishermen catch the crayfish along the shores with special traps. The crayfish is sold to local restaurants that make well-known dishes (tails 'gratin').

In the Camargue, the species was first recorded in 1976. A sampling program was started in 1988 and still continues. The results indicated that *Orconectes* was present in almost all the sites sampled, but the main populations were established in drainage and irrigation canals (Rosecchi & Crivelli 1997).

2.1.3 *Luxembourg*

The River Mosella, on the border between France, Luxembourg, and Germany, has been colonized for a long time by *O. limosus* (Fig. 4). The species has been found

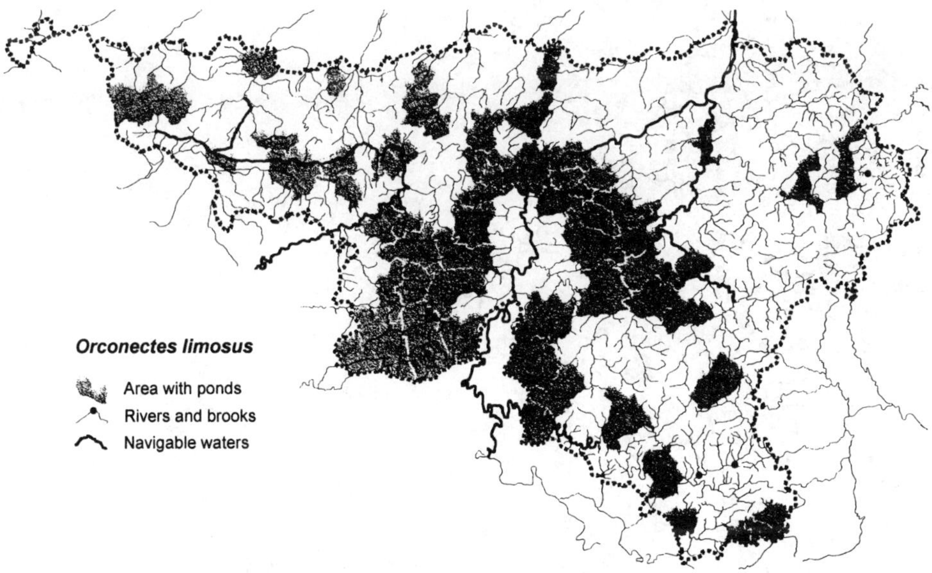

Figure 2. *Orconectes limosus* in Belgium (Wallonia).

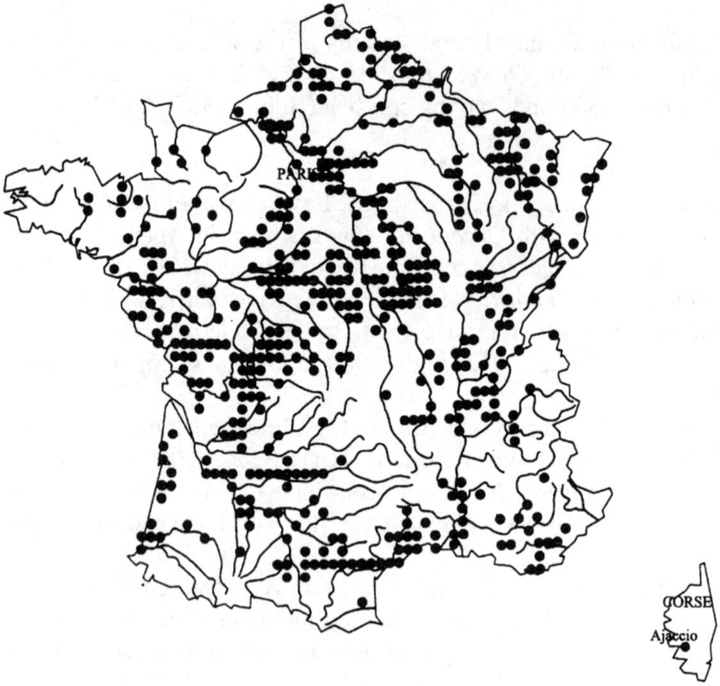

Figure 3. *Orconectes limosus* in France.

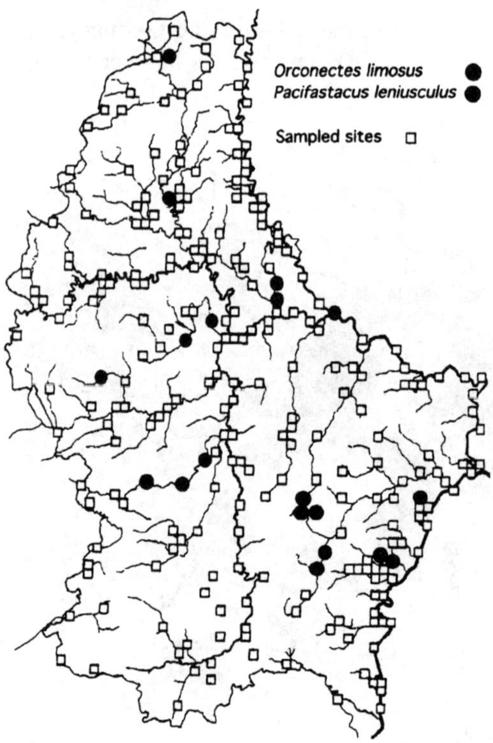

Figure 4. *Orconectes limosus* and *Pacifastacus leniusculus* in Luxembourg.

more recently downstream of the Sûre river, a tributary of the R. Mosella. Recently it was also found in the reservoir of the upper R. Sûre.

2.2 *Pacifastacus leniusculus*

Pacifastacus leniusculus was manually distributed mainly from the hatchery of Simontorp (Sweden) in 1972, but also directly in the central part of France from the Lakes Tahoe and Donner (USA) in 1974. The species is present in all the countries, being considered here in ponds and gravel-pits, in brooks and streams, occasionally in competition with *Orconectes*, *Austropotamobius pallipes*, *Astacus astacus*, and *A. leptodactylus* (see Laurent 1988).

2.2.1 *Belgium*

In Belgium (Fig. 5), *P. leniusculus* was found to be present in 39 ponds out of 388 having crayfish and in 27 out of 120 streams sampled (Herman & Denoel 1997).

2.2.2 *France*

In France (Fig. 6), an inquiry revealed 47 locations (29 in lakes and ponds) in 38 departments. In the central part of the country (Creuse), 27 new locations were occupied between 1992 and 1996 mainly in brooks having small populations of *A. pallipes* (Vigneux 1997). In Lake Leman (582 km^2) the species escaped from an experimental laboratory and has invaded the shores since 1976. The first *P. leniusculus* was caught in October 1983 by a commercial fisherman. Today on the French part of the lake this crayfish accounts for about 10 metric tons of the annual yield of the commercial fishermen. The crayfish is sold to restaurants along the shore, but sometimes it is sent to the central market of Paris (Rungis) (Laurent 1997b).

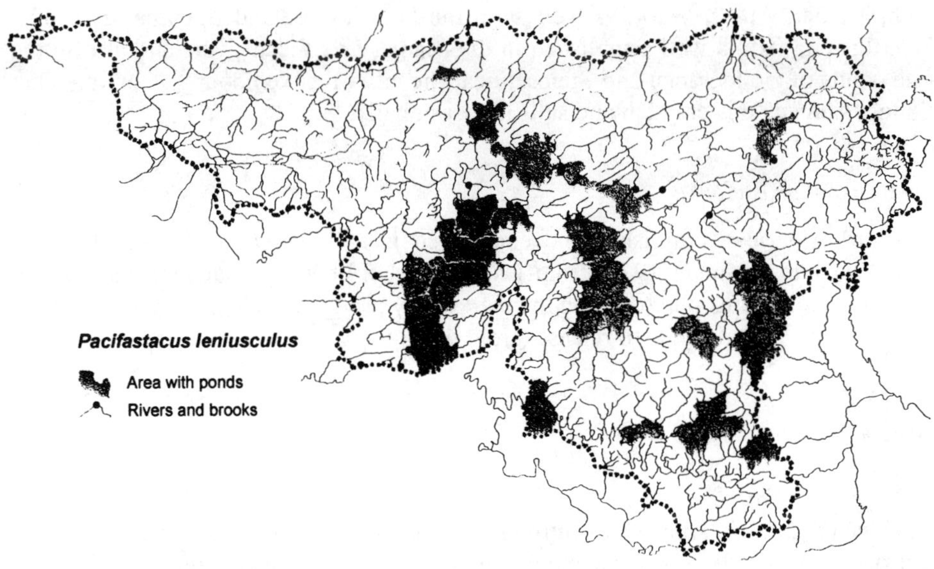

Figure 5. *Pacifastacus leniusculus* in Belgium (Wallonia).

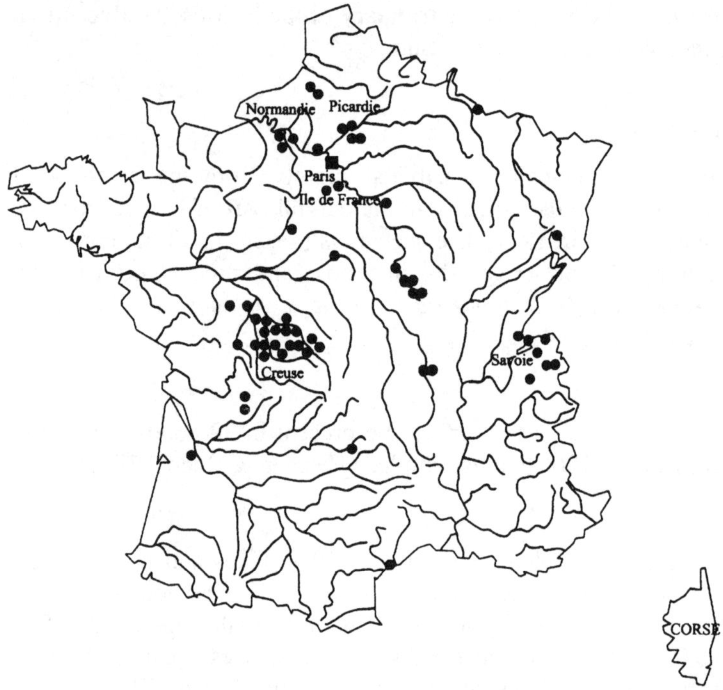

Figure 6. *Pacifastacus leniusculus* in France.

2.2.3 *Luxembourg*

The stocking with this species (Fig. 4) was decided by the Government in order to replace *Astacus astacus* destroyed by two waves of plague. From 1972 to 1978, 40,000 juveniles of *P. leniusculus* from Simontorp were introduced in the rivers Eisch, Mamer, Attert, Wark, as well as in the Pont-Misère and Bavigne reservoirs. Now, this crayfish is well established in the Mamer and Eisch rivers and also in the large ponds of Gonderange and Steinsel near the city of Luxembourg. A private club of anglers breeds this crayfish and sells it occasionally.

2.3 *Procambarus clarkii*

The third American crayfish to be considered is *Procambarus clarkii.* This was introduced into Spain in 1974, from Louisiana (USA). Its propagation has mainly been due to numerous purchases on the markets by private people having for stocking in their ponds. It is also due to releases from the aquaria.

2.3.1 *Belgium*

This crayfish is not yet present.

2.3.2 *France (Fig. 8)*

In France (Fig. 7), *P. clarkii* was introduced in 1974 in a fish farm near the Charente river on the west side of France and subsequently in some others areas. In 1995, the species had invaded 36 departments (Vigneux 1997).

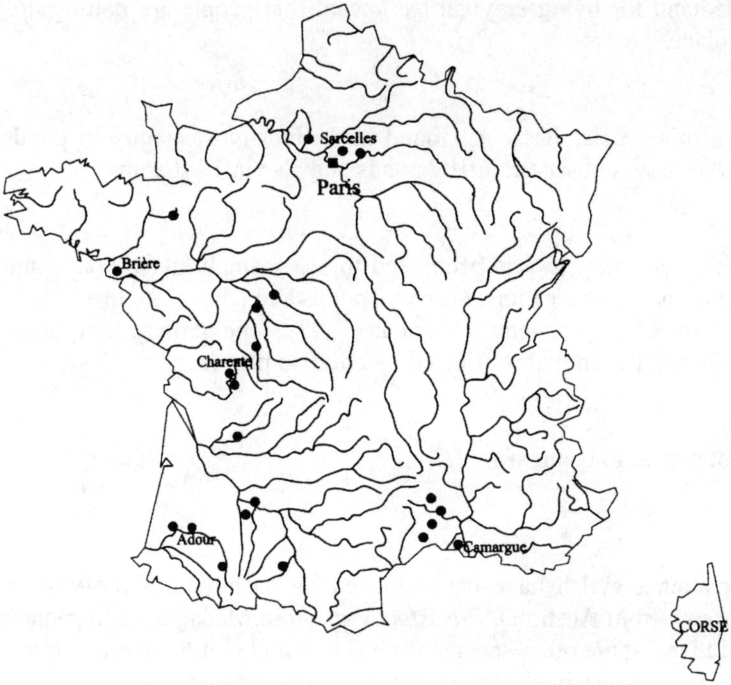

Figure 7. *Procambarus clarkii* in France.

Procambarus clarkii is a very active crayfish and has rapidly colonized rivers, streams, ponds, and marshes of the southern part of France, the marshes in Brittany, and some ponds in the northern part of the country. It spread slowly in the 1980s but more rapidly in the 1990s.

In the Camargue, *P. clarkii* was found in 1995. It remains rare and it has not been proved that this species has established a population. Nevertheless, it may be found in the numerous rice fields, a suitable habitat for proliferating (Rosecchi & Crivelli 1997) as has been found in Spain (see Gutierrez Yurrita et al. this volume).

An example of the rapid acclimatization of this species is found in a pond near Paris (Sarcelles). Seventeen specimens were introduced in 1985; they are now providing a harvest of between 500 and 800 kg/ha/yr (Arrignon 1997).

Procambarus clarkii escaped from a private crayfish farm in 1981 and has colonized a very large part of the swamp of Brière (200 km^2, South Brittany). The eel fishermen quickly learned to catch the crayfish and to sell it to restaurants and private persons.

2.3.3 *Luxembourg*

This crayfish is not yet present.

2.4 *Astacus leptodactylus*

Due to the lack of *Astacus astacus* for gastronomic purposes, *Astacus leptodactylus* has been imported for many years from Turkey. Crayfish trade figures for France are regularly published and show that *A. leptodactylus* dominates imports (Laurent

1997a). The high demand for living crayfish has meant that people are putting this species all over the place.

2.4.1 *Belgium*

In Belgium (Fig. 8), this species has been found in 26 locations, mainly in ponds (Gerard 1986). In 1996, it was observed in 109 ponds and also in 17 streams.

2.4.2 *France*

In France (Fig. 9), *A. leptodactylus* has been used to stock small lakes, ponds and gravel-pits. It can be found in 55 locations in rivers, ponds and lakes (22). In 1995, its presence was noted in 43 departments (Vigneux 1997). The propagation of *A. leptodactylus* is erratic and the animal is affected by crayfish plague.

2.4.3 *Luxembourg*

This species does not appear to be present yet.

2.5 *Other species*

A number of other alien crayfish have been stocked in water bodies of Western Europe, e.g. *Cherax* spp. from Australia, *Parastacus* sp. from Madagascar (Arrignon 1997) and, here and there, some other species from South and North America. None of these species have become established in Belgium, France, or Luxembourg.

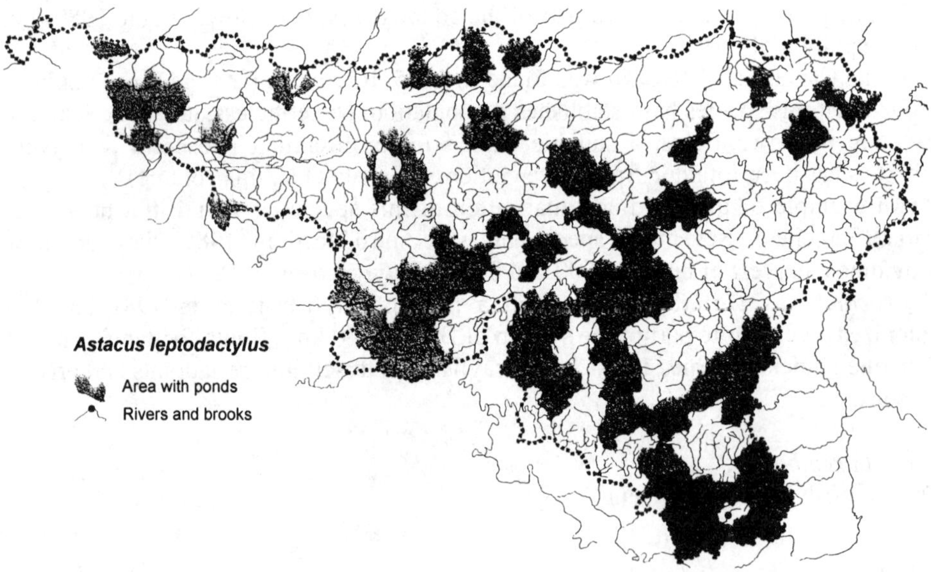

Figure 8. *Astacus leptodactylus* in Belgium (Wallonia).

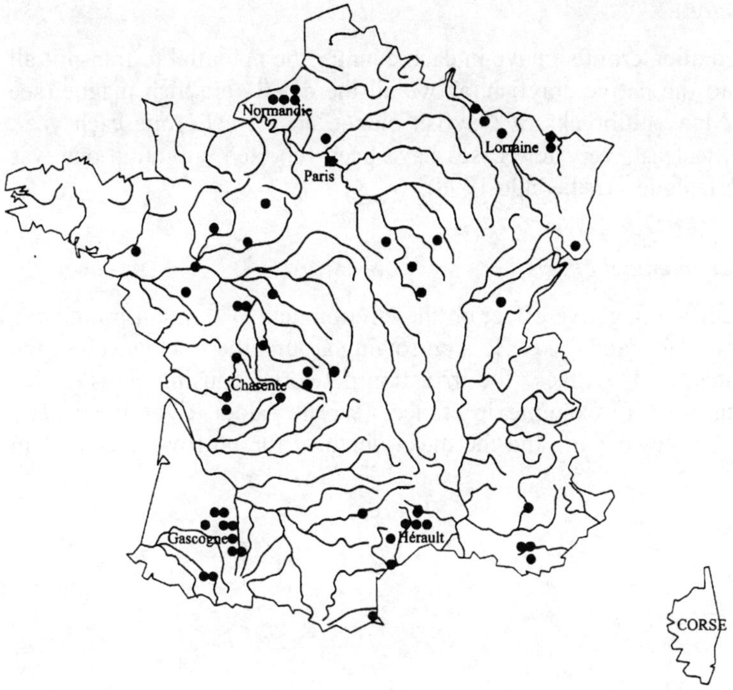

Figure 9. *Astacus leptodactylus* in France.

3 ENVIRONMENTAL EFFECTS

3.1 *Competition*

When they occur, *Orconectes limosus*, *Pacifastacus leniusculus*, *Procambarus clarkii*, and *Astacus leptodactylus* are severe competitors of native crayfish species in Belgium, France, and Luxembourg.

Nevertheless, in France, it was observed than due to intensive gravel extraction in the River Saône, and the presence of a new predator, the catfish, *Silurus glanis*, the population of *O. limosus* has decreased considerably.

3.1.1 *Belgium*

Orconectes limosus and *P. leniusculus* have increased their distribution during the last 10 years (between the two national surveys). On the contrary, *A. astacus* has disappeared during the same period in 67% of its locations (Herman & Denoel 1997).

3.1.2 *France*

From ponds located in the upper basins, *O. limosus* is invading brooks and small streams, dominating over the native *A. astacus* and *A. pallipes*. Laurent (1988) reported that native crayfish in France had disappeared from many waters occupied by *O. limosus*. In Brière, *P. clarkii* is a severe competitor of *O. limosus* that have been present in these marshes for a long time Laurent (1997b).

3.2 *Pathological vectors*

The newly introduced alien crayfish have in each country the potential to transmit all the typical diseases to the native crayfish, above all the deadly crayfish plague (see Vogt this volume). Many outbreaks of crayfish plague occur in Europe each year. However, in France, at least, very few cases have been reported, although one was recently (Machino & Diéguez-Uribeondo 1998).

3.3 *Ecological and economical effects*

Procambarus clarkii has a negative effect on the environment. The main problem is due to its burrowing in dams and dykes. It is burrowing around the foundations of the sluices and water gates. It also attacks the irrigation networks, causing considerable damages to the agriculture. For example, in France, near the Adour River, the density of the burrows is two holes per m^2 and the mean depth of the burrows was 1.21 m (Picard 1991).

4 REMEDIES

4.1 *Eradication*

Manual eradication is not effective for a well-established population of alien crayfish.

Overfishing just limits the population of crayfish in the area being considered but with the risk of spreading the species elsewhere (Arrignon 1997).

The use of poisons (rotenone and others) is strictly forbidden in the running waters of the countries under consideration. Nevertheless, in France, it is not forbidden in the ponds having no relationship with running waters. All the chemical means used in agriculture (weed killers, pesticides) are more effective on the native crayfish than on the aliens. Laurent (1995) and Roqueplo et al. (1995) have carried out experiments with organophosphate insecticides and these are reviewed by Holdich, Gydemo & Rogers (this volume). There is presently no mean to eradicate any of the alien crayfish species.

4.2 *Regulations*

4.2.1 *European regulations*

The Convention for the conservation of European wildlife and natural habitats ('Bern Convention', Appendix III) has *Astacus astacus*, *Austropotamobius pallipes*, and *Austropotamobius torrentium* listed as protected species, but has not forbidden the introduction of alien species. The Convention states that each contracting party shall take appropriate and necessary legislative and administrative measures to ensure the protection of the species specified in Appendix III. The measures to be taken do not include the fight against the introduction of new and competitive species.

The Court of Justice of the European Communities does not forbid the import of living crayfish, included alien species, due to the principle of Free Trade.

Consequently, the national regulations against the import of living crayfish from foreign countries are illegal.

4.2.2 *National regulations*

The measures are different from country to country. In France and Luxembourg, there are national regulations. In Belgium, each Region has its own regulations. In Belgium and Luxembourg, the capture of the native crayfish is often strongly restricted or forbidden and the capture of alien crayfish is allowed. On the contrary, in France the capture of the native crayfish is permitted during the open season without limit of catches, the catching of alien crayfish is allowed, but the transport of living animals is strictly forbidden.

Belgium

The catching of the alien crayfish is authorized during the open season for catching fish. This period differs depending on the categories of the rivers. No legal catching size exists.

France

In the French Act (8/11/85, No. 85-1189), *Orconectes limosus* and *Pacifastacus leniusculus* are considered as two species causing biological disturbances in nature. Consequently, their introduction in wild waters and the transportation of the living animals are forbidden. A special regulation concerns more strictly *Procambarus clarkii*. The catching of this alien crayfish must be immediately followed by the killing of the animals.

Luxembourg

The regulations (31/08/1986) prohibit all catching of crayfish in the state waters (1st category). In private waters (2nd category), the catching of *Orconectes limosus* and *Pacifastacus leniusculus* males is authorized from June 1 to September 30 inclusive. The legal catching size is 8 cm for *O. limosus* and 12 cm *P. leniusculus* (from the eye to the telson). There are special regulations concerning the rivers bordering Germany, where the open season for the two species is from the 15 June to November 1 and the catching size 10 cm for *P. leniusculus* and 8 cm for *O. limosus*.

5 CONCLUSIONS

In order to protect the native crayfish in Wallonian Belgium, France, and Luxembourg, it is necessary to have a good view of the distribution and ecological effects of the native alien crayfish in the neighboring countries, as well as in the countries themselves. In this respect, careful mapping of the species should be carried, as has been done in France (Vigneux et al. 1993). There is also a need to harmonize national regulations and to initiate a set of European regulations concerning alien crayfish which are more restrictive than those in Article 11 (2b) of the Bern Convention, i.e. 'Each Contracting Party undertakes to strictly control the introduction of non-native species'.

On the other hand, we must take into consideration that the four alien crayfish species are now in place and acclimatized, with their economical, ecological, pathological advantages and disadvantages. It seems opportune to consider these animals with pragmatism and, without the ability of a total eradication, to organize the best control and management of their stocks.

The scientific community of astacologists must constantly attract the attention of the European Commission on the problem of alien crayfish and give some useful advice to the European organizations and nations. Consequently, it would be opportune to propose to creation of a European Astacological Committee (see concluding chapter of this volume).

REFERENCES

Arrignon, J.C.V. 1997. Status of the foreign crayfish in France. *Freshwater Crayfish* 11:665-670.

Gérard, P. 1986. Les différentes espèces d'écrevisses en Belgique et leur répartition géographique. *Station de Recherches forestières et hydrobiologiques* – Travaux – Série D.N 54.

Herman, D. & Denoel, P. 1997. *Rapport final d'une convention de recherche sur la distribution des écrevisses en Région Wallonne*. (unpublished).

Laurent, P.J. 1988. *Austropotamobius pallipes* and *A. torrentium*, with observations on their interactions with other species in Europe. In D.M. Holdich & R.S. Lowery (eds), *Freshwater crayfish: biology, management and exploitation*. London: Croom Helm (Chapman & Hall).

Laurent, P.J. 1995. Eradication of unwanted crayfish species for astacological management purposes. *Freshwater Crayfish* 8:121-133.

Laurent, P.J. 1997a. Commerce extérieur des ecrevisses en France en 1996 et coup d'oeil à cette activité en France. *L'Astaciculteur de France* 53:8-11.

Laurent, P.J. 1997b. Crayfish introductions into France and in the world, history and consequences. *Bull. Fr. Pêche Piscic.* 344-345:345-357.

Machino, Y. & Diéguez-Uribeondo, J. 1998. Un cas de peste des écrevisses en France dans le bassin de la Seine. *L'Astaciculteur de France* 54:2-11.

Momot, W.T. 1988. *Orconectes* in North America and elsewhere. In D.M. Holdich & R.S. Lowery (eds), *Freshwater crayfish: biology, management and exploitation*: pp. 262-282. London: Croom Helm (Chapman & Hall).

Picard, B. 1991. L'Ecrevisse *Procambarus clarkii* dans la région de Dax – Landes. *Rapport fait au Conseil Supérieur de la Pêche*. Paris

Roqueplo, C., Laurent, P.J. & Neveu, A. 1995. *Procambarus clarkii* Girard (ecrevisse rouge des marais de Louisiana). Synthèse sur les problèmes poses par cette espèce et sur les essais pour contrôler ses populations. *L'Astaciculteur de France* 45:2-17.

Rosecchi E. & Crivelli, A.J. 1997. Will introduced crayfish proliferate in Camargue (Rhône river delta, southern France)? International Workshop – The introduction of alien species of crayfish in Europe. Florence, Sept. 24-27. Abstract Volume:42.

Vigneux, E. 1997. Introductions of freshwater decapod crustaceans into France. Can we speak of management. *Bull. Fr. Pêche Piscic.* 344-345:357-373.

Vigneux, E., Keith, P. & Nöel, P. 1993. Atlas préliminaire des Crustacés Décapodes d'eau douce de France. Paris: Muséum National d'Histoire Naturelle.

Distribution of crayfish species and legislation concerning crayfish in Switzerland

THOMAS P. STUCKI
Zoological Museum, University of Zürich, Zürich, Switzerland
ERICH STAUB
Swiss Environmental Protection Agency (BUWAL), Fisheries Section, Bern, Switzerland

ABSTRACT

Three native crayfish species live in Switzerland: *Austropotamobius torrentium*, *Austropotamobius pallipes*, and *Astacus astacus*. *A. torrentium* occurs in many brooks in the northeastern part of Switzerland. *A. pallipes* lives in brooks, canals, and ponds in the western part of Switzerland and in the valleys of the alpine region. Many populations of both species have disappeared. The full extent of this decrease is not documented. *A. astacus* can be found in many lakes, ponds, rivers and brooks throughout Switzerland. Cultivated ponds have been restocked after population collapses. Four introduced species live in Switzerland. *Astacus leptodactylus* can be found in many lowland lakes and ponds at present. *Orconectes limosus* forms strong populations in the western part, that are spreading eastwards along the River Aare and can be found in single populations in several parts of Switzerland. *Pacifastacus leniusculus* has been found in a few ponds, in Lake Geneva, in two brooks and in the River Aare. *Procambarus clarkii* lives in four ponds in the northern part of Switzerland. The legislation in Switzerland tries to avoid the appearance of alien species and to protect the native species. Possible ways of introduction and the situation of crayfish species in Switzerland are discussed.

1 INTRODUCTION

Carl (1920) and Bott (1972) described numerous populations of *Austropotamobius torrentium*, *A. pallipes*, and *Astacus astacus* in Switzerland. *Austropotamobius torrentium* was described in lakes and brooks in the north-eastern area of Switzerland, bordering Austria and Bavaria (Germany). According to Carl (1920) and Albrecht (1982) *A. pallipes* was present in lakes, streams, ponds, and brooks in the western part of Switzerland, in Graubünden (eastern Alpine region), and in Ticino. *Astacus astacus* was reported from ponds, rivers and lakes all over Switzerland with the exception of the higher Alpine region. Bott (1972) stated that *Astacus leptodactylus* and the American crayfish species, present in other European countries, had not been found in waters in Switzerland. Büttiker (1987) mentioned the occurrence of *A. leptodactylus* and the American crayfish *Orconectes limosus* in Switzerland and – at that time – the suspected presence of *Pacifastacus leniusculus*. First catch of *O. limosus* was reported from Lake Geneva in 1976 and of *A. leptodactylus* in 1979 (Laurent 1983). The occurrence of a fourth introduced species, *Procambarus clarkii* was described by Stucki (1997).

A great number of water bodies in Switzerland offer splendid conditions for crayfish life. The water collected in small brooks and moorlands in the Alpine region, the Jura Mountain Chain and in the lowland hills, drain into four major drainage systems of Europe: Rhine, Rhône, Danube, and Po. Natural river beds and lake floors offer gravel, rock, clay and sandy habitats with submerged plant cover and reed areas. The high human population density, mainly in the lowland of Switzerland, however, puts great pressure on the ecological quality of watercourses. The water quality of lakes and rivers has improved in the last few decades due to sewage disposal plants. Short, acute water pollution still occurs and leads to death of the local water fauna in restricted geographical areas. A major threat to crayfish and fish populations today are physical modifications of water bodies such as corrections, obstructions, and covering of rivers or brooks, as well as fill-ups in lakes.

A well-known threat to European crayfish species is the crayfish plague fungus *Aphanomyces astaci*, which has caused severe damage to endemic populations all over Europe (Ackefors this volume, Vogt this volume). Carl (1920) described plague outbreaks at the end of the 19th century. Today crayfish plague incidences are reported almost every year somewhere in Switzerland. Outbreaks in small, not cultivated crayfish populations, often remain undiscovered. Büttiker (1987) stated that crayfish plague did not lead to such decimation in Swiss crayfish populations, as it has happened in many other European countries (see Ackefors this volume).

In this paper an outline of the current crayfish distribution in Switzerland and the legislation relevant to crayfish management is given.

2 LEGISLATION CONCERNING CRAYFISH

The three native crayfish species in Switzerland, *Austropotamobius torrentium*, *A. pallipes*, and *Astacus astacus*, are protected on the European level according to the ' Convention on the conservation of European wildlife and natural habitats' (Bern Convention, agreed in 1979). According to the Swiss fisheries legislation, *A. pallipes* and *A. torrentium* are considered as strongly endangered, *A. astacus* as endangered. The catch of native crayfish is restricted by protection periods and minimum size limits fixed in the Federal legislation (Table 1). The cantonal fisheries administrations have the possibility to strengthen these protection measures. In most cantons, the protection period lasts from the beginning of October until the end of July. The

Table 1. Annual protection periods and minimum size limits fixed in the Swiss fisheries legislation.

Species	Protection period	Minimum size limit
Astacus astacus	40 weeks	12 cm
Austropotamobius pallipes	40 weeks	9 cm
Austropotamobius torrentium	40 weeks	9 cm

native species can be transported alive and, within their natural range of distribution, they can be stocked into other waters.

There are no protection measures for the exploitation of alien species in the federal legislation. However, their exploitation is restricted by the prohibition of transporting living alien crayfish species. This restriction was decided in 1991 at the federal level. The prohibition of transporting alien crayfish alive or their stocking in any water body (aquarium, open waters, etc.) is in accordance with the 1987 resolution of the International Symposium of Astacology (IAA 1987). The prohibition is as valid for inland transport as for importation of crayfish, however governmentally – approved exceptions for research and public exhibitions do exist.

The importation of living crayfish is only permitted for the yabby, *Cherax destructor,* from Australia. These crayfish have to be transported cooled and closed in boxes until they arrive at the food supplies, which must be correctly instructed that these animals may only be stored in the refrigerator. As these crayfish are kept refrigerated and never come into contact with an aquatic environment, the risk of escape is considered low. The import of crayfish needs a sanitary certificate by the crayfish supplier in Australia.

3 DISTRIBUTION OF CRAYFISH SPECIES

3.1 *Methods of distribution assessment*

Detailed inventories of crayfish distribution are available from about 40% of the Swiss territory (Cantons Bern, Thurgau, Waadt, and Wallis). This information was collected by local fishermen and was assessed by scientific staff using traps and by making direct observations at night.

Verification of dubious declarations in these inventories, assessments of crayfish presence in potential habitats and sampling in additional areas were carried out by the Zoological Field Station of the University of Zürich and a group of interested people from the Nature Protection Association of Switzerland (Stucki & Jean-Richard in prep.). During these assessments, lake and river populations of crayfish were sampled by scuba diving and by using traps. Specimens caught by nets of commercial fishermen were also examined. Populations in brooks and small ponds were sampled at night by using electric torches.

3.2 *Distribution of native crayfish species*

Because the distribution of *A. astacus* has been strongly modified by human in the past, the question, whether the species is really indigenous, is not finally answered. However, it is considered as a native crayfish species in Switzerland (Büttiker, 1987). The presently known distribution of the native crayfish species in Switzerland is shown in Figure 1.

Austropotamobius torrentium still occurs in many brooks in the northeastern part of Switzerland. Although the presence of this species is possible in several lakes in this area, its occurrence was proved only in Lake Ägeri, in low densities.

Austropotamobius pallipes lives in brooks, canals and ponds in the western part of

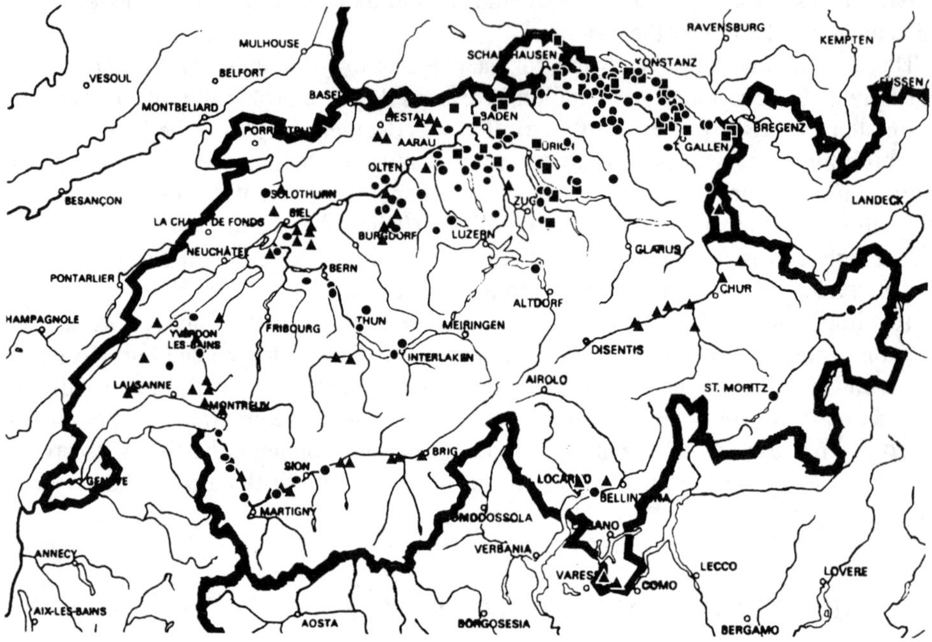

Figure 1. Distribution of native crayfish species in Switzerland. ■: *Austropotamobius torrentium*, ▲: *Austropotamobius pallipes*, ●: *Astacus astacus*.

Switzerland and in the valleys of the alpine region. It can be found in mountain lakes up to an altitude more than 1000 m where it has been introduced. It probably still occurs in some lakes in the western part of Switzerland but is supposed to have disappeared from Lake Neuchâtel, Lake Murten and parts of Lake Geneva.

Astacus astacus can be found in many ponds, lakes, rivers, and brooks in Switzerland. In lakes, it usually occurs in medium densities in restricted areas, often together with the introduced species, *A. leptodactylus*. In cultivated ponds it can reach high densities. After population collapses, populations of *A. astacus* are usually restocked.

3.3 *Distribution of alien crayfish species*

At least, four introduced species are living in Switzerland. There is the European species *Astacus leptodactylus*, and three American species, *Orconectes limosus*, *Pacifastacus leniusculus,* and *Procambarus clarkii* (Stucki 1997). The presently known distribution of the alien crayfish species in Switzerland is shown in Figure 2.

Astacus leptodactylus is present in many lakes and ponds in the lower parts of Switzerland. The density is usually very high. It is supposed that the species has replaced and is still replacing *A. astacus* and probably also *A. torrentium* in many lakes. Two large lake-populations of *A. leptodactylus* (Lake Hallwil and Lake Zug) collapsed in the last couple of years because of the crayfish plague. In both lakes the species has been seen again afterwards.

Orconectes limosus forms strong populations in lakes of the western part of Switzerland and in Lake Sihl and Lake Lauerz. In the lakes of western Switzerland (Lake

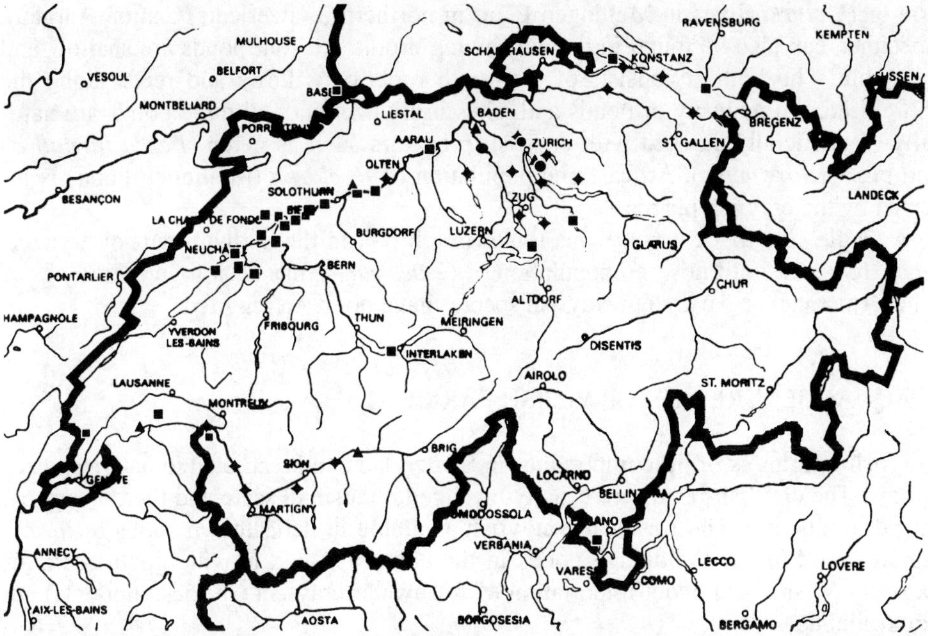

Figure 2. Distribution of alien crayfish species in Switzerland. ◆: *Astacus leptodactylus*, ■: *Orconectes limosus*, ▲: *Pacifastacus leniusculus*, ●: *Procambarus clarkii.* Lacustrine populations are indicated in the center of the lake.

Geneva, Neuchâtel, Murten), it may have replaced populations of *A. pallipes*. *O. limosus* has spread downstream in the River Aare. It has also been found in drainage canals and slow flowing waterbodies alongside the river. The species is also present in the River Rhine north of Basel, in Lake Constance and Lake Lugano. *O. limosus* is known to be very tolerant of unfavourable environmental conditions. The water quality of the locations it colonized in Switzerland seems to be good enough to form possible habitats for native species, as well. *O. limosus* occurs in medium and high densities with large geographical distributions within the shorelines of the lakes.

Pacifastacus leniusculus is found in Lake Geneva and in two ponds in the northern part of Switzerland (Rumensee, Dättwil Pond). The ponds can be described as shallow and eutrophic waters with thick sediment layers covering the substratum. The shoreline consists of roots and patches of reed. In Dättwil Pond, *A. torrentium* was present decades ago. In both ponds, *P. leniusculus* is suspected of carrying the crayfish plague fungus. The species has recently been found in a brook in southwestern Switzerland and together with *O. limosus* in the River Aare.

Procambarus clarkii is reported in three ponds near Zürich (Rumensee, Schübelweiher and Chatzensee). In Schübelweiher, it is thriving, whereas in the other two ponds only a few specimen were found. In Rumensee it forms a sympatric population with *P. leniusculus*. Despite the presence of *P. clarkii* and *P. leniusculus* in brooks draining into Lake Zürich, until now, no colonization of Lake Zürich has been observed. Upstream migration of adult *P. clarkii* in September and October of 1995 and 1996 was observed in the tributary of Schübelweiher (Stucki 1997). A fourth population of *P. clarkii* lives in Mellingen Pond in northern Switzerland (Canton Aargau). Specimen can also be found in the outflowing brook. All four ponds are shallow and eutrophic. The bottom consists of clay with roots of willows and reeds along the shore. Like the majority of ponds and lakes in Switzerland, all three ponds are naturally or artificially stocked with crayfish predators such as perch (*Perca fluviatilis*) and pike (*Esox lucius*). At least one population of *P. clarkii* (Mellingen Pond) is infected by the crayfish plague.

A single *Cherax destructor* was found in a brook in the northern part of Switzerland. However, until now, no population of *C. destructor* has been found. Findings of other American or Australian crayfish species have not been reported.

4 AQUACULTURE AND CRAYFISH FARMING

Crayfish farming is of little importance in Switzerland and restricted to native species by law. The crayfish-ponds are usually managed on a small scale and there is no real inland production. The freshwater crayfish available in delicatessen shops or in restaurants are imported, although ponds in the country would provide good stocks of native crayfish. Data about importation of freshwater crayfish (species, amounts) are not available.

5 DISCUSSION OF THE PRESENT SITUATION

The alien crayfish species in Switzerland have been introduced accidentally or illegally. *Astacus leptodactylus* most probably had escaped or was released from holding facilities in hotels, restaurants and delicatessen shops. *Pacifastacus leniusculus*, which has been imported for food supply as well, could have had the same path of introduction. Its distribution pattern implies that the introductions were more on a private scale. Both species could have been used to stock water bodies by insufficiently informed persons. In former years, *Procambarus clarkii* was available in pet-shops. It may have been introduced into the wild by hobby aquarists. *Orconectes limosus* most probably has been transferred from other European waterbodies into Lake Geneva and other lakes of the western part of Switzerland, from where it colonizes connected water bodies.

Today, the interest in crayfish as food is rather limited in Switzerland. Most people do not even know that there are crayfish species of a suitable size for eating in the country.

The Swiss legislation tries to avoid the appearance of further alien species and to protect the native species. In recent years, the fisheries administration was increasingly confronted with the presence of alien species and outbreaks of the crayfish plague. There are several ways to solve these problems: 1. To increase the knowledge on crayfish distribution and interspecific competition between native and alien species, 2. To stop the import and transport of living alien crayfish species (only allow the sale of frozen crayfish), 3. To improve information and awareness concerning the severe problems linked with illegal transport or release of alien species, and 4. To increase cultivation and management of the native crayfish species. The first two steps of this plan are realised to some extent: Inland transport and importation of living alien crayfish is only permitted for refrigerated *Cherax destructor* and detailed information of crayfish distribution is available from four major cantons. Data of crayfish distribution from the other cantons, an improvement of public information concerning the problem of introduced species and a management of native crayfish species are still needed.

According to the Convention on Biological Diversity (Rio Convention, agreed in 1992), the contracting parties should 'as far as possible and as appropriate...prevent the introduction of, control or eradicate those alien species which threaten ecosystems, habitats or species' (Art. 8 let. h Rio Convention). As *O. limosus*, like in other European countries, is widely spread in Switzerland, eradication seems improbable with the techniques available at the present time. As the distribution of *P. clarkii* and some populations of *P. leniusculus* is restricted to small ponds, eradication or control of these populations can be taken into consideration. However, technical and psychological problems (public acceptance) are not yet solved. It is not known at the moment whether eradication of the less spread species is possible and whether such eradication is sensible in relation to the widespread occurrence of *O. limosus*.

REFERENCES

Albrecht, H. 1982. Das System der europäischen Flusskrebse (Decapoda, Astacidae): Vorschlag und Begründung. *Mitt. Hamb. Zool. Mus. Inst.* 79:187-210.

Bott, R. 1972. Besiedlungsgeschichte und Systematik der Astaciden West-Europas unter besonderer Berücksichtigung der Schweiz. *Revue Suisse de Zoologie* 13:387-408.

Büttiker, B. 1987. Concerning crayfish in Switzerland. *Freshwater Crayfish* 7:2-5.

Carl, J. 1920. Décapodes (écrevisses). *Catalogue des invertébrés de la Suisse*. 12:1-32.

IAA (International Association of Astacology) 1987. Resolution. *Freshwater Crayfish* 7:xi.

Laurent, P.J. 1983. Les écrevisses du Léman. *Bull. de l' Assoc. des Riverains et Usagers du Léman* 12:5-6.

Stucki, T.P. 1997. Three American crayfish species in Switzerland. *Freshwater Crayfish* 11:130-133.

Stucki, T.P. & Jean-Richard, P. (in preparation) Vorkommen von Flusskrebsen (Astacidea; Decapoda) in der Schweiz – Beobachtungen 1987-1998.

German conservation strategies for native crayfish species with regard to alien species

PETER DEHUS & SIMON PHILLIPSON
Fisheries Research Station, Langenargen (Lake Constance), Germany
ERIK BOHL
Bavarian State Office for Water Resources Management, Institute for Water Research, Wielenbach, Germany
BIRGIT OIDTMANN
Institute of Zoology, Fish Biology and Fish Diseases, University Munich, Germany
MAX KELLER
First Bavarian Crayfish Hatchery, Augsburg, Germany
SANDRA LECHLEITER
Fish Health Service, Stuttgart, Germany

ABSTRACT

In Germany three native and five alien crayfish occur regularly in the wild. *Astacus astacus*, *Austropotamobius torrentium*, and *Austropotamobius pallipes* are the native species. The stocks of *A. astacus* were heavily affected by the crayfish plague that spread quickly in the last century beginning in the seventies in southern Germany. Isolated populations of *A. astacus* can still to be found in several lakes and rivers. *A. torrentium* is naturally distributed in the southern part of Germany and it occurs in many small rivers and brooks. *A. pallipes* is only known from the south-west of Germany where a few populations exist. The most widespread alien species is *Orconectes limosus,* which was introduced in the last century from North America. *Astacus leptodactylus* has been imported since the turn of the century from countries of eastern Europe and Asia Minor for the table market. Crayfish farmers first introduced *Pacifastacus leniusculus* in the 1960s and a few populations occur in the wild. *Procambarus clarkii* can be found in a few ponds and recently a population of *Orconectes immunis* was found in a stream. The aquarium trade probably imported both species.

The precise production, catch and import levels in Germany are unknown. A total of around 110 metric tons is imported, mostly *A. leptodactylus*, but also *Cherax destructor* from Australia. German crayfish farmers grow increasing amounts of *A. astacus* for stocking purposes and consumption, about 10 tons is produced for the table market.

Fisheries and conservation laws protect native species in various ways. The catch of *A. astacus* and *A. torrentium* with minimal sizes is allowed, only *A. pallipes* is protected. The release of alien crayfish is illegal, but these rules seem to be ineffective. People are often not aware about the dangers of releasing alien crayfish in the wild. Information and education is therefore important. Seminars for anglers are provided and pamphlets and brochures are produced and distributed to the public.

1 INTRODUCTION

In Germany, three native crayfish species are to be found: the noble crayfish, *Astacus astacus*, the stone crayfish, *Austropotamobius torrentium*, and the white-clawed crayfish, *Austropotamobius pallipes*. Two non-native species, the narrow-clawed crayfish, *Astacus leptodactylus,* and the spiny-cheek crayfish, *Orconectes limosus,* were introduced at the turn of the century and since then can be found in many waters. Over the last decades three other alien species: The signal crayfish, *Pacifastacus leniusculus*, the red swamp crayfish, *Procambarus clarkii*, and the paper shell crayfish, *Orconectes immunis,* were introduced and all of these species have established breeding populations. A few specimens of the white river crayfish, *Procambarus zonangulus,* have also been found in natural waters.

Due to the growing interest in conservation much has been done in Germany to protect the native species. Rivers and streams were renaturalized and the water quality in lakes was improved, to recreate suitable habitats for the crayfish. Noble and stone crayfish are widely accepted as being indicators for clean and intact water bodies. Crayfish were stocked in many waters so populations could become re-established and primarily not for use on a commercial basis. However, imports of alien crayfish have never been so high as today, mostly for the table market but also as pets for aquaria and garden ponds. However the import of live crayfish often leads to these creatures being released into the wild. This increases the danger of the crayfish plague fungus, *Aphanomyces astaci*, being introduced from infected crayfish, which provides a real danger to native species and can lead to the extermination of individual populations. The majority of people who release such crayfish into the wild are unfortunately not aware of the risks involved.

This paper describes the present distribution of native crayfish species in Germany, with regard to alien species. Production and catches of native species are compared with imports of alien ones. The laws and rules governing the catching and stocking of crayfish are explained. The possibilities to protect existing native stocks and prevent the further spread of alien species are discussed from a legal viewpoint. The final section describes the work being undertaken in Germany to inform the public about the possible risks involved in releasing alien crayfish into the wild.

2 DISTRIBUTION OF CRAYFISH SPECIES IN GERMANY

An extensive knowledge of the present crayfish distribution in Germany is required when attempting to assess the possible dangers for the different species. It is also important to have an idea of the various population densities. Each Bundesland is responsible for its own research program, which leads to varying degrees of accuracy in the different Bundesländer when compiling information on crayfish populations and possible dangers to their existence. Over the last 10 years, detailed studies on the distribution of crayfish species have taken place in the south in Baden-Württemberg (Phillipson et al. in prep.) and Bavaria (Bohl 1989b), in the north-west (Lower Saxony, D. Blanke pers. comm.) and in the north in Schleswig-Holstein (Dehus 1990). Studies on fish distribution take place in several Bundesländer on a regular basis, in which the different crayfish species are mapped; this is, however, mostly due to random finds, rather than due to a deliberate search.

Until the end of the last century, the noble crayfish was spread over most parts of the country. This wide distribution is most probably due to the intensive dealing in live crayfish (Albrecht 1983), which were also popular in those times. The invasion of the plague fungus caused severe damage and most stocks were eliminated. Today, the noble crayfish is still widespread in many waters, as can be seen in Figure 1. The populations are mostly small and isolated, which results in the patchy distribution. The stocks existing in lakes and streams are mostly in higher regions, which are not so suitable for noble crayfish (Bohl 1989a). Due to the increasing interest in conservation issues, noble crayfish are being newly introduced into many waters, which leads to the present day increase in distribution. In many regions of Bavaria the noble crayfish was and still is the dominant crayfish species.

Historically, the noble crayfish was widespread in the large rivers such as the Rhine and Oder, as described by such as Fontane in the 19th century in 'Excursions in Mark Brandenburg'. The enlargement, canalization and pollution of the major rivers led to an early reduction of the crayfish populations. Today almost all the major

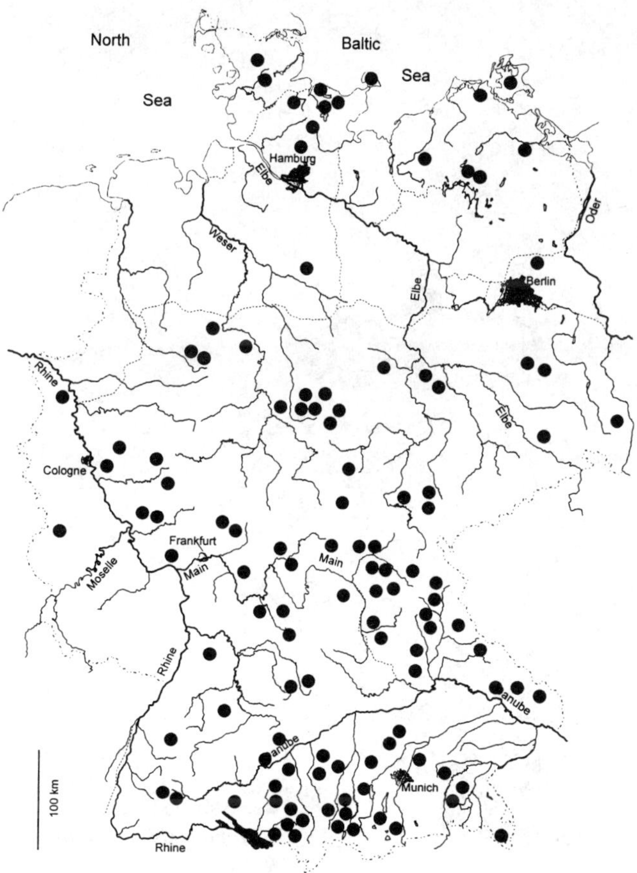

Figure 1. Distribution of *Astacus astacus* in Germany.

rivers have been colonized by the spiny-cheek crayfish, which makes the re-establishment of noble crayfish impossible.

The stone crayfish is only found in the southern parts of Germany. The most northern region in which they can be found is by Cologne and in tributaries of the River Main (Fig. 2). However, they are widespread in the Black Forest and south of the Danube. Most stone crayfish populations exist in the streams and smaller rivers of the higher regions, but they can also be found in several lakes. Due to its small body size, the stone crayfish is of no commercial importance. There are a few known habitats where both stone and noble crayfish coexist, competition does not appear to cause problems in these cases.

The distribution of the white-clawed crayfish is confined to a few places in the southwestern part of Germany (Fig. 2). Klunzinger (1882) reported on its occurrence in western areas of the River Rhine, in Alsace, and recently a few stocks in the Black Forest, i.e. east of the River Rhine, were described (Troschel & Berg 1989, Troschel & Dehus 1993). It is possible that the noble crayfish played a role in driving the white-clawed crayfish out of its habitats in a westerly direction. Populations in the neighboring countries, i.e. Tyrol (Austria) (Füreder & Machino 1995) and Liechtenstein, are known to originate from stocking activities.

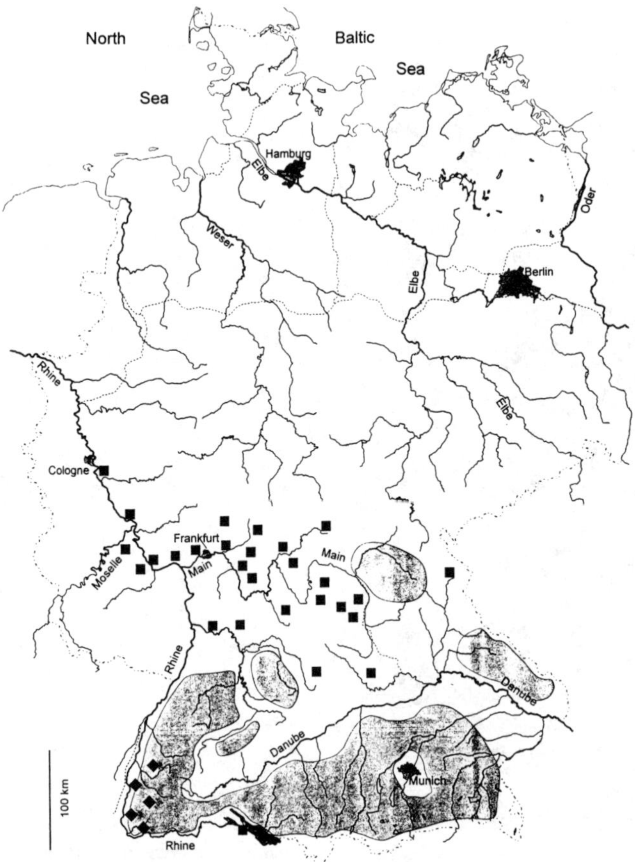

Figure 2. Distribution of *Austropotamobius torrentium* (■ and shaded areas) and *Austropotamobius pallipes* (◆).

As European species, the narrow-clawed crayfish plays a special role within the alien species. It is also susceptible to the plague in the same way as the native species. In the early part of the century it was imported from eastern countries to replace the plague-affected noble crayfish stocks, as it was believed to be less sensitive to the plague. The distribution is presented in Figure 4. *Astacus leptodactylus* occurs only in lakes; attempts to stock in rivers have not yet been successful.

The spiny-cheek crayfish is widespread in Germany in rivers, canals and some lakes (Fig. 3). It was first released in the near of the Oder, north-east from Berlin in the last century. From there, it spread to the western and southern parts of the country, but this spreading was accelerated by the release of crayfish near the towns, where it was sold live on the markets. It is still absent in the higher regions, and small brooks and rivers appear to be unsuitable habitats for the spiny-cheek crayfish.

The signal crayfish was brought to Germany in the early sixties mainly from Swedish hatcheries, but also from Austria. *Pacifastacus leniusculus* was said to be plague resistant, and therefore it was released in some lakes and rivers. German crayfish farmers also began to grow signal crayfish. The animals can easily escape, especially from ponds, and they then develop populations in the neighboring water systems. The stocks known at present (Fig. 4) are not very abundant. In a few running

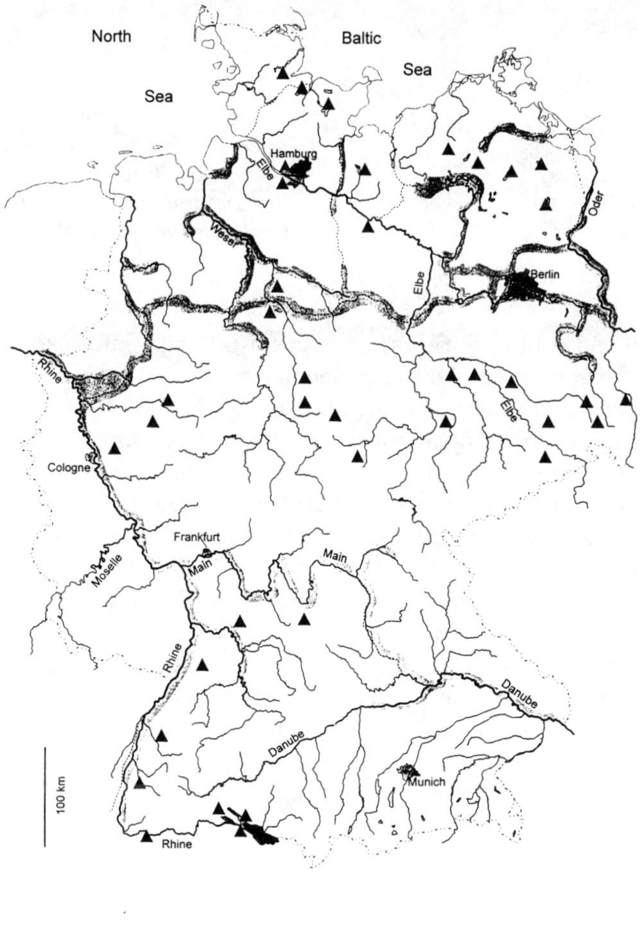

Figure 3. Distribution of *Orconectes limosus*.

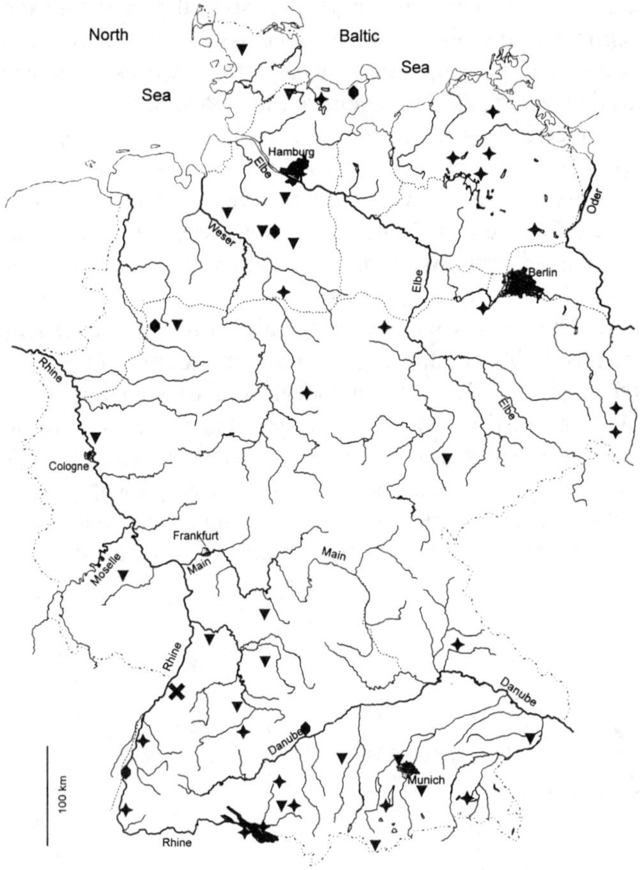

Figure 4. Distribution of *Astacus leptodactylus* (✦), *Pacifastacus leniusculus* (▼), *Procambarus clarkii* (⬧), and *Orconectes immunis.* (✖).

waters, populations of signal and noble crayfish seem to coexist, though information on direct interaction and fluctuations in numbers is not available.

The red swamp crayfish were first imported in small quantities in 1985 for the table market due to the declining Turkish imports of the narrow-clawed crayfish. It is also common in the aquarium trade. Only a few reproducing populations are to be found and it appears that climatic influence is not as important as often thought. One population in northern Germany must tolerate an ice cover for several weeks during the year, but this has obviously no influence on the stock (Dehus 1990). Single specimens have also been found in mountain brooks in northern Bavaria that evidently originate from the food and aquarium trade.

The white river crayfish and the paper shell crayfish were favorite pets in aquaria and in garden ponds some years ago. It is known that many of them were released in natural waters and reproducing populations of the paper-shell crayfish already exist.

Australian crayfish species are imported to Germany for the table market. At present, there is no evidence of living stocks in natural habitats.

3 PRODUCTION, CATCH AND IMPORTS OF NATIVE AND ALIEN SPECIES

During the last century Germany exported large amounts of noble crayfish, mostly from Berlin to Paris. Around 600 tons were sold every year. With the introduction of the crayfish plague and the subsequent loss of most populations, the tradition of catching and the culinary interest were also lost.

Precise data on production, catch and import of crayfish are not available. At present around 10 tons is produced per year for the table market, the majority noble crayfish. Production is spread mostly over smaller pond production units, mainly on a semi-professional basis. However, demand is increasing and more crayfish farmers are now starting to produce crayfish for the table market. The market price is at present DM 50.00-160.00 per kg crayfish, with an individual weight of 80-120 g. By optimizing the farming methods, production in southern Germany could be as high as 300 kg/ha (Keller & Keller 1995).

A few farmers hold signal crayfish in their ponds, producing for the table market. Exact production levels are not known. At least one crayfish farmer has given up with signal crayfish as problems arose when draining the ponds and trying to collect the crayfish and prices were too low. The farmer has now changed to culture noble crayfish (see also Anonymous 1995).

Most farmers in Germany produce crayfish for stocking purposes, juveniles are sold for ca DM 0.50 and crayfish of two summer old for DM 2.30-3.00. The crayfish are sold as a primary stocking basis for ponds, smaller lakes and rivers to enable new populations to become established. At certain times the demand in Germany for such crayfish exceeded the supply, but this is no longer the case.

The harvesting of wild noble crayfish from ponds and lakes is of no commercial importance. Under strict regard of the regulations, it would be illegal to sell noble crayfish that originate from a wild population, as they are a protected species. Narrow-clawed crayfish are harvested from a few lakes, though the amounts are also small. The catch data for Germany in 1994 shows a total harvest of 12.3 tons. This figure is probably derived from catches of spiny-cheek crayfish. Professional fishermen catch this species in waters around Berlin, and in minimal amounts in the River Rhine and in Lake Constance, though financial profits are low.

The official German import and export statistics for freshwater crayfish are inaccurate and invalid. A total of around 110 tons of live crayfish are imported per annum. About 80-90 tons are narrow-clawed crayfish from Turkey, 10-15 tons of *Cherax destructor* ('yabbies') are imported every year from Australia. Signal and red swamp crayfish are of little significance in the table market and imports of the latter are mostly for the aquarium trade.

4 LEGAL POSITIONS

The two most important conservation laws in Germany are the Nature Conservation Act (1993) and the Species Conservation Legislation (1997). Both the noble crayfish and the stone crayfish are in the category 'specially protected species'. This prohibits all activities that could be detrimental to the natural stocks. However, both species may be caught in the wild if then used solely for own consumption, a further sale of these crayfish would be illegal. These rules do not apply to the managed ponds from crayfish farmers. Furthermore these laws also govern and prohibit the release of alien species into the wild. However, when these species manage to reproduce in the wild over several generations, they are then legally seen as being indigenous. When this aspect of the law is taken into consideration, then almost all alien crayfish should be recognized as being indigenous.

Laws more applicable to the practical production of crayfish are the German fisheries laws, under which the crayfish are also affected. Every Bundesland has its own fisheries law and is therefore responsible for its own crayfish stocks. In Baden-Württemberg, for example, the white-clawed crayfish is protected and trapping is forbidden. A minimal catch size and weight is given for noble crayfish in most Bundesländer. Alien species are usually not protected, though a minimal catch size for spiny-cheek crayfish is included in the fisheries law of Berlin.

The release of alien crayfish into the wild is forbidden under most fishery laws. Some Bundesländer do not have a specific law covering this, and in one Bundesland it is actually permitted. As a rule, alien crayfish may be kept under farming conditions though, as the fisheries law has little influence here. These laws also cannot prevent people releasing alien species, e.g. crayfish from an aquarium.

On the basis of the Species Conservation Legislation, Germany banned the import of live crayfish in 1989. Both native and alien species were affected by the ban, which attempted to stop the import of all infected crayfish. The governing body included the possibility for exceptions to this rule. An import was only allowed when the crayfish were sold directly to the end consumer, e.g. restaurants. These also had to comply to certain rules concerning keeping the animals, disinfection of wastes and waste water. Due to these demands, Germany was promptly accused by the Commission of the European Union of not complying to the free trade policies by restricting imports (see also Holdich, Gydemo & Rogers this volume). In 1994, the European Court decreed that this import ban must be lifted (European Court of Justice 1994). It is possible that Germany will be further prosecuted for compensation.

5 PROTECTION POSSIBILITIES FOR NATIVE CRAYFISH SPECIES ON THE BASIS OF LAWS AND DIRECTIVES

Under the Federal Nature Conservation Act (1993) and the Species Conservation Legislation (1997) it is possible to protect the existing crayfish populations. Past experience has shown however that these laws are mostly ineffective. The definition of native species (see above) is unclear in these laws, and a permission from the European Court to impose an import ban is not expected.

In attempts to protect native species in Germany, the EU Directive 91/67/EEC for fish diseases has proved to be in most cases applicable. This directive provides the legal possibility for areas or regions, that have no history of diseases, to become acknowledged disease-free zones with defined import-export rules. Fish may not be brought into these areas either for fish farms or for stocking in natural waters, unless they also come from a recognized disease-free area. This helps prevent the transmission of diseases through the country. Crayfish plague is also covered by this directive, plague-free regions can be protected by only allowing the import of live crayfish into these regions that are certified as plague-free.

The main requirement for such an area is an extensive knowledge of the existing native and foreign crayfish populations. The crayfish plague must also become recognized as a notifiable disease in Germany, which is not yet the case. This must first be agreed upon by the various Bundesländer, which is at present being undertaken. Some Bundesländer could be against this proposal as they have no valuable natural crayfish stocks. Another requirement is a refinement of the diagnostic techniques, so that native and alien crayfish stocks can be quickly and effectively tested for *Aphanomyces astaci*. The Bundesländer involved with such regions must also be prepared to monitor stock movement and when necessary take action to prevent the spread of the disease in the case of an outbreak.

As no 'disease-free' zone exists in the EU at present, it is expected that it will take some time before an application for an area is approved.

A further possibility to create areas in which only disease-free crayfish may be imported would be through the various fisheries laws from the Bundesländer. However a nationwide concept of how these could be created for the whole of Germany would require the cooperation of all the Bundesländer, which is at present improbable.

6 INFORMATION AND EDUCATION

It is widely believed that the laws and regulations available would enable an effective plan to be created to protect the native stocks from alien species. This is however not the case and especially with regard to aquarium crayfish it is extremely difficult to prevent alien species being released into the wild. To prevent this happening several Bundesländer have been trying to inform the public about such dangers. Fisheries authorities and fisheries research institutes regularly provide seminars for the instructors from fishing clubs, who then give this information further to the anglers. The seminars are often organized together with nature conservation clubs, to inform other groups of the importance of these native stocks. An initiative has been started together with the largest aquarium organization in Germany, to inform aquarists about the dangers involved in releasing alien species into the wild. A pamphlet on the subject has been produced and is being distributed through aquarium shops and clubs (Dehus & Lechleiter 1997). Articles on keeping freshwater crayfish in aquariums are often published in aquarist magazines (recently Dost 1995, Werner 1997), but it is rarely mentioned that the release of these animals into the wild could endanger whole populations of native crayfish.

Furthermore, information for fishing clubs and conservationists is being produced and distributed (Dehus 1997). The positive reactions received show a need for this information in the public sector and that the recommendations are being acknowledged. A pamphlet is at present being developed for the table market and is aimed at the large importers of crayfish, who are cooperating with the project and will also distribute the information.

7 CONCLUSIONS

In the 1970s and 1980s more and more stocks of noble, stone and even white-clawed crayfish were registered in Germany during monitoring programs. Years ago, it was presumed that all populations of noble and stone crayfish were more or less extinct. The occurrence of the white-clawed crayfish in southern Germany was unknown. Nowadays our knowledge about the distribution of the different species is quite extensive. Beside the native crayfish species, some populations of alien crayfish have been found, but further stocks of such crayfish certainly exist. The population dynamics are not usually investigated, but this would provide important information, especially in cases where native and alien crayfish occur together in the same river or lake.

Crayfish farmers in Germany are increasingly producing noble crayfish which are sold for stocking purposes and the table market. However, they import signal crayfish, which are also grown and sold to consumers. These crayfish often escape from hatcheries or ponds and enter the natural waters nearby. The crayfish farmers are responsible that no animals escape from the farms.

The import of signal, red swamp and other alien crayfish is not significant when compared to the amount of narrow-clawed crayfish imported. As the species are imported and sold live, they were often released in natural waters. Existing laws are not applicable enough to prevent this happening and have therefore no effect. Since the import ban for living crayfish into Germany was lifted by the EU, German producers should try to increase production as quickly as possible. Germany would then no longer be dependent on imports. This is already the case with juvenile crayfish for stocking purposes, and it should also be possible for the table market.

The existing German and European laws should be used further to protect the native stocks more effectively. It is important that this also occurs on a regional basis, as this ensures a wider acceptance of the problems by the general public. The value of this project is obvious when the people concerned with affected areas are already informed and can react. Several Bundesländer are therefore working intensively on providing information for aquarists, conservationists and fishing clubs and their members.

REFERENCES

Albrecht, H. 1983. Besiedlungsgeschichte und ursprünglich holozäne Verbreitung der europäischen Flußkrebse. *Spixiana* 6:61-77.

Anonymous 1995. German farmer picks his customers. *Fish Farming International* 22(7):49.

Bohl, E. 1989a. Comparative studies on crayfish brooks in Bavaria (*Astacus astacus* L. and *Austropotamobius torrentium* Schr.). *Freshwater Crayfish* 7:287-294.

Bohl, E. 1989b. *Untersuchungen an Flußkrebsbeständen.* Wielenbach: Bayerische Landesanstalt für Wasserforschung Versuchsanlage Wielenbach.

Dehus, P. 1990. Die Verbreitung der Flußkrebse (Decapoda; Astacidae, Cambaridae) in Schleswig-Holstein. *Faun.-Ökol. Mitt.* 6:95-105.

Dehus, P. 1997. *Flußkrebsé in Baden-Württemberg: Gefährdung und Schutz.* 2nd ed. Aulendorf: LVA.

Dehus, P. & Lechleiter, S. 1997. *Ja zu Flußkrebsen im Aquarium, Nein zum Aussetzen von Aquarien-Krebsen.* Langenargen: Fischereiforschungsstelle Baden-Württemberg.

Dost, U. 1995. Flußkrebse im Aquarium. *DATZ* 48:502-508.

European Court of Justice 1994. Case C-131/93, Free movements of goods. Commission of the European Communities V Federal Republic of Germany.

Füreder, L. & Machino, Y. 1995. Record of the white-clawed crayfish *Austropotamobius pallipes* (LEREBOULLET 1858) from Plansee (Tyrol, Austria). *Ber. Nat.-Med. Verein Innsbruck* 82:241-246.

Keller, M.M. & Keller, M. 1995. Yield experiments with freshwater crayfish *Astacus astacus* (L.) in aquaculture. *Freshwater Crayfish* 10:506-511.

Klunzinger, C.B. 1882. Ueber die Astacus-Arten in Mittel- und Südeuropa und den Lereboullet'schen Dohlenkrebs insbesondere. *Jh. Ver. Vaterl. Naturkde Württ.* 38:326-342.

Phillipson, S., Haberbosch, R. & Dehus, P. (in preparation). *Flußkrebse in Baden-Württemberg: Verbreitung, Gefährdung, Schutz*. Langenargen: Fischereiforschungsstelle Baden-Württemberg.

Troschel, H.J. & Berg, R. 1989. Ein Nachweis des Dohlenkrebses (*Austropotamobius pallipes*) Lereboullet (1858) in Baden-Württemberg. *Veröff. Naturschutz Landschaftspflege Bad.-Württ.* 64:283-288.

Troschel, H.J. & Dehus, P. 1993. Distribution of crayfish species in the Federal Republic of Germany, with special reference to *Austropotamobius pallipes*. *Freshwater Crayfish* 9:390-398.

Werner, U. 1997. *Cherax quadricarinatus*, ein australischer Süßwasserkrebs: Biologie und Pflege im Aquarium. *Aquarium Heute* 15:572-577.

The status of crayfish populations in Spain and Portugal

PEDRO J. GUTIÉRREZ-YURRITA, JOSÉ. M. MARTÍNEZ,
MIGUEL Á. BRAVO-UTRERA & CARLOS MONTES
Departamento de Ecología, Universidad Autónoma de Madrid, Spain

MARIA ILHÉU & JOÃO M. BERNARDO
Departamento de Ecologia, Universidade de Évora, Portugal

ABSTRACT

Austrapotamobius pallipes is the only native species of crayfish in the Iberian Peninsula. The continuous reduction of density, distribution range and loss of native crayfish populations in the Iberian Peninsula and a growing commercial demand of this product generated the idea of introducing American plague-resistant species for aquacultural trials. Several species of crayfish have been introduced with commercial purposes with a different degree of success in Spain. Up to the present date, out of six species of crayfish that have been object of introduction, only *P. clarkii*, *P. leniusculus*, and *C. destructor* are naturalized. Native populations of *A. pallipes* were greatly reduced in abundance or become extinct due to habitat alteration, human disturbance, large-scale over-exploitation of natural populations, crayfish plague and the exotic species from North America that are vectors of the *Aphanomyces astaci*. At present, there are few populations of native crayfish in Spain, and relicts populations in Portugal. *P. clarkii* accounts for the largest catch. Nearly, all red swamp crayfish is produced in Guadalquivir marshes. Portugal produces a small, but growing volume. Maximum production of red swamp crayfish in Spain was about 5000 tons during 1986, but this production decreased due to inadequate management of the resource and severe drought. The intention of introducing *P. leniusculus* in Spain was mainly to replace *A. pallipes* in waters devastated by the plague. The catch of *P. leniusculus* is at least 600 kg annually and the catch of *A. pallipes* is less than one ton. However, it must be pointed out that the data on catches available from most regions are estimates. Crayfish production in Iberian Peninsula is growing, especially due to good stocking results with the red swamp crayfish *P. clarkii*. The signal crayfish *P. leniusculus* has proved to be an excellent substitute to the indigenous crayfish for recreational purposes. Nevertheless, both species have caused considerable damage to the native crayfish habitats. We briefly review the status of crayfish populations in Spain and Portugal, with special emphasis on *P. clarkii*, since this species is, by far, the most widespread with wide spread effects. Conservation and management measures for the native and alien species are proposed, with the purpose to make the best of a bad situation.

1 INTRODUCTION

There is only one species of native crayfish in Spain and Portugal (the Iberian Peninsula, IP), the white-clawed crayfish *Austropotamobius pallipes*. This species has been previously referred to as *Astacus fluviatilis torrentium* (Machado 1931) and *Astacus fluviatilis lusitanicus* (Mateus 1934), nowadays a synonym of *Astacus astacus*. Lastly, it has been cited as *Astacus pallipes lusitanicus* (Mateus 1937). This has led to confusion about the occurrence of other different taxa than *Austropotamobius pallipes lusitanicus* in the IP.

During this century, native populations of *A. pallipes* have been greatly reduced in abundance or have become extinct due to habitat alteration and human disturbances. These include engineering operations, e.g. canalization, clearing, dredging and embanking of rivers, changing the river beds, construction of reservoirs, regulation of water levels and stream flows. Other factors are pollution; large-scale overexploitation of natural populations; crayfish plague; and the introduction of alien species from North America, which are vectors of the oömycete fungus *Aphanomyces astaci* Schikora (Gaudioso 1977, Riera 1982, Sánchez 1982, Muez & Muez 1988, Bernardo & Ilhéu 1994a, Diéguez & Rueda 1994, Anastácio & Marques 1995).

The continuous reduction of density, distribution range and loss of native crayfish populations in the IP and a growing commercial demand of this product generated the idea of introducing American plague-resistant species for aquacultural trials. It is worthy to note that crayfish have attracted considerable gastronomic attention for centuries in northern Spain and in other countries, so their fisheries represent a great economic value (see Ackerfors this volume). North American crayfish species have proved highly plague-resistant, with the exception of some cases in Sweden, when signal crayfish, *Pacifastacus leniusculus,* infected with *A. astaci,* have died from the combined effects with other stress factors such as low temperatures (Unestam 1975, Svärdson et al. 1991, Gydemo 1992, Westman & Westman 1992).

Crayfish plague probably arrived in Spain in 1958, when contingent mortalities of native crayfish were detected, though it was always related with stressing factors in water (e.g. Serres et al. 1985, Carral et al. 1993, Diéguez & Rueda 1994, Bolea 1995). These mortalities were associated with crayfish plague, because there was an attempt to introduce the noble crayfish, *A. astacus* in central Spain (River Duero basin) in that period (Gutiérrez- Calderon 1964). Reports of new outbreaks were registered in the Rivers Iregua (La Rioja) and Ucero (Soria) during 1964, 1975-1976, but in this case it was not possible to prove that the fungus *A. astaci* was the pathogenic agent (Gaudioso 1977, Cuéllar & Coll 1983). The first record of *A. astaci* was reported in 1977 (Cuéllar & Coll 1983).

Nowadays, because of the expanded range of the fungus plague in the IP, in association with the red swamp crayfish, *Procambarus clarkii*, and *P. leniusculus*, *A. pallipes* is considered threatened and endangered (Wells et al. 1983, Gaudioso et al. 1987, Carral et al. 1993). It was listed by the International Union for Conservation of Nature (IUCN) as being vulnerable and rare (Rosas et al. 1992, Groombridge 1994, Maitland 1996). In Portugal, natural populations of the native crayfish are almost extinct, occurring only in the small rivers Azibo and Tortulhas, both in Tras-os-Montes region (Bernardo & Ilhéu 1996).

Several species of crayfish (three from North America, two from Europe, and one from Australia) have been introduced for commercial purposes with a different degree of success in Spain. *A. astacus* (introduced in 1963-1964) has had apparently with no success; *P. clarkii* (introduced in 1973-1974) with great success; *Procambarus zonangulus* (introduced in 1974) without success; *P. leniusculus* (introduced in 1974) with considerable success. *Astacus leptodactylus* (introduced in 1975) without success; *Cherax destructor* (introduced in 1983) apparently with success in the area of its introduction. Although some authors had cited that *Orconectes limosus* was introduced in Spain, populations of this species have not been detected in natural environments. Thus, out of six species of crayfish that have been object of introduction, only *P. clarkii*, *P. leniusculus*, and *C. destructor* are naturalized, especially the former, and are now very common in rivers, canals, reservoirs, and rice fields in Spain, Portugal, and some islands (Habsburgo-Lorena 1979, Algarín 1980, Ramos & Pereira 1981, Ocete & Lopéz-Sánchez 1983, Bruxelas 1986, Gaudé III 1986, Almaça 1990, Sall-Reina & Vergara-Martin 1990, Adão & Marques 1993, Carral et al. 1993, Montes et al. 1993, Bernardo & Ilhéu 1994a, 1996, Correia 1994, Correia & Costa 1994, Bolea 1995, Bernardo et al. 1997 among others).

Procambarus clarkii clearly accounts for the largest catch (about 3500 tons annually in Spain, although this varies from year to year depending on the availability of water). Nearly all red swamp crayfish are produced in the Guadalquivir marshes (about 2900 tons). Portugal produces a small but growing volume (about 700 tons). Maximum production of the red swamp crayfish in Spain was about 5000 tons during 1986. However, this production decreased because of the inadequate management of the resource and severe drought in southern Spain during 1992-1995 (Habsburgo-Lorena 1979, 1983a,b, 1986, Algarín & Habsburgo-Lorena 1983, Gaudé III 1986, Soto 1992, Montes et al. 1993, Veiga 1995, Ackefors this volume). From an ecological point of view, this species is the most studied crayfish in Spain and Portugal. It became a key and engineering species in the ecosystems invaded (Algarín 1980, 1981, López & Fernández 1980, Ramos & Pereira 1981, Sánchez 1982, Algarín & Habsburgo-Lorena 1983, Gaudé III 1983, 1986, Ocete & Lopéz-Sánchez 1983, Gállego & Ocete 1985, Bruxelas 1986, Torreblanco et al. 1987, Algarín & Librero 1988, Correia 1990, 1993, Escosa 1990, Soto 1992, Adão & Marques 1993, Anastácio 1993, Ilhéu & Bernardo 1993a, b, 1995, Montes et al. 1993, Bravo et al. 1994, Correia & Costa 1994, Gutiérrez-Yurrita et al. 1994, Anastácio & Marques 1995, Correia & Ferreira 1995, Gutiérrez-Yurrita 1997, Gutiérrez-Yurrita et al. 1997a, Gutiérrez-Yurrita et al. 1998, Gutiérrez-Yurrita & Montes 1998a, Beroiz in press, among others).

The intention of introducing *P. leniusculus* in Spain was mainly to replace *A. pallipes* in waters devastated by the plague, since signal crayfish resists the plague well and have similar habitat requirements. Despite the importance of *A. pallipes* in the ecosystems where it occurs, few studies have been carried out on its ecology. Most of the studies were focused on its aquacultural potential, description of its habitats (Celada & Gaudioso 1985, Temiño et al. 1986, Celada 1987, Coll-Morales 1988, Almaça 1989, 1990, Álvarez et al. 1989, Peixoto et al. 1996), and its fishery (Pardo 1951, Gaudioso & Vigil 1976, Gaudioso 1977, Muñoz 1978, Fuente 1982). The catch of *P. leniusculus* is at least 600 kg annually and the catch of *A. pallipes* is less than 1 ton. However, it must be pointed out that the data on catches available from most regions are estimates, more detailed statistical data does not exist

This study briefly reviews the status of crayfish populations in Spain and Portugal, with special emphasis on *P. clarkii*, since this species is by far the most widespread and is increasing its range in IP. It is also the species having the highest incidence in the ecosystems, is the most important in culture trials and has the greatest fishery potential. Finally, conservation and management measures for the native and alien species are proposed, with the purpose of getting the best of a bad situation.

2 AUSTROPOTAMOBIUS PALLIPES

The natural distribution range of *A. pallipes* is restricted to the north and central part of the IP. Just a few populations occur in the southeast (Sierra Nevada) close to Granada, Jaen, and Murcia and in southeastern Córdoba. Its habitat comprises slow-flowing rivers, or even ponds, of clear and cold water (16-20°C); the largest populations are found in small isolated lakes with a high calcium carbonate and potassium content (Torre-Cervigon & Rodríguez-Marques 1964, Muñoz 1978, Serres et al. 1985, Bolea 1995, Gil & Alba-Tercedor 1998). Apparently, the high calcium requirement of this species has restricted it to the karstic area of the IP (Torre-Cervigon & Rodríguez-Marques 1964, Gaudioso & Vigil 1976). Natural occurrences of *A. pallipes* in Silicean regions such as Badajoz, Lugo, and Orense in Spain (Torre-Cervigon & Rodríguez-Marques 1964), is due to several introductions and translocations of the species, or to the geographical origin of this rivers in karstic areas that maintain high calcium levels in water.

Substantial reduction of *A. pallipes* populations began in the late 1950s when it increased in its gastronomic and recreational importance. Some years later, Torre-Cervigon & Rodríguez-Marques (1964) reported that the large-scale over-exploitation, habitat destruction, water contamination, and river canalization might lead this resource to a critical situation. Crayfish plague has greatly accelerated the native crayfish extinction process (Bolea 1993, 1995, Bernardo et al. 1997). Figure 1a shows the distribution ranges of *A. pallipes* in the 1960s in Spain (Torre-Cervigon & Rodríguez-Marques 1964) and in the mid-1970s in Portugal (Almaça 1990).

Present native crayfish distribution is restricted to the highest reaches of the rivers, even at the altitudes in which they had never been found before (more than 1100 m). This is the case of three populations in Aragón (Bolea 1995) and one population in Lake Enol in Asturias, both in Spain (pers. obs.). Bolea (1995) has described crayfish populations whose maximum densities were 4 animals/m^2 in Matarraña River basin, where the species was introduced by the beginning of the century. He also pointed out that this basin has probably the best conditions for the conservation of *A. pallipes* in Spain. The highest number of river sections settled by *A. pallipes* is found in Teruel, Spain, in spite of the 1979-1980 aphanomycosis (Bolea 1995). Reduction of *A. pallipes* distribution range in Aragón are more related to habitat loss and water pollution rather than crayfish plague (Bolea 1993, 1995). In the same way, *A. pallipes* presence is restricted to the highest sections of small rivers and streams in other areas of Spain, such as the País Vasco, Castilla-León, Castilla-La Mancha, La Rioja, Córdoba and Granada (Muñoz 1978, Habsburgo-Lorena 1979, Fuente 1982, López-Fernández & Infante-Miranda 1982, Riera 1982, Muez & Muez 1988, Aldabe et al. 1991, Carral et al. 1993, Diéguez & Rueda 1994, Gil & Alba-Tercedor 1998, F.

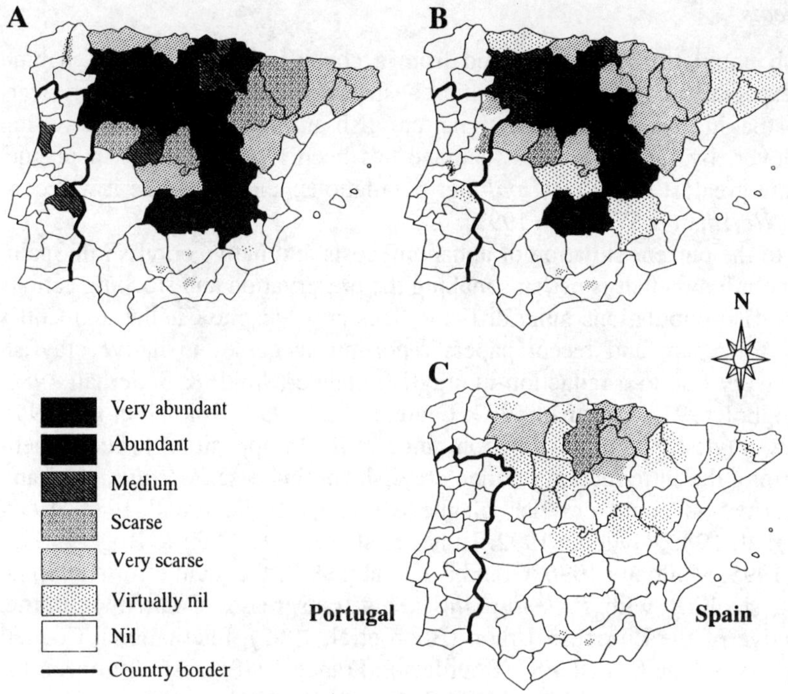

Figure 1. Distribution of populations of *Austropatamobius pallipes* on the Iberian Peninsula. A) in the 1960s, B) from 1979-1986, C) in the 1990s.

Alonso pers. comm., L. Bolea pers. comm.). Figure 1b shows the areas with the greatest populations of *A. pallipes* between 1979-1986 (Riera 1982, Almaça 1990). Figure 1c shows the present distribution range of *A. pallipes* in IP (Bernardo et al. 1997, L. Bolea personal communication).

Native crayfish populations in Portugal are in a worse situation than in Spain. Severe drought, over-fishing of the larger size classes, and human activity are probably the main causes of *A. pallipes* disappearance in this country. Several populations in the northeast rivers disappeared in 1986. The mass mortality occurred in a short time period and according to local people 'it was as if crayfish had been poisoned'. This description suggests that crayfish plague was the cause of the observed mortality. Previously, the Direcção Geral das Florestas had promoted several stocking actions from 1977 to 1979 (Douro, Mondego and Tagus river basins). After 1986, restocking was undertaken in 1993 (Douro river basin), after a period of several years in order to guarantee plague free rivers, and in an attempt to restore crayfish population in its traditional distribution area. However, the distribution is now confined to a few sections in the small Rivers Azibo and Tortulhas (Douro water basin, Trás-os-Montes, northeastern Portugal). The distribution range of native crayfish is now limited to a few small river sections in Trás-os-Montes (Azibo and Tortulhas rivers, NE-Portugal) (Bernardo & Ilhéu 1996, Bernardo et al. 1997) (Fig. 1c.).

2.1 *Major threats*

Native crayfish populations have suffered from a considerable amounts of habitat modification especially due to pollution and construction work and also from overfishing, good stocking results with the signal crayfish and the spread of red swamp crayfish. However, by far the greatest damage has been from the crayfish plague, which is still the greatest threat to *A. pallipes* populations, and looks like it will continue to be so (Westman & Westman 1992).

As regards to the current situation of aphanomycosis and native crayfish in Spain, there is a great confusion that is perhaps making the preservation and stocking actions with native crayfish populations more difficult. This is so despite Cuéllar and Coll's (1983) review for Spain and recent papers reporting successes in native crayfish population recovery due to a reduction in crayfish plague (Smith & Söderhäll 1986, Skurdal & Taugbol 1991, Alderman 1993, Laurent et al. 1993, Holdich et al. 1995). Figure 2 shows crayfish plague distribution range in IP. In opposition to recommendations concerning the introduction of alien crayfish (in this case *P. leniusculus*) and to current information about crayfish plague (Cerenius & Söderhäll 1984, 1992, Smith & Söderhäll 1986, Gydemo 1992, Huner et al. 1992, Holdich & Rogers 1995, Holdich et al. 1995, Mailtland 1996, Grandjean et al. 1997), the main efforts in Spain are focused in stocking with *P. leniusculus*. This transgresses Council of Berne, Habitats Directive of the European Union (Boon et al. 1992), International Council for the Exploration of the Sea (ICES) (Bourdeaux, France 1988), FAO European Inland Fisheries Advisory Commission (EIFAC), International Union for the Conservation of Nature (IUCN), and International Association of Astacology (IAA) recommendations. These documents advise that introducing or translocating crayfish should be subject to tighter controls or not made (Westman & Westman 1992). There are two main reasons why these statements are not followed in Spain. The first lies in the belief that the presence of the signal crayfish may stop the expansion of the red

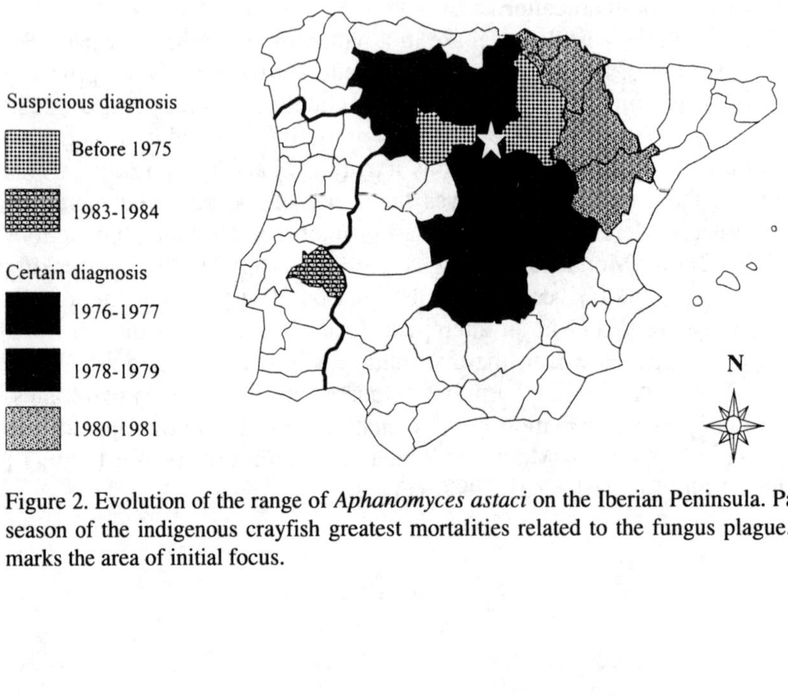

Figure 2. Evolution of the range of *Aphanomyces astaci* on the Iberian Peninsula. Patterns show the season of the indigenous crayfish greatest mortalities related to the fungus plague. The white star marks the area of initial focus.

swamp crayfish, which is a vector of crayfish plague and a great competitor for habitat and other life resources of native crayfish. This of course ignores the fact that the signal crayfish has not been proven to be a hindurance to the spread of red swamp crayfish (Goldman 1973, Lowery & Holdich 1988, J. Arrignon pers. comm.). Rather it has the same devastating effects, including acting as a vector of crayfish plague (Unestam 1973, Persson & Söderhäll 1983, Lowery & Holdich 1988, Gydemo 1992, Diéguez & Rueda 1994, Holdich this volume). The second reason arises from the ancient belief that signal crayfish is an 'ecological substitute' of native crayfish (Abrahamsson 1973, Goldman 1973), just because it can occupy similar habitats (Vandermeer 1972, Chesson 1991).

Although Abrahamsson (1973) and Goldman (1973) pointed out that it is very important to do ecological studies prior to the introduction or translocation of exotics, in Spain there was no previous work concerning the ecology of the rivers where *P. leniusculus* was introduced. So, it was not possible to evaluate the effect of introducing the signal crayfish on the relationship between biodiversity and ecosystem processing (Lawton 1989, 1990, Covich 1996). Other studies are needed to determine how this relationship might vary when other crayfish dominates the ecosystem, for instance, what would happen on the cascade food-web effect if we remove a key or a driver species? (Schindler 1990, Lawton & Jones 1994, Harris 1995, Guitay et al. 1996, Holdich and Nyström this volume). Every successful introduction may change the ecosystem's processes, ability to withstand or recover following extreme events (Vitousek 1990, Lodge 1993, Williamson 1996, Gutiérrez-Yurrita & Montes 1998a). Following Hutchinson´s (1959) definition of niche as an abstract n-dimensional hypervolume in an n-dimensional space, it is clear that although *A. pallipes* and *P. leniusculus* could coexist, they occupy different but overlapping ecological niches. Thus they have several marked differences. 1. The population dynamics vary (see Lowery & Holdich 1988, Laurent 1988, Blake & Hart 1993). 2. The disease resistance differs, because signal crayfish harbor parasites not naturally found in native crayfish, e.g. *Psorospermium haeckeli* (Hilgendorf) (Alderman & Polglase 1988, Diéguez et al. 1995, Henttonen et al. 1997), and is a vector of aphanomycosis. 3. The physiological efficiencies are different (Moshiri et al. 1971, Taylor & Wheatly 1981, Mackêviciene 1993, Whiteley et al. 1995). 4. The feeding preferences and ecological impact on aquatic macrophyte beds vary (Matthews et al. 1993, Nyström & Strand 1996, this volume) where *P. leniusculus* has proved in most cases to be more efficient, aggressive, and to has a higher growth rate than *A. pallipes* (see Holdich this volume).

Finally, Laurent (1988) reported that when both species coexist, they will interact directly and will mate, strongly reducing the reproductive efficiency of wild populations of *A. pallipes*, since offspring do not develop longer than embryonic state and mating occurs between male *P. leniusculus* and female *A. pallipes*. Critics of translocations emphasize the contamination of local genetic stock (Grandjean et al. 1997, Hodder & Bullock 1997). Therefore, the use of alien, or even non-local stocks, is often seen as negative in itself, with little regard for any positive or negative impacts on biodiversity. In addition, there is evidence that genetic mixing between genetically differentiated populations has the potential to introduce locally maladaptive traits (Bullock & Hodder 1997, Hodder & Bullock 1997).

Besides, the lack of disease control measures on the individuals used for stocking is extending aphanomycosis. Diéguez (cited in Diéguez & Rueda 1994) detected the presence of *A. astaci* in crayfish for stocking; the presence of zoospores is maintained in the environment promoting the phenomena known as repeated zoospore emergence (Cerenius & Söderhäll 1984). At present, it is known that probabilities for successful stocking with native crayfish are very high, if infested rivers are left without vector crayfish for a prudent time (Smith & Söderhäll 1986, Holdich & Reeve 1991, Skurdal & Taugböl 1991, Gydemo 1992, Alderman 1993, Maitland 1996).

In summary, if introduction and translocation of alien species are avoided and European regulations followed, it actually would be possible to restore native crayfish populations.

2.2 *Yield*

There are few and confusing data on the actual number and size of the native crayfish populations, not only in Spanish Public Administration but also in Autonomous Communities Governments. Just a few communities have specialized human resources to update evaluations of crayfish populations (Castilla-León, País Vasco, Castilla-La Mancha, Aragón and Navarra). Estimations of the total annual catch were made only for a short period in Spain (from 1974 to 1980) by the Ministerio de Agricultura, Pesca y Alimentación (MAPA), formerly Ministerio de Agricultura, Pesca y Caza, when Spain had probably the most important European production of white-clawed crayfish. Nowadays, neither MAPA nor Ministerio del Medioambiente have registered all the captures coming from the wild, because these Ministries are not interested in species with a low economic value. Table 1 shows native crayfish produc-

Table 1. Native crayfish (*Austropotamobius pallipes*) annual production at different seasons in Burgos Province (Spain) and in Spain*.

Year	Place	Mt	No animals (thousands)	Source
1935	Burgos	31.5	2,100.0	1
1962		900.0	60,000.0	2
1981		105.0	7,000.0	3
1962	Spain	1472.9	98,191.7	1
1974		855.0	27,600.0	4
1975		1057.0	34,100.0	4
1976		837.0	27,000.0	4
1977		778.0	25,100.0	4
1978		643.0	20,800.0	4
1979		519.0	31,800.0	4
1980		200.0	6,460.0	4
1981		540.0	36,000.0	3
1988		5.0	333.3	5
1996		< 1.0	< 66.7	6

(mt = metric tons). [1]Pardo 1951, [2]Torre-Cervigon & Rodríguez-Marques 1964, [3]Fuente 1982, [4]Ministerio de Agricultura 1982, [5]FAO 1989, [6]J.M. Carral pers. comm. *Animal conversion to weight assuming a crayfish is 30g.

tion data in Spain for different seasons. Portugal does not have these estimates, since native crayfish is not subject to a commercial fishery. Burgos Province production data are detailed because is by far the largest producer in Spain.

Total profits of this activity generated in 1962 in Spain were more than 51 million Pesetas at that time (Torre-Cervigon & Rodríguez-Marques 1964), and the most productive provinces were Burgos (25 million Pesetas), Guadalajara (6), Albacete (5), Cuenca (3.3), Ciudad Real (2.5) and Segovia (2). Later Servicio Nacional de Pesca y Caza from the Ministerio de Agricultura (Spain) reported in 1964 that Burgos and Palencia were the provinces with higher fishing potential, followed by Ciudad Real, Guadalajara, Cuenca, Soria, Zamora, Segovia, Valladolid, León, Ávila, Navarra, Álava, Logroño, Zaragoza, Teruel, Salamanca, Madrid and Toledo. From the 1970s until now Burgos is the most productive province (Fuente 1982, C. Temiño pers. comm.). Spain only exported crayfish to France from 1962 to 1965, but prices were lower than Italian exports, probably, because the majority of the crayfish exported were used for restocking rivers in Herault, south France, and not for human consumption (Vigneux 1980, Laurent 1988).

Statistics shown in Table 1 indicate that after the confirmed outbreak of crayfish plague in 1977-1978, yield of native crayfish declined from 519 (in 1979) to 200 (in 1980) tons. The yield increased slightly in 1980 because of the fishery increase, since the government issued a greater number of fishing licenses. Mean size of individuals has decreased notably, from 30.9 g average weight in 1978 to 15.0 g in 1980.

In summary, there is a long history of crayfish consumption in Spain, and though *A. pallipes* is less exploited than other species at this time, the domestic importance of this species is very high in the north and to a lesser extent in central Spain.

2.3 *Present situation*

The effects of over-fishing, habitat loss, hydraulic management, and pollution were evident when the crayfish production decreased from 1962 to 1978. Furtive fishing and crayfish plague have a synergistic effect whose manifestation started in the late 1970s and last until today, driving *A. pallipes* populations almost to extinction in IP. Public Administration has dealt with this problem in several ways:

1. Reduction in the number of fishing licenses, i.e. in Duero region 153,067 licenses were issued in 1977, but just five years later only 87,377 licenses were issued in that region (Fuente 1982).

2. Disease measures control since January 1981, to prevent and stop the spread of crayfish plague (Martin 1982).

3. Development of production farms of *A. pallipes* yearlings for restocking, e.g. at present there are two farms for the culture of native crayfish in Castilla-La Mancha. One (Rillo de Gallo, Guadalajara) produces about 35,000-45,000 0+ crayfish (E. Llorente pers. com.) and the other one (El Chaparrillo, Ciudad Real) produces 200,000 0+ crayfish a year (F. Alonso pers. comm.). Yearly production of *A. pallipes* individuals, 0+ or summerlings, for restocking could be about 80,000 in Spain. Several Autonomous Communities (AC) have promulgated laws regulating fishing of native crayfish. Fishing seasons vary greatly by AC but in the majority of cases closed seasons are from June 20 to the first week of October; fishing is allowed only in two days a week, mainly weekends and in festivals. Other regulations concern the

legal size of animals (≥ 8 cm total length); type and number of gear (lift net) and mesh size are variable according to AC; daytime crayfishing is restricted to dawn and nightfall and the maximum permitted catch is 80 animals per day, in most cases. However, fishing is prohibited by law (Decreto 91/1994 which regulates Freshwater Fishing) and native crayfish cannot be commercialized (Real Decreto 1118/89) in some AC, e.g. Castilla-La Mancha.

4. Efforts are being made in order to include native crayfish in the Red Book of Spain as threatened, endangered or vulnerable species (Rosas et al. 1992).

Portuguese Public Administration is acting as follows:

– The Instituto de Conservação da Natureza promoted in 1994 a survey in order to know the status of *A. pallipes* populations and to develop the adequate conservation strategies (Bernardo & Ilhéu 1994a, Bernardo et al. 1997), because several sites of the white-clawed crayfish were included in the NATURA 2000 proposal.

– Several dispositions of the present legislation intend to preserve the native crayfish populations: 1. Fishing is banned from September to March, 2. 9 cm of minimum size, 3. The only legal fishing method is baited traps ('balança' or 'ratel'), 4. Fishing days are Thursday, Sunday and National holidays, and 5. 40 crayfish/fisherman per day is the maximum catch.

– Crayfish farming and stocking programs are being prepared by the Direcção Geral das Florestas, the Portuguese Forestry Department. Fishing will be totally banned for a number of years until populations are restored.

2.4 *Proposals for action*

Action proposals to preserve and increase native crayfish populations in IP are similar in both countries and focus on the following:

1. Fishing could be banned completely, for a specific period each year; fishing methods and equipment could be subjected to restrictions, controls, and only a certain size class of the crayfish population can be targeted. It is very important to declare this species officially as 'threatened' or 'vulnerable' and to adopt measures for protection.

2. Ecological, immunological, genetic and aquacultural research of *A. pallipes* should be carried out.

3. Native crayfish farming should be increased.

4. Stocking with *A. pallipes* is the most important method for restoring plague-devastated waters to their productivity for men but, in order that they succeed, it is also important to monitor them.

5. Although there are no reliable methods to eradicate crayfish plague, especially from larger water bodies and rivers, it is possible to improve its control by stricter manipulation of stocking operations, by using only plague-free stocking material, and by effective disinfecting of traps and other equipment.

6. Mapping the areas with active plague infestation and preventing uncontrolled transport of crayfish, traps and equipment from these areas.

7. Prohibition of translocation and introduction of alien crayfish species in natural water bodies. In respect, the current legislation must include the establishment of large-scale no-go areas where future crayfish restocking and/or farming developments will be banned as has been done in Great Britain (see Holdich, Rogers & Rey-

nolds this volume)

8. Environmental education programs aimed at the public and fishermen in particular should be developed. It is also very important to encourage the public to collaborate in *A. pallipes* preservation and restoration campaigns.

Interest for preserving this species is growing among scientists, crayfish farmers, fishermen, crayfish consumers, and the general public. However, the extinction rate of native crayfish is faster than Public Administration can act to preserve it, so, if legal procedures do not speed up, there will be no species to protect.

3 THE INTRODUCTION OF CRAYFISH INTO THE IBERIAN PENINSULA

It can be seen in Figure 3 that with the exception of crayfish of the genus *Procambarus*, introduced into the southwest of the Iberian Peninsula, introductions of crayfish have taken place in the center and north of Spain.

The first attempts to acclimatize various species of European and American crayfish took place in the 1960s, with varying success (Habsburgo-Lorena 1978, Lowery & Holdich 1988). In 1962, the noble crayfish, a species typical of Central Europe and Finland, was tried in a crayfish farm, but without success.

In 1973 a batch of *P. clarkii* was imported from Louisiana (USA) into the province of Badajoz. Its acclimatization was so successful that it has now become naturalized. Subsequently, in 1974, white river crayfish *P. zonangulus* were also introduced into the rice fields of the Lower Guadalquivir in the province of Seville, together with another batch of *P. clarkii*. This species apparently did not prosper (Habsburgo-Lorena 1979, 1986).

In 1974, unlike other alien species whose batches were imported from within their natural distribution range, the signal crayfish *P. leniusculus*, a species typical of the West Coast of North America, was imported from Sweden. This species achieved considerable success in acclimatizing itself (Carral et al. 1993). This species is currently being introduced in places where the native crayfish has disappeared in such a

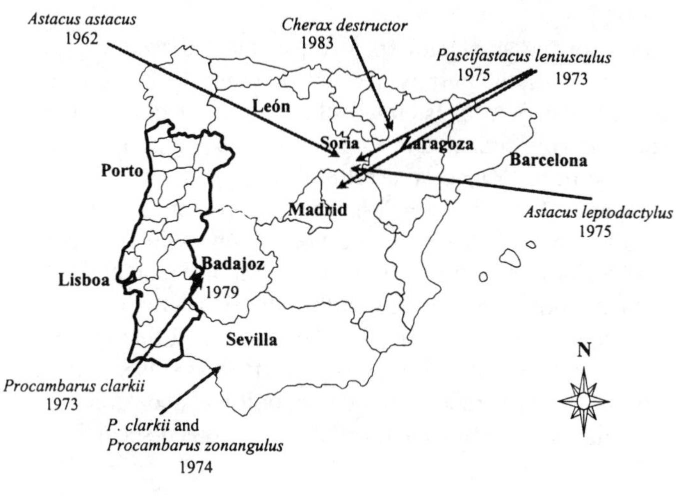

Figure 3. Introductions of exotic crayfish on the Iberian Peninsula.

way that its range of distribution has considerably increased. Populations of *P. leniusculus* have been described in the following Autonomous Communities: The País Vasco, Castilla-León, Castilla-La Mancha, Navarra and La Rioja. Likewise, the number of farms cultivating them for the purpose of restocking have increased: There are currently three in operation. Fishing of this species is permitted and well-regulated in some AC, such as Castilla-León, Navarra and the País Vasco, while its fishing and sale is prohibited in Castilla-La Mancha (Decreto 91/1994 and R.D. 1118/89, respectively). It is estimated that its production reached 1 ton in 1988 (A. Habsburgo-Lorena personal communication), and that about 0.6 ton for human consumption, and 1.5 million second stage juveniles for restocking were produced on farms during 1994. It is important to point out that these figures could be underestimated due to the type of fishing and the heavy poaching of this species, which ensures that the total catch is not always declared to the authorities.

Finally, there was an also unsuccessful attempt in 1975 to introduce a batch of crayfish narrow clawed crayfish, *A. leptodactylus,* from Turkey into various ponds on a farm in the province of Valladolid. This species is typical of the Ponto-Cáspian Basin (Black Sea, Caspian Sea, Turkey). In 1983, a batch of Australian yabby, *C. destructor,* were brought to Cataluña for aquacultural purposes, but because permission was not granted to set up the farm, the businessmen moved the animals to the province of Zaragoza. As they did not have permission for a crayfish farm there either, they decided to release the animals into a lake in the province, where the population has since prospered (Bolea, 1995).

3.1 *The case of Procambarus clarkii*

There are several reasons why *P. clarkii* is the most widely distributed species of crayfish in the world. It has long been known that the red swamp crayfish is the most ecologically plastic of the entire Decapoda. For this reason it has been the center of attention for a large number of aquacultural projects in its area of origin. This has resulted in the creation of a flourishing multi-million dollar industry in the state of Louisiana (USA) (50,000 ha under cultivation), related with its extensive and intensive production, processing, consumption and distribution in national and international markets (Huner 1985, 1988).

There are three basic reasons for the artificial expansion of the red swamp crayfish. First, they are imported for recreational purposes as pets in domestic aquaria and are subsequently released into the aquatic ecosystems of the country of introduction. This is the origin of the introductions in France and England (Hobbs III et al. 1989). The second reason arose from the need to feed aquacultural important species, such as the case of the bullfrog in Japan in 1918 (Penn 1954). The third reason is related to the attempt to create commercially exploitable breeding populations. This has been the fundamental motive for the majority of introductions, including in Spain and subsequently in the south of Portugal, although it is known that the first individuals arrived in Portugal as a result of the natural expansion of the population that became naturalized in Badajoz (Correia 1993). In any case, as Gaudé III (1983) has pointed out, each introduction features a series of personal, economic, political and ecological aspects that ultimately explains the origin and execution of a determined species importation project.

The result of this behavior has been the widespread distribution of this species throughout the world (see Holdich this volume). It is currently present on every continent with the exception of Australia and Antarctica. Its distribution continues to extend year by year, undoubtedly encouraged by the tendency to importation, towards zones with warm climates characterized by mild autumns and winters and hot summers (Huner 1985, 1988, Hobbs et al. 1989).

3.1.1 *The introduction of P. clarkii into the Iberian Peninsula*

Unlike the introduction of other animal and plant species into Spain in general, and of astacoid crayfish in particular, the importation of the red swamp crayfish has been thoroughly documented and dealt with by various authors (Habsburgo-Lorena 1979, 1983a,b, 1986, Algarín 1980, 1981, Gaudé III 1983, 1986, Ocete & Lopéz-Sánchez 1983).

The favorable conditions for the introduction of this species into Spain were the result of the optimal international conditions that existed in the 1970s for the introduction of American species resistant to the crayfish plague. This decimated the astacoid populations in continental Europe where the crayfish is in great demand by the market as it is considered to be a high quality foodstuff (Westman & Westman 1992, Anastácio & Marques 1995, Ackefors this volume).

These conditions unleashed a series of failed introductions of other European and American crayfish taxa at the beginning of the 1960s and 1970s. This led to the consideration of a species with a broad ecological spectrum, and suitable for a country more than 60% of whose territory is characterized by a Mediterranean climate (Montes et al. 1993). In this context, the first experiment concerning the acclimatization of *P. clarkii* (100 kg) originally taken from New Orleans and placed in the rice fields of a farm in Badajoz was carried out in June 1973 (Habsburgo-Lorena 1986). Subsequently, a larger scale introduction was carried out, with institutional help, to occupy a zone of potentially greater economic benefit on account of its area (28,000 ha): the rice fields of the upper marsh of the Guadalquivir (Algarín 1980). All the legal procedures were followed and observed, and permission, help, and assessment were provided by ICONA and Dr James Avault, who had previously visited Spain to identify different zones appropriate for the introduction (Habsburgo-Lorena 1986). Thus, 500 kg of red swamp crayfish and white river crayfish were brought to Seville on 10 May 1974, although the latter species appears not to have become established (Huner 1988). Given that the native crayfish had never been present in nor was suited to this zone, it was thought that there would be no problems of competition or elimination due to the transport of the crayfish plague.

Although the majority of the crayfish died in transit, 100 kg of crayfish were released into an eel pond belonging to Mr Grau, who financed the operation, in the municipality of Puebla del Río, Seville. The absence of filters allowed the crayfish to escape and to begin the natural colonization of the nearby ditches and canals. In 1975, expansion was accelerated by fishermen. Because of the high prices of crayfish, they began to distribute animals throughout the entire Guadalquivir marsh zone, including the Doñana National Park and its area of influence. The idea was not only to break Grau's commercial monopoly, but also to establish zones for their storage during the winter and spring when the rice fields are dry.

By 1976 there was commercially exploitable production in the rice fields and the canal systems, and the first individuals were caught in the Rocina stream and Las Nuevas Channel within the National Park (Algarín 1980). By 1977, the Rocina had become a fishing site which by 1979 had extended along the entire Lower Guadalquivir, including the different natural aquatic environments and dams of the provinces of Seville, Huelva and Cádiz. Figure 4 illustrates the expansion of the red swamp crayfish in the Iberian Peninsula.

The knowledge of the economic benefit this species was providing in the lower Guadalquivir and the ease of obtaining live material (this is the form in which it is sold in the national markets) encouraged different owners of rice fields to attempt its transfer to other regions of the country. This was the case of the rice fields of Valencia in 1978 and the Ebro Delta in 1979 (Fig. 4).

A consequence of the highly active behavior of the red swamp crayfish is that its rate of natural dispersal is very high. As a result of this and the geographical location where the first Spanish introduction took place, it was discovered in Portugal at the end of the 1970s. Ramos & Pereira (1981) state that *P. clarkii* arrived in Portugal in 1979, specifically in the River Caia, near the Spanish border. Abundant populations were detected in the mid-1980s in the hydrographic region of the Tagus (Ferreira 1985, Reiner 1985, both cited in Correia 1994). In 1987, large numbers of individuals were observed in the River Mondego, central Portugal (Marques et al. 1992). Adão (1992) reported that by the middle of the 1980s the red swamp crayfish was found throughout the entire hydrographic region of the Guadiana (SE) and by the beginning of the 1990s it had reached the River Sado (SW). Although there are no data concerning its arrival, Adão and Marques (1993) mentioned its occurrence in the River Douro (N). This initially natural expansion was seen to be favored by the diverse introductions and translocations that Spanish fishermen made around the south of Portugal, with the aim of increasing their fishing area. A massive translocation of *P. clarkii* has been undertaken in Spain and Portugal since 1980 (Fig. 4).

The expansion of the crayfish through Spain and Portugal was not only restrained to the continent. Stable crayfish populations have now been detected in Lagoa do Peixe, São Miguel Island, Azores, Portugal (Correia & Costa 1993, Costa et al. 1996), in the Balearics and in San Andrés steep river bank, Tenerife Island, Canary

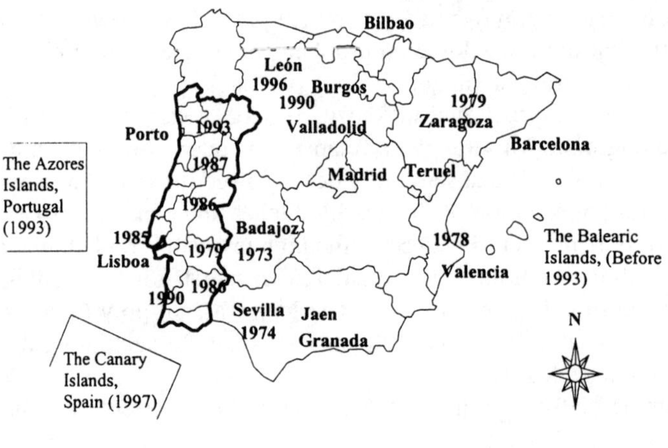

Figure 4. Expansion of the red swamp crayfish *P. clarkii* in Spain and Portugal.

Islands, Spain (Gutiérrez-Yurrita & Martínez unpublished data, Viceconsejería de Medio Ambiente de Canarias personal communication).

3.1.2 *Ecological impact*

P. clarkii is a key species in the function of aquatic ecosystems (Jones & Momot 1983, Momot et al. 1978, Momot 1984, 1995, Holdich 1988, Gutiérrez-Yurrita et al. 1996, Gutiérrez-Yurrita 1997, Lindqvist & Huner and Holdich this volume). The direct impact on the ecosystem after its introduction should be considered from three perspectives. The first is its potential for altering the total production of the ecosystems (Huner & Barr 1991, Momot 1995, Gutiérrez-Yurrita et al. 1998). The second concerns its feeding habits. Although it is an important carnivorous predator, while it searches for animals it eats large quantities of vegetation and detritus at the same rate that reduces the animal protein (Gutiérrez-Yurrita et al. 1998). At high crayfish densities, foraging may be so extensive as to modify the habitat totally, destroying the microbial and the macrophyte beds (Momot et al. 1978, Feminella & Resh 1986, 1989, Lodge & Lorman 1987, Olsen et al. 1991, Beroiz in press). Finally, the role of crayfish as the prey of other invertebrate and vertebrate animals must be considered (Hernando 1978, Adrián & Delibes 1987, Beja 1996, Elvira et al. 1996, Elvira 1998). In this respect, it is very important to notice that *P. clarkii* may accumulate toxins and heavy metals in its body and this could lead to a transfer of the toxins further up the food chain (López & Fernández 1980, Torreblanco et al. 1987, Lirås et al. 1998; Gutiérrez-Yurrita & Montes 1998b).

Several studies have revealed *P. clarkii* as a key species in many freshwater ecosystems. As a major consumer, red swamp crayfish may perform a significant role ingesting large quantities of detritic organic material (Ilhéu & Bernardo 1993b, 1995, Gutiérrez-Yurrita et al. 1998). This consumption promotes the acceleration of the decomposition process, transforming the coarse particulate organic matter into fine particulate organic matter, thus providing food for collector-gathering groups and primary production is stimulated by the increase of nutrient turnover rates (Bernardo & Ilhéu 1994b). If the availability of vegetal detritus is low, then crayfish trophic behaviour shifts to macrophyte grazing (Ilhéu & Bernardo 1994a, 1995, Gutiérrez-Yurrita et al. 1998). Moreover, it has been demonstrated that it is an engineer species of the ecosystem, physically transforming its environment and altering the availability of resources for other species (Gutiérrez-Yurrita 1997). These changes have been achieved principally by structural modification of the marsh and other aquatic ecosystems (Gutiérrez-Yurrita & Montes 1997). The crayfish bring this about by reducing or totally eliminating the aquatic macrophyte beds (Gutiérrez-Yurrita 1997, Beroiz in press). Also, they have a habit of burying themselves during the breeding season and building burrows, which affects the physical composition of the ecosystem (Montes et al. 1993, Bravo et al. 1994, Gutiérrez-Yurrita et al. 1994, Gutiérrez-Yurrita 1997; Gutiérrez-Yurrita & Montes 1998b). The most important ecological consequence of the reduced aquatic vegetation cover is the enriching of the water column with nutrients, which favor the development of phytoplankton communities, changing the ecological balance from fresh to turbid water, as Figure 5 shows (Scheffer 1990, Montes et al. 1993, Scheffer et al. 1993, Bouffard & Hanson 1997, Gutiérrez-Yurrita 1997).

On the other hand, it has to be borne in mind that due to this species' importance

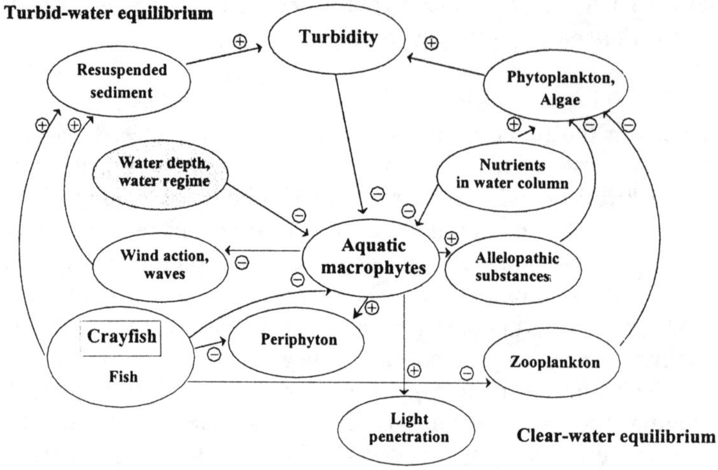

Figure 5. General feedback loops thought to be responsible for the existence of alternative equilibria in freshwater marsh ecosystems. The qualitative effect of each route in the diagram can be determined by multiplying the signs along the way. Both the aquatic macrophytes and the turbidity-state are self-reinforcing. Qualitative effect of management measures discussed in the text can be checked in the same way if the influence of a manager acts with either a positive or negative effect indicated by the arrows.

as a fishing resource in diverse regions of the Iberian Peninsula, it also brings with it an impact on the ecosystems due to its exploitation. This impact manifests itself in the disruption caused by fishermen when they set and recover their traps, the fishing techniques employed and the form and the sites where fishing takes place (Molina-Vázquez 1984, Domínguez 1987, Green 1994, Gutiérrez-Yurrita et al. 1997a, b,). This impact shows itself in the destruction of riparian vegetation by fishermen and by the high mortality of aquatic vertebrates (fish, amphibians, and reptiles) and birds, due to the type of trap employed. For example, more than 42,000 birds and more than 750,000 aquatic vertebrates of various species died in the traps in the Doñana National Park and the Brazo del Este Natural Park, in the 1989 season. Two of these bird species (*Marmaronetta angustirostris* and *Porzana puzilla*) are featured in the Red Book for Spain as being in danger of extinction (Gutiérrez-Yurrita et al. 1997a).

The presence of the red swamp crayfish in the marsh zones not only has repercussions in the natural areas that it colonizes, but also in the rice fields. Various authors have demonstrated the considerable damage that the crayfish can cause to the agricultural infrastructure and the technical difficulties of its minimization (Algarín 1980, Ocete & López-Sánchez 1983, Gállego & Ocete 1985, Adão & Marques 1993, Anastácio & Marques 1995).

3.1.3 *Socio-economic impact, or how to make the best of a bad situation*

Although the red swamp crayfish has been introduced into many places on the planet, its commercial exploitation is basically confined to Louisiana (USA), its area of origin, the People's Republic of China, Kenya, and the marshes of the Lower Guadalquivir (Huner 1989, Huner et al. 1992, Ackefors this volume). The reasons for this are cultural and sanitary. In most places where it has been introduced, neither this nor any other similar crustacean was traditionally consumed. This is also the case in African and American countries, with the exception of the USA, where large quantities of *P. clarkii* are consumed (Huner & Barr 1991), and Mexico, where other crayfish species have been eaten since pre-hispanic times (Gutiérrez-Yurrita & Latournerié-Cervera 1992). In Japan, as well as being considered a foodstuff of low quality, it also acts as a vector of certain intestinal parasites, which deters its consumption (Penn 1954).

In Spain, only the Lower Guadalquivir has an industry capable of channeling the production of crayfish to secure their placement in wholesale markets. Fishing in other regions can be considered a sport limited to private consumption, or a minority market, there being no fishermen's associations or other business organizations based on the crayfish, as is the case in Valencia and in the Ebro delta (Escosa 1990). On the other hand, it could be considered as without economic importance, as occurs in the Valencian Albufera (lagoon) (Jover et al. 1990, M. Jover personal communication).

With respect to Portugal, the exploitation of this species has been very limited, and has basically been confined to poaching by Spanish fishermen when, because of the drought in the few remaining water bodies and in the marshes of the Guadalquivir, there were no crayfish (various fishermen personal communications). In 1995, as a result of the drop in production of crayfish in Spain, a Dutch company, based in Seville, showed an interest in installing a crayfish processing factory in the Mondego Valley to commercialize it as a frozen and tinned food and as paté in North European markets. The estimated current yield in this region for export to Spain is 700 tons, but the potential yield for 1998 could be 3000 tons (Veiga 1995).

In 1975, numbers of red swamp crayfish were caught in the nets of eel fishermen of the Lower Guadalquivir. They regarded the new species as being of great economic potential as a substitute for the eel, which was in evident decline. They began to disperse part of their captures, generally the smaller crayfish, among the old eel reserves that were assigned to them. The first years' crayfish catches were an overwhelming success for the collective (Algarín 1981). The annual Spanish crayfish production from the time of its introduction to the present is shown in Figure 6. The greatest catch of 5000 tons was achieved in 1987. Due to the heavy demand in markets in the north of the Peninsula, crayfish is sold for good prices. By virtue of the existence of the reserves, the fishermen maintained total control of production. Wholesalers came to the marsh and went away with the merchandise without incurring any additional costs (Soto 1992). Figure 7 shows the evolution of the number of crayfish fishermen in the Guadalquivir marshes over time.

With the incessant increase in production, the prices dropped and the competition between the different producers and sellers increased. Figure 8 shows the evolution of the prices of the crayfish in Spain. This situation ended up in a very disorganized state. The red swamp crayfish fishing industry changed from being the economic hope of the region to being a marginal and seasonally exploitable resource – combined with an alternation with agriculture, the fishing of other species, and sporadic work in areas away from the primary sector (Montes et al. 1993). Due to this and the severe drought in this area, the annual production of crayfish in the Guadalquivir marshes at the beginning of the 1990s was practically nil. The quantities that have been reported have been largely dependent on the poaching carried out in Portugal, as the captured animals were transported to Spain, where they were processed and mar-

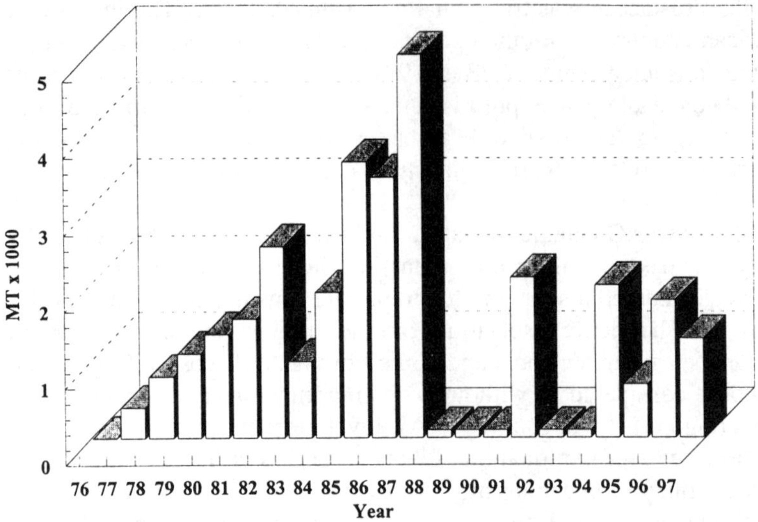

Figure 6. Yearly production of red swamp crayfish (*P. clarkii*) in the Guadalquivir marshes.

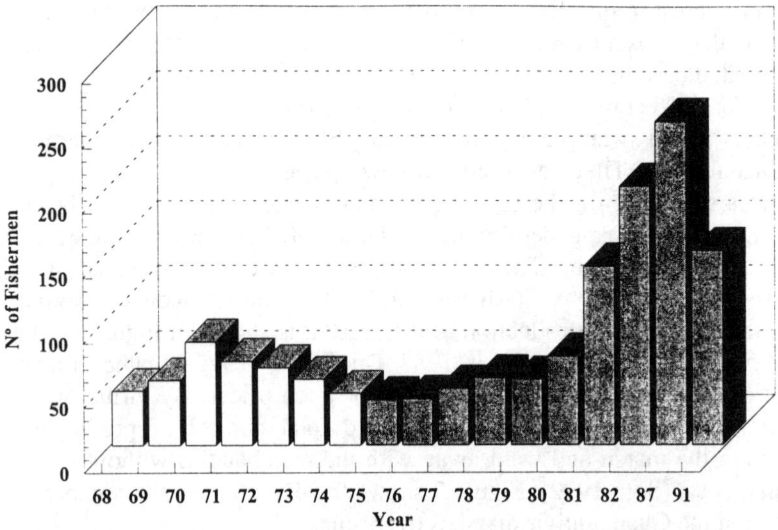

Figure 7. Yearly distribution of number of fishermen in the Guadalquivir marshes.

keted (P.J. Gutiérrez-Yurrita personal observation, various fishermen personal communications). It is currently thought that with good management measures of this resource and with years of average precipitation (540 mm), the Guadalquivir marshes could yield about 3000 tons (Soto 1992). This level is about to be achieved, as during the 1997 campaign production has risen to 2900 tons.

The primary productive reserve, in the El Rocío marsh, changed from being a protected space with the broadening of the limits of the Doñana National Park in 1978. Crayfish fishing was legally practiced there until 1984 when henceforth it was

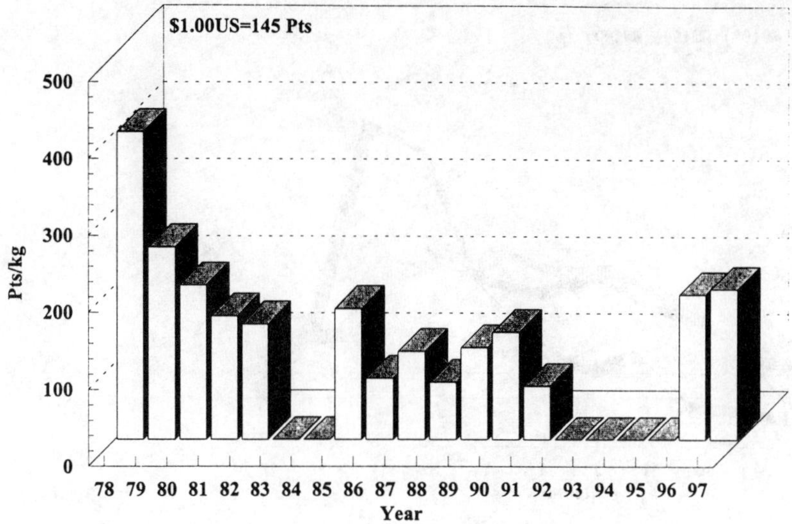

Figure 8. Yearly average price for *P. clarkii* in Spain.

prohibited due to the increased environmental impact on the Park's ecosystems (Molina-Vázquez 1984, Domínguez 1987).

In 1985 the Junta de Andalucía entrusted, through the Fishery General Direction (Dirección General de Pesca), the carrying out of a sociological analysis of the fishermen's collective (DOPP 1986), which is the most recent investigation. This portrayed a still unresolved and situation panorama, with important accusations between the different sectors with respect to production. The parties demanded external regulation by the public bodies with the aim of them somehow counteracting the trend of a few processing plants towards a monopoly, and the unequal sharing of profits. The most recent reviews concerning this matter in this zone have been those realized by Algarín and Librero (1988), wherein new data were presented concerning both production and the development of the fishermen's collective, and a sociological analysis carried out by Soto (1992) for the 1991 campaign. It is notable that the economic problems of the fishermen are largely due to the fact that the production of *P. clarkii* is strongly affected by its seasonality (more than 97% of the production comes from captures in natural areas, whereas less than 3% comes from farms). It has not proved possible to maintain the stability of production throughout the year. Figure 9 illustrates the monthly production of *P. clarkii*, while Figure 10 shows the monthly price per kg ($1.00 US Dollar ≈145 Ptas). Crayfish prices in Portugal are lower than in Spain. For instance, while fishermen sell crayfish from 135 to 270 Ptas/kg in Portugal (P. Anastácio personal communication), the crayfish prices in Spain were about 180 Ptas/kg in Lower Guadalquivir (pers. obs.) and from 350 to 400 Ptas/kg in Valencia (M. Jover personal communication) from July to early August. During this period, fishermen received the lowest price because there were a lot of crayfish in the market. Furthermore, prices went down to 25 Ptas/kg at the end of August and early September due to the great production and the small size of crayfish in the Lower Guadalquivir. Prices of crayfish were higher than 800 Ptas/kg (maximum price re-

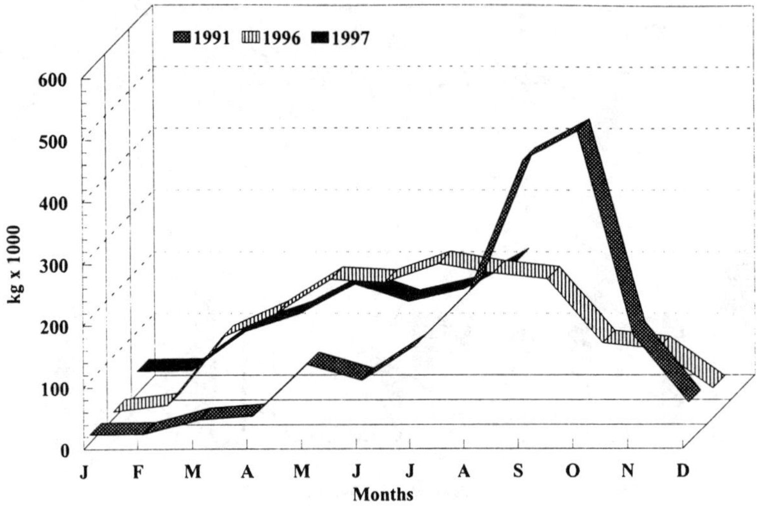

Figure 9. Evolution of the monthly production of in the Guadalquivir marshes.

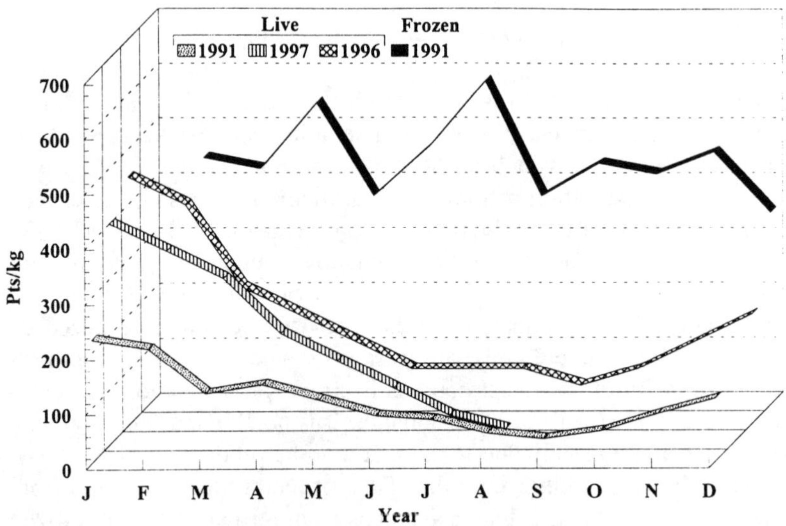

Figure 10. Monthly average price for live and frozen *P. clarkii*.

corded was 1300 Ptas/kg), at least 1000 Ptas/kg, and as high as 630 Ptas/kg in Madrid, in Canary Islands and in Portugal, respectively, despite going down in price in fishing places.

The market for *P. clarkii* is illustrated in Figure 11. It is distributed basically in the central and north regions, Madrid, and Valladolid being the main provinces in which it is consumed. Figure 11a shows the European market for this species. The estimated exportation is 301 tons, of which 78.1% is sent to France, 13.6% to Sweden, 7.6% to Belgium and 0.7% to Denmark.

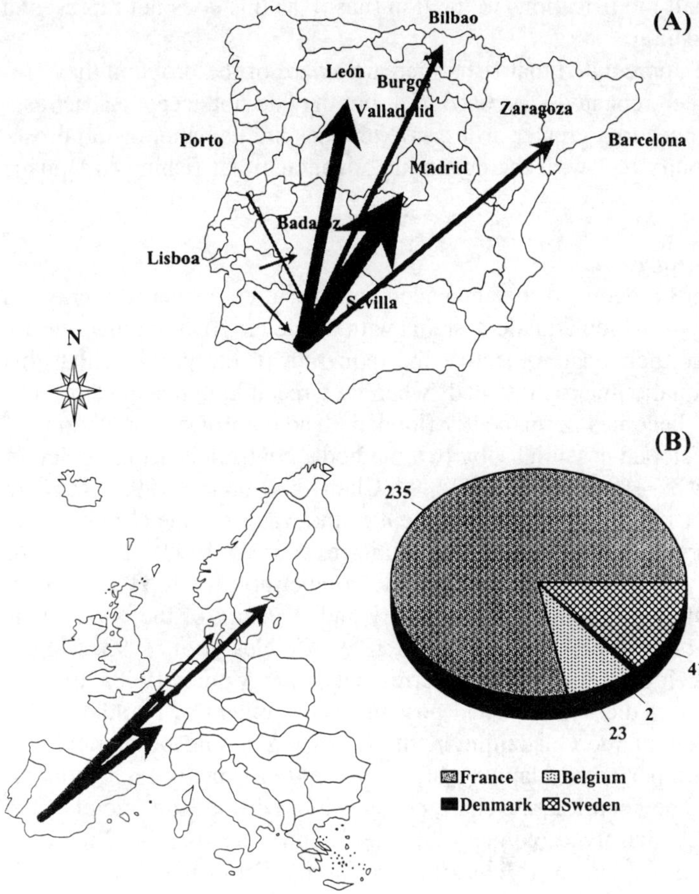

Figure 11. Market distribution of *P. clarkii* in Spain and Portugal (A) and export markets of whole frozen crayfish (B). Arrow width indicates importance of the market. Data expressed in metric tons.

3.1.4 *Environmental awareness of fishermen*

The degree of environmental awareness shown by the fishermen's collective is quite low. Fishing is an extractive activity involving little or no intervention in the medium, except for the environmental impact to which we have made reference. Thus, it is difficult for a fisherman to appreciate why his presence in particular zones could disturb or be dangerous to a natural system from which he has traditionally made his living. Almost 70% of the fishermen fish in El Rocío, in the Doñana National Park, this term being used to refer jointly to the marsh of El Rocío and la Madre as far as the Doñana Biological Reserve. The majority of those who do not practice this live on the left bank and fish in another protected zone, the Brazo del Este Natural Park.

Indeed, 80% of fishermen believe that poaching not only causes no damage to the Park, but on the contrary that their activity is beneficial as it controls the spread of the species in some sense. Those who thought that poaching did have a harmful effect on the Park pointed out that if fishing were regulated within the Park, then the damage would be much less. On being questioned about the effect that the nets had on the birds, 69% of the fishermen replied that this damage was minimal and that it could be avoided by making small modifications to the trap (nasa), as this does not represent a major impediment to fishing.

In general, it can be appreciated that fishermen are aware of the problem they created, since they have been repeatedly accused of it, but they do not accept a strict ban. In fact, poaching continues to a greater or lesser extent despite its banning and prosecution. The reasons finally rest with the economic advantage that fishing in Doñana provides.

3.1.5 *Management measures*

From the above, it can be deduced that management measures concerning crayfish populations, or the way in which aquatic systems with stable crayfish populations are managed, must have as their basic objective the reduction of the pressure that this species exerts on the aquatic macrophyte beds when they reach high densities of individuals and the system becomes permanently flooded . The control of populations of *P. clarkii* could be developed essentially by two methods: controlled fishing (Momot 1991, 1993, Huner 1992, Huner et al. 1992, McClain & Romaire 1995, Romaire 1995, Gutiérrez-Yurrita 1997) and the management of the water regime (Huner 1988, 1995, Whatley & Eversole 1993, Jones 1995, Gutiérrez-Yurrita 1997). The goal of both methods consists of reducing the intraspecific competition for food and cover, while adequately maintaining the numerical density and structure of the populations for each management objective (Lutz & Walters 1989, Geddes et al. 1993, Momot 1993, McClain 1995, Yeh & Rouse 1995, Gutiérrez-Yurrita & Montes 1998a, b).

As important aspects of the dynamic and possible management of populations of *P. clarkii* in the marshes of the Guadalquivir, the following should be pointed out. Crayfish populations are perfectly adapted to the daily and seasonally highly fluctuating aquatic systems. They can tolerate sharp changes in environmental factors such as temperature (5-38°C), dissolved oxygen (2-14 mg/l) and water regime. Figure 12 compares the different life cycles of *P. clarkii* in the Iberian Peninsula. *P. clarkii* adjusts its life cycle to the period of permanency of the water in the ecosystems, from a cycle of annual reproduction (in systems with < 4 months of flooding) to one of almost continuous reproduction (in systems with 12 months of flooding). In this way, the populations of crayfish could be correctly managed with an adequate control of the flooding regime (Gutiérrez-Yurrita 1997).

The introduction of *P. clarkii* has brought with it a new economic resource to part of the Lower Guadalquivir and has served as a complementary source of income to improve the standard of living of a considerable number of families. Nevertheless, it has caused very important environmental changes. One can speak of a before and after the introduction in many types of aquatic environments, including various sectors of the Doñana National Park. The economic benefits of the introduction of the crayfish in the Lower Guadalquivir are obvious, but does it compensate the environmental costs it has generated?

Since the introduction of *P. clarkii* in the Lower Guadalquivir, only 25 years ago, this species has become a key species in the functioning of aquatic ecosystems. Nowadays, its biology, ecological impact and the way to produce and culture it are better known. However, it is important that all this knowledge can be developed in a region where ecological and economical interest in natural and protected areas come together. In this context, the Environmental Department of the Andalusian Govern-

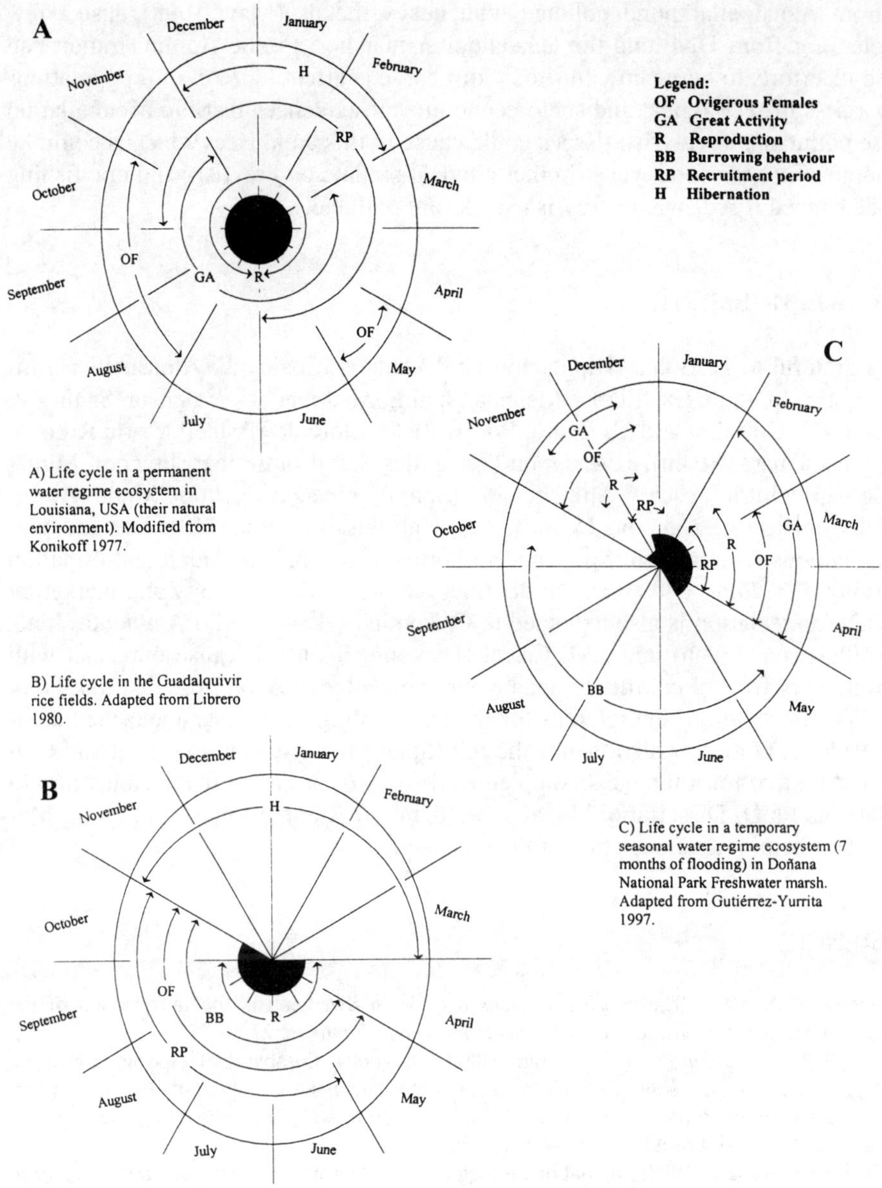

A) Life cycle in a permanent water regime ecosystem in Louisiana, USA (their natural environment). Modified from Konikoff 1977.

B) Life cycle in the Guadalquivir rice fields. Adapted from Librero 1980.

C) Life cycle in a temporary seasonal water regime ecosystem (7 months of flooding) in Doñana National Park Freshwater marsh. Adapted from Gutiérrez-Yurrita 1997.

Figure 12. Life cycles of *P. clarkii* in the lower Guadalquivir marshes and in its natural environment in the USA.

ment together with the Ecology Department of the University of Madrid are developing a new general project to implement crayfish culture in transformed rice-fields, and a new Fishery Plan in order to evaluate, control and reduce ecological impact of red swamp crayfish in the Lower Guadalquivir and protected natural areas (Doñana Natural Park, Doñana National Park). Furthermore, due to the water and sludge outflow from Aznalcóllar pond polluted with heavy metals (May, 1998), just a few kilometers far from DNP and the Guadalquivir marshes, Public Administration had focused its efforts to know how this toxic discharge is affecting *P. clarkii* populations and to assess the ecological and socio-economic consequences that bio-accumulation of these pollutants in crayfish tissues could cause in this area. Heavy metals could be incorporated into the food webs if other animals predate on crayfish, and crayfishing could be banned if red swamp crayfish stocks are polluted.

ACKNOWLEDGEMENTS

We are grateful to L. Bolea (Diputación Gral. Aragón, Spain), F. Alonso (Serv. Invest. Agraria, Junta de Castilla-La Mancha, Spain), M. Jover (U. Valencia, Spain), P. Anastácio (U. Coimbra and U. Évora, Portugal), E. Llorente (Astacifactoría Rillo de Gallo, Guadalajara, Spain), J. Arrignon (Acad. des Sci. d'outre-mer, France), Ministerio de Agricultura, Pesca y Alimentación (Spain), Viceconsejería de Medio Ambiente del Gobierno de Canarias (Spain), A.S. Habsburgo-Lorena, M. Ferrer and several anonymous fishermen in Spain and in Portugal, for their invaluable information concerning *A. pallipes* and *P. clarkii* distribution, production, ecology and marketing in the IP. Appreciation is also extended to C. Temiño (Cons. Medio Ambiente, Junta de Castilla-León, Spain) and J.M. Carral (U. León, Spain), because our talks with them were very fruitful in order to update our knowledge of *A. pallipes* distribution in Spain. We also thank to the referees for their interesting comments about the manuscript. Pedro J. Gutiérrez also thanks the ICI (Spain) for a subvention grant and CONACyT (México) for a financial complement of the former grant that enabled him to complete his Ph.D. Dissertation. Finally, we thank to Arantxa López and Philip Mason for their help in the translation of the manuscript.

REFERENCES

Abrahamsson, S.A. 1973. The crayfish *Astacus astacus* in Sweden and the introduction of the American crayfish *Pacifastacus leniusculus*. *Freshwater Crayfish* 1:27-40.

Adão, H. 1992. *Procambarus clarkii* (Girard 1852) (Decapoda, Cambaridae) espécie exotica em Portugal. Consideraçoes sobre a ecobiologia e aspectos sobre sua biologia populacional na barragem de Monte Novo (Alentejo, Portugal). Provas de Aptidao Pedagogica e Capacidade Cientifica (Trabalho de sintese), Universidade de Évora.

Adão, H. & Marques, J.C. 1993. Population biology of the red swamp crayfish *Procambarus clarkii* (Girard, 1852) in Southern Portugal. *Crustaceana* 63:336-345.

Adrián, M.I. & Delibes, M. 1987. Food habits of the otter (*Lutra lutra*) in two habitats of the Doñana National Park SW Spain. *J. Zool. Lond.* 212:399-406.

Aldabe, J., J.I. Esquisabel, M.I. Fernández et al. 1991. Estudio de distribución y abundancia de las especies de cangrejos presente en los ríos navarros. Servicio de Medio Ambiente del Gobierno de Navarra. Negociado de Getión Cinegética y Piscícola.

Alderman, D.J. 1993. Crayfish plague in Britain. the first twelve years. *Freshwater Crayfish* 9:266-272.

Alderman, D.J. & Polglase, J.L. 1988. Pathogens, parasites and commensals. In D. M. Holdich & R. S. Lowery (eds) *Freshwater crayfish: biology, management and exploitation*, pp. 167-212. London: Croom Helm.

Algarín, S. 1980. Problemática y perspectivas de la introducción del cangrejo. *El Cangrejo Rojo de la Marisma*. España: Junta de Andalucía, Conserjería de Agricultura y Pesca: 25-31.

Algarín, S. 1981. Estudio sobre la viabilidad a corto, medio, y largo plazo de la puesta en práctica de las conclusiones acordadas en las Jornadas de Estudio sobre el cangrejo rojo de las marismas, celebradas en Sevilla durante el mes de Junio de 1980.

Algarín, S. & Habsburgo-Lorena, A.S. 1983. El cangrejo de las marismas en España. Jornadas de Estudio sobre el Cangrejo de Río. Diputación Provincial de Burgos.

Algarín, S. & M. Librero. 1988. Biología general de los cangrejos rojos de marismas *Procambarus clarkii* Girard, en las Marismas del Guadalquivir. Primeras Jornadas de Acuicultura de Castilla-La Mancha.

Almaça, C. 1989. On the biology of north-eastern Portuguese populations of *Austropotamobius pallipes* (Lereboullet, 1858). *Arquivos do Museo Bocage, Nova Série* 1(29): 419-428.

Almaça, C. 1990. Recursos animais e sua conservação: as populações portuguesas do lagostim-de-rio, *Astacus pallipes* Lereboullet, 1858. *Museu Nacional de História Natural*. 3-22.

Álvarez, E., Ausán, I. & Cabañas, C. 1989. Repoblación con cangrejo señal (*Pacifastacus leniusculus*) en los territorios de Álava, Vizcaya y Guipúzcoa. Informe técnico 5pp.

Anastácio, P.M. 1993. Ciclo biologico e produção do lagostim vermelho da Louisiana *Procambarus clarkii* Girard na região do Baixo Mondego. M. Sc. thesis, Universidad de Coimbra, Portugal. 109 pp

Anastácio, P. & Marques, J.C. 1995. Population biology and production of the red swamp crayfish *Procambarus clarkii* (Girard) in the Lower Mondego River Valley, Portugal. *Journal of Crustacean Biology* 15:156-168.

Beja, P.R. 1996. An analysis of otter *Lutra lutra* predation on introduced American crayfish *Procambarus clarkii* in Iberian streams. *Journal of Applied Ecology* 33:1156-1170.

Bernardo, J.M. & Ilhéu, M. 1994a. Distribuiçao em Portugal do lagostim-de-patas-brancas *Austropatomobius pallipes* (Lereboullet), caracterizaçao das suas populaçoes e medidas de conservaçao. Instituto de Conservaçao da Natureza, Universidade de Évora.

Bernardo, J.M. & Ilhéu, M. 1994b. Red swamp crayfish (*Procambarus clarkii*): contribution to material cycling. *Verh. Internat. Verein. Limnol.* 25:2447-2449.

Bernardo, J.M. & Ilhéu, M. 1996. Present status of *Austropotamobius pallipes* (Lereboullet) in Portugal. *Freshwater Crayfish* 11:671-674.

Bernardo, J.M., Ilhéu, M. & Costa, A.M. 1997. Distribution, population structure and conservation of *Austropotamobius pallipes* in Portugal. *Bull. Fr. Pêche Piscic.* 347:617-624.

Beroiz, B. (in press). Estudios bioenergéticos de dos poblaciones, Doñana y El Alentejo, de cangrejo rojo (*Procambarus clarkii*) alimentadas con vegetación natural macrofítica del Parque Nacional de Doñana y su posible impacto en los ecosistemas. Tesis de licenciatura. Depto. de Ecología, Universidad Autónoma de Madrid.

Blake, M.A. & Hart, P.J.B. 1993. Habitat preferences and survival of juvenile signal crayfish, *Pacifastacus leniusculus* – the influence of water depth, substrtum, predatory fish and gravid females crayfish. *Freshwater Crayfish* 10:318-332.

Bolea, L. 1993. *Establecimiento de las bases para la gestión de una especie autóctona amenazada: el cangrejo de río común Austropotamobius pallipes*. España: Diputación General de Aragón.

Bolea, L. 1995. *El cangrejo de río común en Aragón*. España: Diputación General de Aragón.

Boon, P.J., Morgan, D.H.W. & Palmer, M.A. 1992. Statutory protection of freshwater flora and fauna in Britain. *Freshwater Forum* 2:91-101.

Bouffard, S.H. & Hanson, M.A. 1997. Fish in waterfowl managers' perspective. *Wildlife Society Bulletin* 25:146-157.

Bravo, M.Á., Duarte, C. & Montes, C. 1994. Environmental factors controlling the life history of *Procambarus clarkii* (Decapoda, Cambaridae) in a temporary marsh of the Doñana National Park (SW. Spain). *Verh. Int. Verein. Limnol.* 25:2450-2453.

Bruxelas, S. 1986. Contribução para o estudo do lagostim vermelho da Louisiana (*Procambarus clarkii*, Girard) no sul de Portugal. Relatorio de estágio do curso de engenheiro silvicultor, I.S.A., Lisboa.

Bullock, J.M. & Hodder, K. 1997. Reintroductions: challenges and lessons for basic ecology. *Trend. Ecol. Evol.* 12:68-69.

Carral, J.M., J.D. Celada, J. González, M. Sàez-Royuela, V.R. Gaudioso, R. Fernàndez & C. Lòpez-Baisson 1993. Wild freshwater crayfish populations in Spain: current status and perspectives. *Freshwater Crayfish* 19:158-162.

Chesson, P.L. 1991. A need for niches? *Trend. Ecol. Evol.* 6:26-28.

Celada, J.D. 1987. Desarrollo de los huevos, producción y supervivencia de juveniles en el cangrejo de río (*Pacifastacus leniusculus* Dana) bajo condiciones controladas. Resumen de Tesis doctoral. Universidad de León. 29pp.

Celada, J.D. & Gaudioso, V.R. 1985. Localización y evaluación ecológica de las poblaciones y hábitats del cangrejo de río (*Austropotamobius pallipes* Lereboullet) en la provincia de León. *Boletín de la Estación Central de Ecología* 27:99-105.

Cerenius, L. & Söderhäll, K. 1984. Chetomotaxis in *Aphanomyces astaci*, an asthropod parasitic fungus. *Journal of Invertebrate Pathology* 43:278-281.

Cerenius, L. & Söderhäll, K. 1992. Crayfish diseases and crayfish as a vectors for important diseases. *Finnish Fisheries Research* 14:125-133.

Coll-Morales, J. 1988. El cultivo controlado de los cangrejos de río. *Rev. Lat. Acui.* 35:25-41.

Correia, P.A.C.M. 1990. Estudio da dinâmica populacional de *Procambarus clarkii* (Decapoda, Cambaridae) em quatro habitats naturais. Relatório de estágio em recursos faunísticos e ambiente. Fac. de Ciències de Lisboa, Portugal. 138 pp.

Correia, P.A.C.M. 1993. Length-weight relationships for two populations of red swamp crayfish, *Procambarus clarkii* (Decapoda, Cambaridae) from Portugal. *Freshwater Crayfish* 9:442-450.

Correia, P.A.C.M. 1994. Situación de l'acclimatation de l'ecrevisse rouge des marais *Procambarus clarkii* au Portugal. *L'Astaciculteur de France*, 35:2-9.

Correia, P.A.C.M. & Costa, A.C. 1994. Introduction of the red swamp crayfish *Procambarus clarkii* (Crustacea, Decapoda) in São Miguel, Azores, Portugal. *Arquipélago*, 12A:67-73.

Correia, P.A.C.M. & Ferreira, O. 1995. Burrowing behavior of the introduced red swamp crayfish *Procambarus clarkii* (Decapoda; Cambaridae) in Portugal. *Journal of Crustacean Biology* 15:248-257.

Costa, A.C., Correia, A.M. & Rodrigues, M.L. 1996. Monitoring a population of *Procambarus clarkii* (Decapoda: Cambaridae) in São Miguel (Azores, Portugal). *Freshwater crayfish* 11:213-218.

Covich, A.P. 1996. Stream biodiversity and ecosystem processes. *Bulletin NABS*, 13(2):294-301.

Cuéllar, L. & Coll, M. 1983. Epizootiology of the crayfish plague (aphanomycosis) in Spain. *Freshwater Crayfish* 5:545-549.

DOPP, 1986. Análisis sociológico de los pescadores del Cangrejo Rojo de las Marismas del Guadalquivir. Junta de Andalucía. Conserjería de Agricultura y Pesca. Dirección General de Pesca.

Diéguez, J. & Rueda, A. 1994. Nuevas esperanzas para el cangrejo de río autóctono. *Quercus* 97:8-12.

Diéguez, J., Pinedo-Ruiz, J., Cerenius, L. & Söderhäll, K. 1995. Presence of *Psorospermium haeckeli* in a Spanish population of *Pacifastacus leniusculus*. *Freshwater Crayfish* 10:286-288.

Domínguez, L. 1987. Impacto de la pesca del cangrejo rojo americano (*Procambarus clarkii*) en el Parque Nacional de Doñana, durante la campaña de la temporada de 1987. Informe Técnico del ICONA. 26 pp.

Elvira, B. 1998. Impact of introduced fish on the native freshwater fish fauna of Spain. In I.G. Cowx (ed.), *Stocking and introduction of fish*. pp. 186-190. London: Blackwell Science.

Elvira, B., Nicola, G.G. & Almodóvar, A. 1996. Pike and red swamp crayfish: a nw case on predator-prey relationship between aliens in central Spain. *Journal of Fish Biology* 48:437-446.

Escosa, R. 1990. L'ecrevisse rouge (*P. clarkii*) problemees dans le Delta de L'Ebre. *L'Astaciculteur Francais* 23:2-4

FAO 1989. Annual production of fisheries report. Food and Agriculture Organisation (FAO). Rome, Italy.

Feminella, J.W. & Resh, V.H. 1986. Effects of crayfish grazing on mosquito habitat at Coyote Hills Marsh. *Proc. Calif. Mosq. and Vector Contr. Assoc.* 54:101-104.

Feminella, J.W. & Resh, V.H. 1989. Submersed macrophytes and grazing crayfish: an experimental study of herbivory in a California freshwater marsh. *Holartic Ecology* 12:1-8.

Fuente, F. 1982. El cangrejo se extingue. *Vida Libre* (Junio):21.

Gállego, L. & Ocete, M.E. 1985. Problemática de la introducción de *Procambarus clarkii* Girard (Crustacea: Decapoda) en las marismas del Guadalquivir. *Trazos -Trabajos Zoológicos- Nº 2. Depto. de Zoología y Biología Marina, Facultad de Ciencias, Universidad de Baleares, Palma de Mallorca*:1-30.

Gaudé III, A.P. 1983. The Louisiana red swamp crawfish (*Procambarus clarkii*) in Coto Doñana National Park (Sevilla, Huelva; Spain). Ms.C. Thesis. University of Southern Louisiana.

Gaudé III, A.P. 1986. Ecology and production of Louisiana red swamp crayfish *Procambarus clarkii* in southern Spain. *Freshwater Crayfish* 6:111-130.

Gaudioso, V. 1977. El cangrejo de río en España. *Rev. Ibérica*, 183 (Nov.):370-375.

Gaudioso, V. & Vigil, E. 1976. El cangrejo de río en régimen de laboratorio: mantenimiento y reproducción. *Vida Silvestre* 20:256-265.

Gaudioso, V.R., Celada, J.D., Carral, J. & Rodriguez, P.L. 1987. El cangrejo de río en León. Universidad de León. Spain.

Geddes, M.C., Smallridge, M. & Clarke, S. 1993. The effect of stocking density food type and shelters on survival and growth of the Australian freshwater crayfish *Cherax destructor*. *Freshwater Crayfish* 9:57-69.

Gil, J.M. & Alba-Tercedor, J. 1998. El cangrejo de río autóctono en la Provincia de Granada. *Quercus* 144:14-15.

Goldman, Ch. R. 1973. Ecology and physiology of the California crayfish *Pacifastacus leniusculus* (Dana) in relation to its suitability for introduction into European waters. *Freshwater Crayfish* 1:105-120.

Gouyon, P.H. 1990. Invaders and disequilibrium. In F. Di Castri, A.J. Hansen & M. Debussche (eds), *Biological Invasions in Europe and the Mediterranean Basin*. pp. 365-369. The Netherlands: Pub. Kluwer Academic.

Grandjean, F., Souty-Grosset, C., Raimond, R. & Holdich, D.M. 1997. Geographical variation of mitochondrial DNA between populations of the white-clawed crayfish *Austropotamobius pallipes*. *Freshwater Biology* 37:493-501.

Groombridge, B. (ed.). 1994. 1993 *List of Threatened Animals*. Cambridge: IUCN.

Green, A. 1994. Medidas urgentes necesarias para reducir el impacto producido por las nasas cangrejeras sobre la Cerceta pardilla y otras aves acuáticas en el Brazo del Este. Estación iológica de Doñana, CSIC.

Guitay, H., Wilson, J.B., & Lee, G.W. 1996. Species redundancy: a redundant concept?. *Journal of Ecology* 84:121-124.

Gutiérrez-Calderón, E. 1964. Información sobre la aclimatación en España del cangrejo edelkrebs. *1 Boletín Técnico, Serie Piscícola. Min. Agric. España*. 26 pp.

Gutiérrez-Yurrita, P.J. 1997. El papel ecológico del cangrejo rojo, *Procambarus clarkii* en los ecosistemas acuáticos del Parque Nacional de Doñana. Una perspectiva ecofisiológica y Bioenergética. Ph. D. Tesis, Depto. Ecología, Fac. Ciencias, Universidad Autónoma de Madrid. 348 pp. (In Spanish but English summary).

Gutiérrez-Yurrita, P.J. & Latournerié-Cervera, J.R. 1992. Preliminary studies on exploitation of two native crayfish species of the genus *Procambarus* (Crustacea; Decapoda) by Purépecha people in Michoacán, México. *Proc. of the III Intenational Congess of Ethnobiology*: 61.

Gutiérrez-Yurrita, P.J. & Montes, C. 1997. El cangrejo americano es una especie clave en la ecología de la marisma dulce del Parque Nacional de Doñana. *Zones Humides i Meditérranies (RED-HUM)*,1 (Noviembre):11-12.

Gutiérrez-Yurrita, P.J. & Montes, C. 1998a. Studies of the environmental factors controlling the crayfish *Procambarus clarkii* activity in the Doñana National Park temporary freshwater marsh, SW-Spain. *Comp. Biochem. Physiol.* (in press).

Gutiérrez-Yurrita, P. J. & Montes, C. 1998. Lé ecrevisse rouge (*Procambarus clarkii*) dans le Parc National de Donana. *Bulletin L'Astaciculteur de France*.55 (Juin):2-6.

Gutiérrez-Yurrita & J. M. Martínez (1998). Informe preliminar sobre la población del cangrejo rojo americano localizada en el barranco del Cercado de San Andrés (Tenerife): impacto y control. Informe para la Viveconsejería de Medio Ambiente, Consejería de Política Teritoril y Medio Ambiente del Gobierno de Canarias. Depto. de Ecología, Universidad Autónoma de Madrid. 47 pp.

Gutiérrez-Yurrita, P.J., Bravo-Utrera, M.Á., Jordá, J.R., Baltanás, Á. & Montes, C. 1994. Análisis preliminar de la tasa metabólica estándar en el cangrejo rojo, *Procambarus clarkii* (Decapoda: Cambaridae), en el Bajo Guadalquivir (S. España). *Limnetica* 10:123-128.

Gutiérrez-Yurrita, P.J., Ilhéu, M., Montes, C. & Bernardo, J.M. 1996. Morphometrics of red swamp crayfish from a temporary marsh (Doñana National Park, SW. Spain) and a temporary stream (Pardiela stream, S. Portugal). *Freshwater Crayfish* 11:384-393.

Gutiérrez-Yurrita, P.J., Green, A., López- Mª.A. Romero, Bravo-Utrera, M.Á. & Montes, C. 1997a. Diseño de un nuevo modelo de trampa para reducir el impacto de la pesca del cangrejo rojo (*Procambarus clarkii*) sobre la población de vertebrados en las marismas del Guadalquivir. *Doñana Acta Vertebrata* 24(2):51-66.

Gutiérrez-Yurrita, P.J., Green, A., Bravo-Utrera, M.Á. & Montes, C. 1997b. Impacto de la pesca del cangrejo rojo (*Procambarus clarkii*) en los vertebrados del Parque Nacional de Doñana y su entorno. In H. Bigaré, H. da Cruz, J. A. Pascual y T. Purcell (eds.), *The conservation of wetlands in a north- south perspective: The east Atlantic Migration Flyway*. pp. 215-227. España: Friends of the Earth Int. Wetlands Campaign y Junta de Andalucía.

Gutiérrez-Yurrita, P.J., Sancho, G., Bravo-Utrera, M.Á., Baltanás, Á. & Montes, C. 1998. Natural diet of the red swamp crayfish *Procambarus clarkii* in the Doñana National Park freshwater marsh (SW, Spain). *Journal of Crustacean Biology* 8(1):120-127.

Gydemo, R. 1992. Crayfish diseases and management-the need for knowledge. *Finnish Fisheries Research* 14:119-124.

Habsburgo-Lorena, A.S. 1979. Present situation of exotic species of crayfish introduced into Spanish continental waters. *Freshwater Crayfish* 4:175-184.

Habsburgo-Lorena, A.S. 1983a. Some observation on crawfish farming in Spain. *Freshwater Crayfish* 5:549-551.

Habsburgo-Lorena, A.S. 1983b. Socioeconomic aspects of the crawfish industry in Spain. *Freshwater Crayfish* 5:552-554.

Habsburgo-Lorena, A.S. 1986. The status of the *Procambarus clarkii* population in Spain. *Freshwater Crayfish* 6:131-136.

Harris, P.M. 1995. Are autoecologically similar species also functionally similar?. A test of pond communities. *Ecolgy* 76:544-552.

Henttonen, P., Huner J.V., Rata, P. & Lindqvist, O.V. 1997. A comparison of the known life forms of *Psorospermium* spp. in freshwater crayfish Arthropoda, Decapoda) with emphasis on *Astacus astacus* L. (Astacidae) and *Procambarus clarkii* (Girard) (Cambaridae). *Aquaculture*, 149:15-30.

Hernando, J.A. 1978. Estructura de la comunidad de peces de la marisma del Guadalquivir. Tesis Doctoral. Universidad de Sevilla. España.

Hobbs, H.H. Jr, Jass, J.P. & Huner, J.V. 1989. A rewiew of global crayfish introductions with particular emphasis on two North american species (Decapoda: Cambaridae). *Crustaceana* 56:299-316.

Hodder, K. & Bullock, J.M. 1997. Translocations of native species in the UK: implications for Biodiversity. *Journal of Applied Ecology* 34:547-565.

Holdich, D.M. 1988. The dangers of introducing alien animals with particular reference to crayfish. *Freshwater Crayfish* 7:xv-xxx.

Holdich, D.M. & Reeve, I. 1991. Distribution of freshwater crayfish in the British Isles, with particular reference to crayfish plague, alien introductions and water quality. *Aquatic Conservation*, 1:139-158.

Holdich, D.M. & Rogers, W.D. 1995. A report on the status of freshwater crayfish on Schedule 9 of the wildlife and countryside Act 1981, Contract Report (N° F71-12-293) to the Joint Nature Conservation Committee and English Nature, Peterborough.

Holdich, D.M., Rogers, W.D. & Reader, J.P. 1995. Crayfish conservation, R&D project records 378/10 N&Y for tne National Rivers Authority, UK.

Huner, J.V. 1985. Crayfish Culture in Europe. *Aquaculture Magazine* (March/April):48-53

Huner, J.V. 1988. *Procambarus* in North America and elsewhere. In D.M. Holdich & R.S. Lowery (eds), *Freshwater crayfish: biology, management and exploitation*. pp. 239-261. London: Croom Helm.

Huner, J.V. 1989. Overview of International and domestic freshwater crayfish production. *Journal of Shelfish Research* 8:259-265.

Huner, J.V. 1992. Experiences with the supplemental stocking of red swamp, *Procambarus clarkii*, and white river, *Procambarus zonangulus*, crayfishes for growout purposes. *Finnsh Fisheries Research* 14:53-59.

Huner, J.V. 1995. Ecological observations of red swamp crayfish, *Procambarus clarkii* (Girard, 1852), and white river crayfish, *Procambarus zonangulus* (Hobbs y Hobbs, 1990), as regards their cultivation in earthen ponds. *Freshwater Crayfish* 10:456-468.

Huner, J.V. & Barr, J.E. 1991. *Red swamp crawfish: biology and exploitation*. Sea Grant Publ. LSU Center for wetland resources, Louisiana.

Huner, J.V., Holdich, D.M. & Westman, K. 1992. Crayfish management strategies and management methods in Europe: a summary. *Finnish Fisheries Research* 14:157-159.

Hutchinson, G.E. 1959. Homage to Santa Rosalia, or why are there so many kinds of animals? *American Naturalist*, 93:145-159.

Ilhéu, M. & Bernardo, J.M. 1993a. Experimental evaluation of food preference of red swamp crawfish, *Procambarus clarkii*: vegetal versus animal. *Freshwater Crayfish* 9:359-364.

Ilhéu, M. & Bernardo, J.M. 1993b. Aspects of trophic of red swamp crayfish (*Procambarus clarkii* Girard) in Alentejo, South of Portugal. *Actas VI Congreso Español de Limnología, Granada*: 417-423.

Ilhéu, M. & Bernardo, J.M. 1995. Trophic ecology of red swamp crayfish *Procambarus clarkii* (Girard)- Preferences and digestibility of plant foods. *Freshwater Crayfish* 10:132-139.

Jones, C.M. 1995. Production of juvenile redclaw crayfish, *Cherax quadricarinatus* (von Martens) (Decapoda, Parastacidae). III. Managed pond production trials. *Aquaculture*, 138:247-255.

Jones, P.D. & Momot, W.T. 1983. The bioenergetics of *Orconectes virilis* in a Photole Lake. *Freshwater Crayfish* 5:192-209.

Jover, M., Rodrigo, E. & Fernández, J. 1990. Resultados preliminares del crecimiento del cangrejo (*Procambarus clarkii*) alimentado con pienso de diferente contenido en proteínas y carbohidratos. *Actas del III Congreso Nacional de Acuicultura*: 349-354.

Konikoff, M. 1977. Study of the life history and ecology of the red swamp crayfish *Procambarus clarkii* in the lower Atchafalaya Basin Floodway. Final report for US Fish and Wildlife Service. Dept. Biology, U. of Southwestern Louisiana, Lafayette, USA.

Laurent, P.J. 1988. *Austrapotamobius pallipes* and *A. torrentium*, with observations on their interaction with other species in Europe. In D.M. Holdich & R.S. Lowery (eds), *Freshwater crayfish: biology, management and exploitation*: pp. 341-364. London: Croom Helm.

Laurent, P.J., Nicolas, J. & Paris, L. 1993. Five years of actions in Lorraine and Morvan (France) to restore the noble crayfish, *Astacus astacus*. *Freshwater Crayfish* 9:380-389.

Lawton, J.H. 1989. Food webs. In J.M. Cherret (ed.), *Ecological concepts*. Oxford: Blackwell Sci. Pub.

Lawton, J.H. 1990. Species richness and population dynamics of animal assemblages: patterns in body size, abundance, space. *Phil. Trans. Roy. Soc.* (London). ser. B. 330:283-291.

Lawton, J.H. & Jones, C.G. 1994. Linking species and ecosystems: Organisms as ecosystem engineers. In C.G. Jones & J.H. Lawton (eds), *Linking species and ecosystems*. pp. 141-150. New York: Chapman & Hall.

Librero, M. 1980. Biología y pesca del cangrejo. Jornadas de estudio del cangrejo. Junta de Andalucía:27-35.

Lirås, V., Lindberg, M., Nyström, P., Annadotter, H., Lawton, L.A. & Graf, B. 1998. Can ingested cyanobacteria be harmful to the signal crayfish (*Pacifastacus leniusculus*)? *Freshwater Biology* 39:233-242.

Lodge, D.M. 1993. Biological invasions. Lessons for ecology. *TREE* 8(4):133-137.

Lodge, D.M. & Lorman, J.G. 1987. Reductions in submersed macrophyte biomass and species richness by the crayfish *Orconectes rusticus*. *Can. J. Fish. Aquat. Sci.*, 44:591-597.

López, A. & Fernández, J.M. 1980. Contaminación por insecticidas organoclorados en cangrejo rojo de río de las marismas del Guadalquivir. *Hygia Pectoris* 17(II):39-52.

López-Fernández, A. & Infante-Miranda, F. 1982. Residuos de insecticidas organoclorados en algunos ecosistemas acuáticos del SE de la Provincia de Córdoba. III. Niveles en el cangrejo de río *Austropotamobius pallipes* Lereb). *Archivos de Zootecnia* 31(119):73-90.

Lowery, R.S. & Holdich, D.M. 1988. *Pacifastacus leniusculus* in North America and Europe, with details of the distribution of introduced and native crayfish species in Europe. In D.M. Holdich & R.S. Lowery (eds), *Freshwater crayfish: biology, management and exploitation*: pp. 283-308. London: Croom Helm.

Lutz, C.G. & Wolters, W.R. 1989. Estimation of heritabilities for growth, body size, and processing traits in red swamp crayfish, *Procambarus clarkii* (Girard). *Aquaculture* 78: 35-54.

Mackeviciene, G. 1993. Respiration during postembryonic development of the freshwater crayfish *Pacifastacus leniusculus* (Dana). *Freshwater Crayfish* 9:183-188.

Machado, A. 1931. Notas de Zoología. A existencia en Portugal de aqua doce *Astacus fluviatilis*. *Ann. Soc. Sci. Porto.*, 16:187-192.

Maitland, P.S. 1996. The North American signal crayfish, *Pacifastacus leniusculus* (Dana), established in the wild in Scotland. *Aquatic Conservation: Mar. Freshwater ecosystems* 6:107-110.

Marques, J.C., Adão, H. & Anastácio, P. 1992. Lagostim vermehlo da Louisiana *Procambarus clarkii* (Girard) espécie exotica em Portugal: bioecologie, impacto e controlo. *Actas da 3 Conferencia Nacional sobre Qualidade do Ambiente. Aveiro*: 303-312.

Martin, C. 1982. Medidas higiénico-sanitarias. *Vida Libre* (Junio): 20.

Mateus, A. 1934. O *Astacus* de S. Martinho de Angueira Tras-os-Montes. *Trab. da Assoc. da Filosofia Natural* 1(1):33.

Mateus, A. 1937. L'Astacus de Tras-os-Montes (Portugal). *XII Congrès International de Zoologie, Lisbonne* 3:1564-1566.

Matthews, M.A., Reynolds, J.D. & Keatinge, M.J. 1993. Macrophyte reduction and benthic community alteration by the crayfish *Austropotamobius pallipes* (Lereboullet). *Freshwater Crayfish* 9:289-299.

McClain, W.R. 1995. Management strategies for the production of large crawfish. *Proceedings of the International Congress of Aquaculture, California, USA*:171-172.

McClain, W.R. & Romaire, R.P. 1995. Management considerations for the production of large procambarid crawfish. *Journal of Shelfish Research* 4:553-559.

Molina-Vázquez, F. 1984. La pesca del cangrejo rojo americano y su influencia en el entorno del Parque de Doñana. *Revista de estudios Andaluces* 3:151-160.

Momot, W.T. 1984. Crayfish production: A reflection of community energetics. *Journal of Crustacean Biology* 4:35-54.

Momot, W.T. 1991. Potential for exploitation of freshwater crayfish in coolwater systems. Management, guidelines and issues. *Fisheries* 16:14-21.

Momot, W.T. 1993. The role of exploitation in altering the processes regulating crayfish populations. *Freshwater Crayfish* 9:101-117.

Momot, W.T. 1995. Redefining the role of crayfish in aquatic ecosystems. *Reviews in Fisheries Sciences* 3:33-63.

Momot, W.T., Gowing, H. & Jones, P.D. 1978. The dynamics of crayfish and their role in ecosystems. *Aerm. Midl. Nat.* 99:10-35.

Montes, C., Bravo-Utrera, M.Á., Baltanás, Á., Duarte, C. & Gutiérrez-Yurrita, P.J. 1993. Bases ecológicas para la gestión del cangrejo rojo de las marismas en el Parque Nacional de Doñana, España. ICONA, España. 270 pp.

Moshiri, G.A., Goldman, G.L., Mull, D.R. et al. 1971. Respiratory metabolism in *Pacifastacus leniusculus* (Dana) (Crustacea:Decapoda) as related to its ecology. *Hydrobiol.* 37:283-295.

Muez, M.A. & Muez, V. 1988. El cangrejo de río. *Navarra Agraria* (31-32):35-46.

Muñoz, M. 1978. Notas sobre la cría del cangrejo de río. *Vida Silvestre* 27:160-163.

Ministerio de Agricultura. 1982. Boletín de pesca del Ministerio de Agricultura, Pesca y Caza de España.

Nyström, P. & Strand, J.A. 1996. Grazing by a native and an exotic crayfish on aquatic macrophytes. *Freshwater Biology* 36:673-682.

Ocete, M.E. & López-Sánchez, S. 1983. Problemática de la introducción de *Procambarus clarkii* (Crustacea: Decapoda) en las Marismas del Guadalquivir. *Actas del I Congreso Ibérico de Entomología*: 515-523.

Olsen, T.M., Lodge, D.M., Capelli, G.M. & Houlihan, R.J. 1991. Mechanisms of impact of an introducted crayfish (*Orconectes rusticus*) on littoral congeners, snails, and macrophytes. *Can. Fish. Aquat. Sci.*, 48:1853-1861.

Pardo, L. 1951. Astacicultura continental. Ministerio de Agricultura. Spain.

Peixoto, M.C., Bruxelas, S. & Maia, A. 1996. Contribução para a recuperação das populaçoes de lagostim de pés brancos (*Austropotamobius pallipes* Lereboullet) na bacia hidrografica do Sabor. *Iternal Repport, Instituto florestal*, 16 pp.

Penn, G.H. 1954. Introductions of American crawfish into foreign lands. *Ecology* 35:296.

Persson, M. & Söderhäll, K. 1983. *Pacifastacus leniusculus* Dana and its resistance to the parasitic fungus *Aphanomyces astaci* Schikora. *Freshwater Crayfish* 5:292-298.

Ramos, M.A. & Pereira, T.M.G. 1981. Um novo Astacidae para a fauna portuguesa: *Procambarus clarkii* (Girard,1852). *Boletim do Instituto Nacional de Investigação das Pescas (Lisboa)* 6:37-47.

Riera, G. 1982. Réquiem por el barrendero de nuestras aguas. *Vida Libre* (Junio):18-20.

Romaire, R.P. 1995. Harvesting methods and strtategies used in commercial procambarid crawfish aquaculture. *Journal of Shellfish Research* 14:545-551.

Rosas, G., Ramos, M.A. & Garcia-Valdecasas, A. 1992. Invertebrados españoles protegidos por Convenios Internacionales. ICONA. Colección Técnica.

Sall-Reina, E. & Vergara-Martin, J.M. 1990. Cría del cangrejo de río (*Procambarus clarkii*) en un sistema de circuito cerrado. *Actas III Congreso Nac. Acuicult.*:35-360.

Sánchez, M.A. 1982. Aportación al estudio de *Procambarus clarkii* (Girard) y su aprovechamiento como conserva. Tesis de Licenciatura. Fac. de Veterinaria, Universidad de Santiago de Compostela. España,110 pp.

Scheffer, M. 1990. Multiplicity of stable states in freshwater systems. *Hydrobiologia* 200/201:475-486.

Scheffer, M., Hosper, S.H., Meijer, M.L., Moss, B. & Jeppesen, E. 1993. Alternative equilibria in shallow lakes. *Trend Ecol. Evol.* 8:275-279.

Schindler, D.W. 1990. Experimental perturbations of whole lakes and tests of hypotheses concerning ecosystems structure and function. *Oikos*, 57:25-41.

Serres, J.M., González, C. & Parrondo, J.L. 1985. Problemática del cangrejo de río en España. ICONA, España.

Skurdal, J. & Taugbol, T. 1991. Crayfish management in Norway. *Finnish Fisheries Research*, 14:33-39.

Smith, V.J. & Söderhäll, K. 1986. Crayfish pathology: an overview. *Freshwater Crayfish* 6:199-221.

Soto, J. 1992. Estudio económico y ambiental del cangrejo rojo (*Procambarus clarkii*) en las marismas del Guadalquivir. Informe inédito. Universidad de Córdoba, España.

Svärdson, G., Fürst, M. & Fjälling, A. 1991. Population resilence of *Pacifastacus leniusulus* in Sweden. *Finnish Fishery Research* 12:165-177.

Taylor, E.W. & Wheatly, M.G. 1981. The effect of long-term aerial exposure on heart rate, ventilation, respiratory gas exchange and acid base status in the crayfish *Austropotamobius pallipes*. *Journal of Experimental Biology* 92:109-124.

Temiño, C., J.D. Celada, J.M. Carral & Fernández, R. 1986. Estudio de las poblaciones astacícolas en los ríos de la provincia de Burgos. Perspectivas. *Jornadas de Estudio de Cangrejo de Río, Gobierno Vasco, Vitoria*: 87-109.

Torre-Cervigon, M. & Rodríguez-Marques, P. 1964. El cangrejo de río en España. Dirección General de Montes Caza y Pesca fluvial, Ministerio de Agricultura. Madrid.

Torreblanco, A., Díaz-Mayans, J., Del Ramo, J. & Núnez, A. 1987. Oxygen uptake and gill morphological alterations in *Procambarus clarkii* (Girard) after sublethal exposure to lead. *Comp. Biochem. Physiol.* 86(c):219-224.

Unestam, T. 1973. Significance of diseases on freshwater crayfish. *Freshwater Crayfish* 1:135-150.

Unestam, T. 1975. The dangers of introducing new crayfish species. *Freshwater Crayfish* 2:557-561.

Vandermeer, J.H. 1972. Niche theory. *Annu. Rev. Ecol. Syst.* 3:107-132.

Veiga, P. 1995. Arroz de lagostim. *Correio Agricola* (Junho): 42-43.

Vigneux, D. 1980. Enquête sur les écrevises en France. Conseil Superieur de la Pêche, Paris.

Vitousek, P.M. 1990. Biological invasions and ecosytem processes: towards an integration of population biology and ecosystem studies. *Oikos* 57:7-13.

Wells, S.M., Pyle, R.M. & Collins, N.M. 1983. *The IUCN invertebrate red data book*. Gland, Switzerland: IUCN.

Westman, K. & Westman, P. 1992. Present status of crayfish management in Europe. *Finnish Fisheries Research* 14:1-22.

Whatley, M. H. & Eversole, A.G. 1993. Extension of the crayfish season by delaying harvests and supplemental feeding. *Freshwater Crayfish* 9:18-27.

Whiteley, N.M., Naylor, J.K. & Taylor, E.W. 1995. Extracellular acid-base status in the freshwater crayfish *Austropotamobius pallipes* between 1 and 12°C. *The Journal of the Experimental Biology* 198:567-576.

Williamson, M. 1996. *Biological invasions*. London: Chapman and Hall.

Yeh, H.S. & Rouse, D.B. 1995. Effects of water temperature, density, and sex ratio on the spawning rate of red claw crayfish *Cherax quadricarinatus* (von Martens). *J. of the World Aquaculture Soc.* 26(2):160-164.

Crayfish introductions in the Nordic and Baltic countries

JOSTEIN SKURDAL, TROND TAUGBØL,
Eastern Norway Research Institute, Lillehammer, Norway
ALOYZAS BURBA,
Institute of Ecology, Vilnius, Lithuania
LENNART EDSMAN, BJÖRN SÖDERBÄCK
Institute of Freshwater Research, Fisheries Board of Sweden, Drottningholm, Sweden
BJARNE STYRISHAVE,
Department of Life Science & Chemistry, University of Roskilde, Roskilde, Denmark
JAANUS TUUSTI,
Sagadi Training Centre, Sagadi, Estonia
KAI WESTMAN
Finnish Game and Fisheries Research Institute, Helsinki, Finland

ABSTRACT

The noble crayfish, *Astacus astacus,* is the only native species in the Baltic and Nordic countries. Crayfish plague struck the native stocks in the various countries under consideration during the period 1890-1971 and reduced crayfish abundance and distribution. Later, man made pollution and habitat changes have also reduced both crayfish production and the potential for crayfish production. Many of the remaining stocks of noble crayfish in Norway, Sweden, and Finland result from stocking into new areas during this century. To substitute the loss of the noble crayfish, the crayfish plague resistant, North American, signal crayfish, *Pacifastacus leniusculus,* was introduced into Sweden in 1960. Later the signal crayfish was introduced into Finland, Latvia, Lithuania, and Denmark. The narrow-clawed crayfish, *Astacus leptodactylus,* has expanded, due to a combination of stocking and further natural dispersion, from its former area of distribution in southeastern Europe, and is now found in Lithuania, Latvia, Finland, and Denmark. The North American, introduced, spiny-cheek crayfish, *Orconectes limosus,* has expanded widely from the first introduction in the Oder River in Germany and it has recently invaded Lithuanian rivers and lakes. At present, Norway and Estonia are the only countries without introduced crayfish. The interest for conservation of the noble crayfish has increased in all Nordic and Baltic countries and measures are planned and partly implemented to improve protection of remaining stocks and restore former crayfish localities. The expansion of the introduced North American crayfish species do, however, represent a threat since they both out-compete the noble crayfish and are carriers of the crayfish plague fungus. Also the expansion of the more competitive European narrow-clawed crayfish represent a threat to the native crayfish. Present catch of noble crayfish from the wild, in the Baltic and Nordic countries, is less than 200 tons, compared to historic records of more than 2000 tons.

1 INTRODUCTION

Freshwater crayfish are a major component of the freshwater systems in large parts of the Nordic and Baltic countries. The Nordic countries represent the northern limit for crayfish distribution in Europe, and at present crayfish has extended their distribution north to 67°30'N (Westman 1991). Currently there are two European (*Astacus astacus* and *A. leptodactylus*) and two North American (*Pacifastacus leniusculus* and *Orconectes limosus*) crayfish species in the region. Freshwater crayfish are subject to great concern due to a large decline in the native populations. The native crayfish species of Europe are considered vulnerable and special care is recommended in the management (exploitation and conservation) of these species, as recommended in Bern convention and the EC's Habitats Directive. In addition, crayfish has a high economical, as well as ecological, social, and cultural value, and there is a considerable interest to increase harvest from wild and cultured stocks (see Ackefors this volume).

In this article, we present the distribution, historic development, and current situation for the native noble crayfish in the Nordic and Baltic countries. Then we show the crayfish introductions into the region. Three crayfish species have been introduced and are currently increasing their area of distribution. We present a broad overview of the biology of the four crayfish species in the Nordic and Baltic countries. The extent of crayfish aquaculture in the region is briefly described. We also include a broad overview of the legislation and the efforts to conserve and restore the native crayfish. The future for the native crayfish is at present very uncertain, since the general decline in distribution and abundance seems to continue. However, the change in policy is promising and we hope that an increased and harmonized effort to protect and restore noble crayfish in all countries will yield positive results.

2 FRESHWATER CRAYFISH SPECIES IN THE NORDIC AND BALTIC COUNTRIES

At present, four species of freshwater crayfish are found in the Nordic and Baltic countries. The noble crayfish is the only species native to the area. The European narrow-clawed crayfish has come to the region due to a large number of human transplants over the last century and further natural spread from the sites of introduction (see Holdich this volume). Two North American crayfish species have also been introduced successfully into the region, the spiny-cheek crayfish was introduced into Europe in 1890 and the signal crayfish in 1960. In addition, the North American species, *Orconectes virilis,* was introduced into Sweden in 1960, however, without success. Most management regulations deal with crayfish as a group and do not include specific rules designed for each of the individual species. It is, however, important that the management account for the differences between the species. To strengthen our arguments for a species specific management, we have included a short description of the life history and biology of the four species in the region.

2.1 *General biology of the noble crayfish*

Noble crayfish are widely distributed in Europe and its distribution extends from France and the northern part of Spain in south-east to Russia in the east, and from Italy and Greece in the south to Scandinavia in the north (Cukerzis 1988). Noble crayfish has also been introduced to England, Portugal, and Morocco. The noble crayfish prefers streams, rivers, and lakes with an excess of shelters. This species is sensitive to pollution and physical changes of the environment. Due to high economic value, noble crayfish is heavily exploited throughout its area of distribution. Aquaculture experiments and enterprises are developing in Denmark, Norway, Sweden, Finland, Germany, and France (Westman et al. 1990, 1992, Ackefors this volume). In Lithuania and Estonia limited experiments with aquaculture have been undertaken partly as research projects.

In its northern distribution area, noble crayfish typically reach sexual maturity at a size of 6-8 cm total length, corresponding to an age of 3-7 years. Testes and ovaries mature between July and September. After the first maturation, males normally remature every year while females are reproductively inactive every second or third year. Mating takes place in September and October. Extrusion and fertilization of eggs occur 1-6 weeks after mating. The female carries the eggs attached to the pleopods (normally some 100-200 in number) until they are hatched the following summer in June or July. There is a positive correlation between the body size of the female and the egg number, even though the individual variations are large. Hatching time is largely determined by temperature; a cold spring/early summer leads to late hatching. If hatching occurs too late in the summer, the growing season will be too short for juveniles to survive the winter, thus temperature is a main factor limiting distribution. Newly hatched juveniles remain with their mother until the second molt, 18-20 days after hatching. The third stage juveniles adopt a solitary life style.

In Scandinavian waters, adult noble crayfish (> 6-7 cm total length) normally molt once a year and the growth is slow. Temperature and food greatly influence molting frequency and, where conditions are favorable, two molts each year have been recorded as normal in adult males and reproductively inactive females. Berried females normally molt once per year, after releasing their young. Crayfish continue to grow during their entire life-span and noble crayfish may reach sizes of 17-19 cm total length, corresponding to a weight of 250-300 g. There are no definite records of maximum age, but there are examples from Norway indicating that noble crayfish may live at least 20 years (based on recapture of crayfish in localities with no recruitment).

2.2 *General biology of the signal crayfish*

The signal crayfish is native to northwestern North America. Due to transplantation some 100 years ago, it has established large populations in south-western states such as California. In the 1960s, signal crayfish were also stocked into Sweden (see Ackefors and Holdich this volume). Since then, signal crayfish have been stocked in most European countries (Lowery & Holdich 1988).

Signal crayfish was introduced into Sweden to substitute losses of noble crayfish. The selection of signal crayfish was based on this species' close resemblance to the noble crayfish regarding ecology, appearance and taste; compared to the noble crayfish, the signal crayfish is however more aggressive, grows faster and is also more fecund.

At first, experimental stocking and studies were performed in closed systems under the control of the authorities. From 1969 the Fisheries Board (Fiskeristyrelsen) did however permit stocking with signal crayfish in plague stricken areas, and now signal crayfish is established in more than 1200 lakes, 400 streams and rivers, and 1300 ponds and aquaculture facilities in Sweden (Fiskeriverket 1993). The controversy over the introductions of alien species and the protection of native species is, however, increasing and at present there is an increased interest in protecting the native noble crayfish.

The introduction of North-American species is especially complicated since these species are resistant carriers of the crayfish plague fungus. Successful establishment of any of these species thus also results in risk of the chronic establishment of the crayfish plague fungus.

Signal crayfish are at present spread to most European countries. Norway and Estonia are exceptions (Holdich & Lowery 1988, Fiskeriverket 1993). Like the noble crayfish the signal crayfish is a long-lived species. Despite the difficulty of comparison, it is obvious that the signal crayfish grows faster than many other temperate zone crayfish species, with the exception of some from Australia, and it may achieve a larger final size (Lowery 1988). The fecundity of signal crayfish is high, but the egg size is small. There is a positive correlation between the body size of the female and the egg number which ranges from 200-400. Molting frequency decreases with size and age and adult animals only molt once or twice a year. The breeding cycle of signal crayfish is typical of a temperate zone species. Mating occurs in autumn and the eggs are carried over winter to hatch as soon as the temperature is high enough in summer. The recently hatched juveniles remain with their mother until around the second molt, when they start an independent life. Size at maturity is 6-9 cm total length at the age of 2-3 years.

2.3 *General biology of the narrow-clawed crayfish*

The narrow-clawed crayfish is easily recognized by its long chelae. It has a southeastern distribution and was originally distributed in an area corresponding more or less to Turkey, Ukrania, Belarus, and southwestern Russia. However, it has expanded both in the western and northern direction, partly due to transplants and partly by additional natural spreading, and is now found in Poland, Germany, Lithuania, Latvia, and Finland. The species has also been introduced into Denmark, the Netherlands and England, and in addition France and Italy have stocks of this species (see other papers of this volume). The narrow-clawed crayfish have a wide habitat preference and is found both in shallow and deep lakes, in rivers and streams, and on a wide range of bottom substrates. Shelter is probably not as important for this species as it is for noble crayfish, since they are more active during the night and day. The narrow-clawed crayfish can also live in brackish water (Holdich et al. 1997). Compared to the other native European species, i.e. noble crayfish, white-clawed crayfish and stone crayfish, the narrow clawed crayfish is less sensitive to pollution and the temperature tolerance is also wider.

The narrow-clawed crayfish is exploited in many countries like Russia, Belarus, and Turkey (Köksal 1988). In Turkey, France, Italy, Poland, Belarus, and Russia, experimental aquaculture is taking place (Köksal 1988, Westman et al. 1990, 1992). The narrow-clawed crayfish probably is a more temperate species compared to noble crayfish, but the distribution is currently extending to the north. The length of its reproductive cycle varies according to the climate of the habitat in which it lives. Mating occurs during October and November when the water temperature is 7-12°C and egg laying takes place 4 to 6 weeks later. Incubation of eggs continues through winter and spring. Hatching takes place in late May and June. After hatching, the young remain with their mother for about 20-25 days. During this period the juveniles go through one molt after which they leave their mothers and start an independent life in shallow waters. Males and females reach sexual maturity for the first time in their third year and at a size of 7-8 cm total length. The number of eggs produced per female is generally 200-400. There is a positive correlation between the body size of the female and the egg number, even though the individual variations are large.

As in the other European crayfish species, the molting frequency of narrow-clawed crayfish is size dependent. While the sub-yearlings molt 8 to 9 times during their first year, adult males molt twice a year and females with eggs only once a year. The first molting period of males occurs in mid-spring at which time the females still carry their offspring. The second molting period of males occurs in early autumn, at which time the females also molt for the first time. For this reason males become larger than females and may reach a size of at least 17 cm total length.

2.4 *General biology of the spiny-cheek crayfish*

The spiny-cheek crayfish was introduced into Europe in the last century. The first introduction was in 1890 when it was released into a tributary of the Oder River in Germany (see Dehus et al. this volume). The aim of the introduction was to substitute for losses of noble crayfish due to crayfish plague. It is smaller than the noble crayfish and is exploited to a minor extent. The size rarely exceed 10-11 cm. It has higher tolerance to eutrophic and polluted water than the European species, and this crayfish has spread to a considerable area (Holdich this volume). The present distribution covers central-Europe from Spain and France in the west to Poland and Russia in the east, including Lithuania (Lowery & Holdich 1988, Westman et al. 1990).

The typical orconectoid crayfish life span is about three to four years. Females, which move into shallow waters in early spring, produce only one brood per year. Average fecundity is very high; 200 up to 579 eggs have been recorded in Lithuania. The young are usually attached for one to three weeks and, depending on temperature, hatch in May or June. The young remain attached to the female through two molts and leave after the second molt. The young-of-the-year then frequent the littoral zones of lakes or ponds or the riffle areas of rivers and streams. Most *Orconectes* spp. reach maturity early at the end of the second growing season.

Most of the growth is achieved during the juvenile stage; once mature, adults grow very slowly. Because of the additional summer molt, adult males are usually larger than females of the same age.

Mating usually occur in late summer or early spring. The male transfer sperm to a receptacle in the abdomen of the female using a specially modified pair of abdominal legs, called copulatory stylets. The shape of these stylets differs depending on whether the male crayfish is in the sexually competent Form I or non-breeding quasi juvenile form II. Sexual maturity of females is detected by the presence of the sperm plugs, but greater accuracy is achieved through actual inspection of the gonads. *Orconectes* spp. are able to withstand very high exploitation rates of the adult stock without serious depletion (Momot 1988 and references therein).

3 HISTORY AND CURRENT SITUATION OF NATIVE CRAYFISH

There are five native European crayfish species (Henttonen & Huner this volume), of which only the noble crayfish is native in the Nordic and Baltic countries. The noble crayfish is still widespread in the region; however, the distribution is fragmented and discontinuous as it has disappeared from many of the former important and productive localities.

Crayfish plague first hit the region in Lithuania in 1890 and reached Norway in 1971, and has been a major cause for the decline. Crayfish plague affected most of the area before 1920, but is still active and each year hundreds of localities are affected.

Anthropogenic pollution, habitat deterioration, and introductions of crayfish and predators have contributed significantly to the decline and reduced the potential for native crayfish distribution and production.

Presently, the noble crayfish is mainly confined to less productive areas in small lakes and brooks, at higher elevations and further north. Many of these locations are the result of the stocking practice well known in the area and especially during this century in Finland, Sweden, and Norway. The world's northernmost population of self-reproducing crayfish is the noble crayfish in Lake Leppäjärvi, Finland at 67°30' N. Current harvest of noble crayfish in the region is less than 5% of the pre-plague harvest (Table 1), in spite of the extensive stocking which has increased the area of distribution. The situation and the future perspectives for the native noble crayfish is at present very uncertain, since the general decline in distribution and abundance seems to continue. A presentation of the situation in each country is given below.

Table 1. Pre- and post-plague crayfish harvest in the Nordic and Baltic countries.

Country	Plague	Catch in tonnes	
		Pre-plague	Post-plague
Norway	1971, 1987	> 40	10 (1990)
Denmark	1907	?	?
Sweden	1907	600-1000	30-50 (1995)
Finland	1893	> 650	110 (1990)
Estonia	1896	> 31	1 (1997)
Latvia	1890s	?	?
Lithuania	1886	250	4 (1970)
Total		> 2110	~174

3.1 *Norway*

Norway has only one species of crayfish, the native noble crayfish, which has been part of the Norwegian fauna for at least 250 years. The Norwegian harvest was negligible before the demand increased after the plague struck Sweden in 1907. The harvest has been stable throughout the first half of the century and was estimated at 40 tons in 1966, but has declined to some 10 tons at present due to pollution, siltation, physical habitat deterioration, and crayfish plague.

Most, maybe all, crayfish localities in Norway are a result of stocking, and the stocking activities have continued until recently. The distribution of freshwater crayfish has been studied in 1918, 1969, 1978 and 1995 (Taugbøl & Skurdal 1996). Crayfish are distributed in the south-eastern part of Norway with some scattered populations along the coast as far north as Trondheim. Information has been collected on 577 crayfish localities, of which 365 localities in 74 municipalities still harbour crayfish (Taugbøl & Skurdal 1996). With the exception of a few localities in the counties of Hordaland and Sør-Trøndelag, crayfish are limited to the counties of Østfold, Akeshus, Oslo, Hedmark, Oppland, Telemark, Vestfold and Telemark. Only in 13 of the municipalities the crayfish stocks are characterised as good, whereas in the majority, 46 municipalities, the stocks are characterised as sparse (Fig. 1).

Crayfish plague first appeared in Norway in 1971-1973 in two small watersheds on the border with Sweden. In 1987 crayfish plague reappeared and the crayfish stocks in the Glomma river system were wiped out. The border Lake Store Le was affected by crayfish plague in 1989, and in August 1989 the plague also spread to the nearby Halden water course. No plague incidents have been reported after 1993.

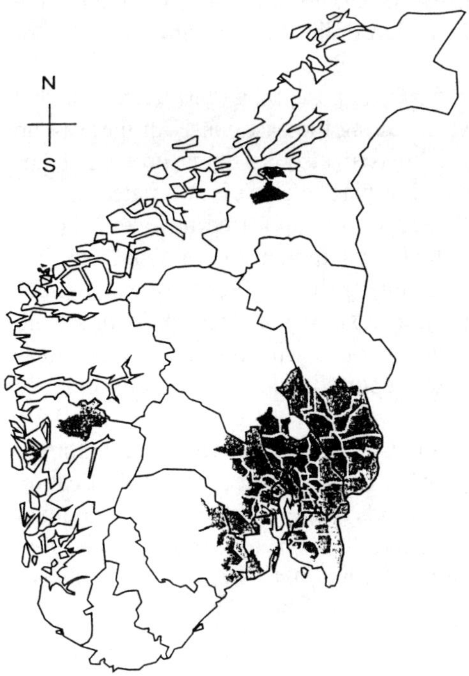

Figure 1. The distribution of noble crayfish *Astacus astacus* in Norway.

3.2 *Denmark*

Denmark has only one native species of crayfish, the noble crayfish. The noble crayfish was formerly a common animal in all parts of Denmark. Within the last century, however, stocks have been greatly reduced due to eutrophication and drainage of lakes and rivers for agricultural purposes.

Crayfish plague is not very well documented in Denmark and there are some discussions if and when the plague affected the native crayfish stocks. Larsen (1990) indicates that some stocks were affected in the period of 1907-1910. However, stocks are still found in most major river systems and in all parts of the country, including the island of Bornholm.

Denmark is separated into 31 regions with respect to river systems. The noble crayfish is found in all 31 regions. On the Jutland peninsula, the largest stocks are found in the River Skals, R. Simested, in the Gudenå river system and in the Skanderborg lakes. On the island of Funen, the largest stocks are probably associated with the Odense river system, and on the island of Zealand, major stocks are found in the Lindved, Tude, and Suse river systems (Larsen 1990).

3.3 *Sweden*

Fossil crayfish has not been found in Sweden and whether the crayfish is a natural immigrant or were stocked has been discussed (Fiskeriverket 1993). However, this is a discussion which only has academic interest and noble crayfish is regarded as a native species by the authorities. Noble crayfish was earlier limited to the southern part of the country, with the border of occurrence stretching from the middle of the county Värmland, through the south-eastern part of the county Dalarna and the southern parts of the county Hälsingland. Even within this original range, the occurrence was sparse, especially in watersheds that drain directly into the Baltic or into the sea on the west coast.

Through extensive stocking during the last 500 years, the range has been extended northwards to also include mainly running waters along the east coast, all the way up to the Finnish border. Production probably increased also as dams prevented upstream migration of eels in many river systems. Pre-plague catch is estimated at 600-1000 tons at the turn of the century. Crayfish plague first appeared in Lake Mälaren in 1907 and spread rapidly. The plague arrived with a shipment of crayfish from Finland to Stockholm. At the end of the 1920s and during the 1930s, many water sheds in southern Sweden were struck by the plague, and since then periods with a high frequency of outbreaks have been alternated with periods with few outbreaks. However, still in 1995, 90 new plague outbreaks were reported from 19 different drainage areas.

The present catch of noble crayfish is 30-50 tons (see also Ackefors this volume). The main reason for the decline in catch is probably crayfish plague, but acidification of the water, water pollution, changes and fluctuations in the water regime in connection to power plants have contributed to the decline. In a survey conducted in 1996, noble crayfish was reported from 1596 localities of which 943 were lakes, 356 were running waters and 297 were ponds (Fig. 2).

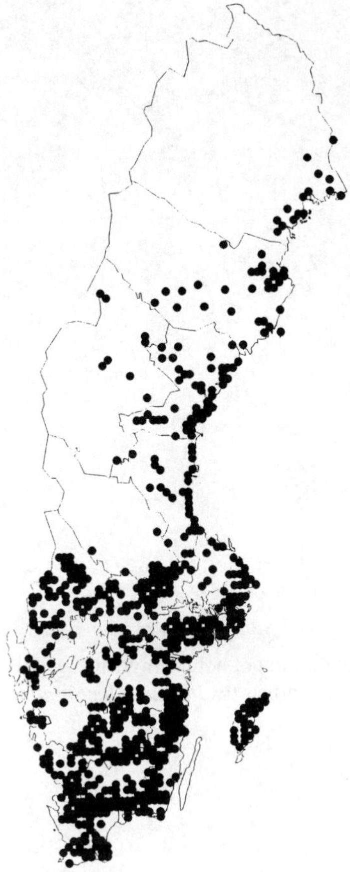

Figure 2. The distribution of noble crayfish *Astacus astacus* in Sweden.

3.4 *Finland*

Although the origins of the noble crayfish in Finland are a matter for speculation, there is a lot of evidence to suggest that it is native to Finland and not introduced by man. The first written description of crayfish dates back to 1558. Fossil remains of crayfish have also been found in Southern Finland in deposits originating from the Littorina Period (c. 3000-1000 BC). The first data on noble crayfish distribution was published in 1859. At that time, noble crayfish were registered north to 62° N. The distribution did not change markedly to the turn of the century.

The spread of noble crayfish has been promoted by stocking as long ago as the early 1500s. Towards the end of the 1800s, their distribution extended beyond the original northern limit. This was a result of increased stocking, as commercial catching became more important in Finland. Stocking peaked in the 1920s and 1930s, due to increased market prices and rewards from the State for successful stocking (Westman 1991, 1995). At present, the continuos area of distribution extends north to 65° N in Eastern Finland and to about 67° N in Northern Finland, i.e. the Polar Circle (Fig. 3). The northernmost population of self-reproducing noble crayfish is found in Lake Leppäjärvi at 67°30' N (Westman 1991).

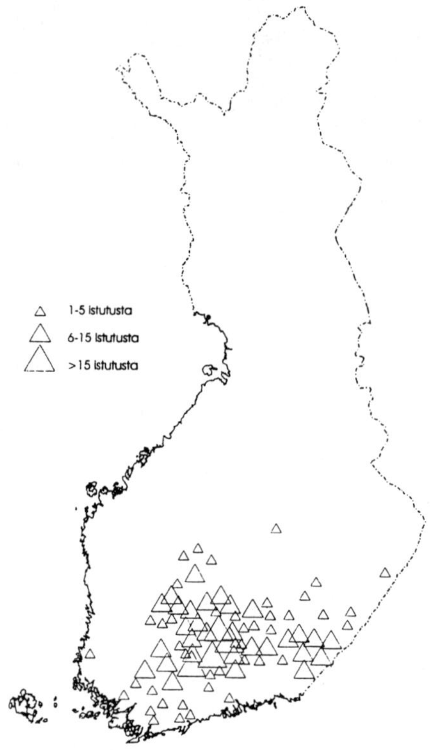

Figure 3. The distribution of noble crayfish *Astacus astacus* and signal crayfish *Pacifastacus leniusculus* in Finland.

Crayfish plague spread from Russia in 1893 and rapidly destroyed a large proportion of the best noble crayfish stocks. Of the 74 major watercourses in Finland in which noble crayfish have lived, less than 20 remain uninfected. In total, the noble crayfish has vanished from thousands of lakes. At present, the distribution is non-continuous and noble crayfish are in many areas confined to a large number of small headwater localities. About 10-20 plague outbreaks are recorded annually in Finland and, since the plague may be active for many years, some hundred plague episodes are active in Finland at any given time (Nylund & Westman 1992).

Commercial catching of noble crayfish dates back to 1840 and it increased throughout the last century, peaking at 20 million crayfish or 850 tons at the turn of the century, of which 15.5 million was exported. The loss due to the plague represents tens of millions FIM of export value a year. The harvest declined sharply due to the crayfish plague. In recent years noble crayfish catch has ranged 2.5-4.8 million specimens or 80-160 tonnes annually. Crayfish plague is the main reason for the decline in crayfish harvest and production, however, habitat deterioration, including acidification, pollution, siltation, and hydropower development, has also contributed.

3.5 *Estonia*

Noble crayfish is the only crayfish species in Estonia (Tuusti et al. 1993, Tuusti

1994), and the species probably migrated into Estonian lakes and rivers concurrently with the end of the last glaciation and the elevation of the land. At the turn of the century, Estonian lakes and rivers were very rich in crayfish.

The plague struck in Estonia for the first time in 1896-1903 in the watersheds of lakes Peipsi and Võrtsjärv. During the following 40 years, the plague wiped out nearly all crayfish stocks on the mainland. The last plague outbreak was registered in 1985-1986, when crayfish disappeared in eight good crayfish waterbodies. Only the crayfish stocks on the islands Saaremaa, Hiiumaa, and Muhu have not been affected by the crayfish plague. Pollution, dredging and poaching have also contributed to the decline in crayfish abundance, as well as to the difficulties in re-establishing noble crayfish in mainland lakes and rivers.

Commercial harvest of crayfish in Estonia topped at 31 tons in 1929. Since 1940, the catch has declined and at present the harvest is only about 1 ton. The biology, distribution and stocks of noble crayfish have been studied in detail in 1952-1956. Information on crayfish stocks and distribution have been collected annually since 1971. At present, the noble crayfish is found in more than 250 waterbodies.

3.6 *Latvia*

The noble crayfish is native to Latvia and were registered in 117 lakes in 1967 (Yurane 1973). We have not been able to obtain updated information on the situation in Latvia, but the development and situation is probably much the same as in the surrounding countries.

3.7 *Lithuania*

The noble crayfish is the only native crayfish in Lithuania (Burba 1994, Taugbøl et al. 1998). As early as in 1578, the Livonian Chronicle stated that there was a lot of crayfish in Lithuania. The noble crayfish was widely distributed, and in the period 1890-1914 about 250 tons of crayfish were harvested annually and exported into other countries. Official catches normally exceeded 100 tons prior to 1940; in the 1950s and early 1960s the catches declined and ranged from 20-30 tons. Presently the catches are very small.

Lithuania was truly one of the most important centres in Europe for crayfish export, since the crayfish from other countries also were exported through Lithuania. Due to crayfish plague, pollution, physical encroachment and eel stocking, the crayfish decreased tremendously throughout this century.

In 1997, the noble crayfish has been confirmed to exist in at least 214 localities in 36 of the 44 regions in Lithuania (Fig. 4). However, in many localities were crayfish have been registered previously, data is lacking on the present situation. It is therefore likely that the noble crayfish is even more widely distributed.

Crayfish plague first occurred in 1886 in the small Duobo Eris Lake in the Rokikis district. Later the crayfish plague spread and in 1893 the Rivers Nemunas and Neris were affected. It is estimated that crayfish plague has occurred in 31 lakes altogether (Cukerzis 1970, 1989).

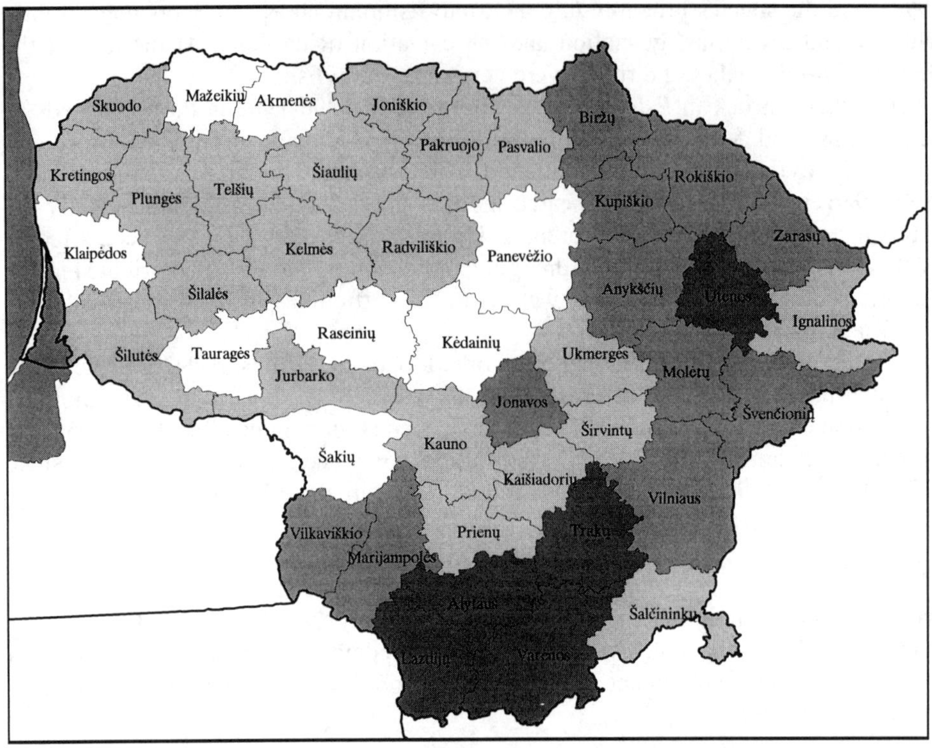

Figure 4. The distribution of noble crayfish *Astacus astacus* in Lithuania. Darker shading indicate more crayfish localities.

4 HISTORY AND CURRENT SITUATION OF INTRODUCED CRAYFISH

The loss of noble crayfish, as well as other native crayfish throughout Europe, strongly influenced demand and hence crayfish became a very valuable resource. The many failures to re-establish the native crayfish after plague incidents increased the search for alternatives, such as the introduction of alien crayfish.

The introductions of crayfish in the Nordic and Baltic are mostly due to deliberate introductions for various purposes like research, aquaculture, ornamental use or stocking in nature which have been approved by the authorities. However, some accidental introductions have also occurred when the crayfish either have escaped or when people are not informed on the regulations and the effects. Natural spreading after initial introduction is also important in spite of the low mobility, and thus spreading ability, for freshwater crayfish. Below, we briefly present the history of introductions and the current situation in the Nordic and Baltic countries.

4.1 *Norway*

There has been some interest importing crayfish into Norway and a few applications have been handled by the authorities; among them, an application to import redclaw crayfish (*Cherax quadricarinatus*) for aquaculture purpose. All applications have, however, been turned down. The only import of foreign crayfish species has thus been in connection to the aquarium industry. Today it is no longer allowed to import or hold crayfish in an aquarium. However, due to the lack of proper control, crayfish are still observed in aquarium shops from time to time.

Due to the legal and illegal spread of signal crayfish in Sweden, this species is found in river systems draining from Norway and in other waterways close to the border on the Swedish side. Sweden has adopted stricter rules to avoid the spread of signals into Norway, but there is a great concern in Norway.

4.2 *Denmark*

In the early 1970s, at least two alien crayfish, the narrow-clawed crayfish and the signal crayfish, were introduced into Denmark in unknown quantities (Bruun 1992). Signal crayfish was introduced partly to replace the noble crayfish in small scale crayfish farming and partly, because signal crayfish with its higher reproductive potential and faster growth, has a greater potential than noble crayfish for large scale commercial production. It soon became evident, however, that this import could potentially threaten the remaining stocks of noble crayfish through competition from signal crayfish and by spreading of crayfish plague. Signal crayfish farming was however already established and is still found in all parts of the country.

No registration of signal crayfish introductions has been made, and therefore a map on the distribution of signal crayfish introductions in Denmark is not available. Signal crayfish are also found in the wild, but no efforts have been made to map the distribution. On the Jutland peninsula signal crayfish has been found in the Ribe river system, and on the island of Zealand, signal crayfish has been found in the Tude river system. These signal crayfish have been documented to be carriers of the crayfish plague fungus, and this is the first documentation of crayfish plague in Denmark. When a large stock of noble crayfish suddenly disappeared from the nearby Bonderup Bog in 1992, it was speculated that this was due to the plague. In Skovse stream associated with the Tude river system, females with hatchlings have been caught. This is the only clear evidence for signal crayfish reproducing in the wild in Denmark.

The narrow-clawed crayfish, was introduced to Denmark primarily from Poland and Turkey in the early 1970s for use in crayfish farming. As a result, stocks are found in most parts of the country (especially in the Jutland peninsula and in the Island of Zealand) in privately owned, isolated lakes and ponds. The distribution of the narrow-clawed crayfish in Denmark in the wild has never been investigated. However, the Danish Crayfish Association has two records on the occurrence of wild populations of the narrow-clawed crayfish in Denmark, both in the southern part of the Jutland peninsula. The narrow-clawed crayfish has been found in a pond associated with the Ribe river system, and in Ferup Lake close to the City of Vejle. The narrow-clawed crayfish is found in all seven districts associated with Zealand and surrounding islands (National Forest and Nature Agency 1997), but has not yet been reported in the wild.

4.3 *Sweden*

Signal crayfish were introduced into lakes in Sweden as a replacement for the noble crayfish, lost mainly due to plague outbreaks. The first introduction was in 1960, and both signal crayfish and *Orconectes virilis* were stocked in Swedish lakes. For unknown reasons, the test with *O. virilis* failed, and signal crayfish was further tested in several lakes. After a trial period in a few lakes in the 1960s, a more extensive stocking programme started in 1969, both with imported adult animals and with juveniles produced by a hatchery in Simontorp, southern Sweden (see Ackefors this volume). The stocking of signal crayfish into lakes continued to be extensive for the next 20 years. This was partly due to the fact that it was more or less the rule to stock signal crayfish into all lakes when they had been struck by the crayfish plague. In a survey conducted in 1996, signal crayfish was reported from 2697 localities of which 1042 were lakes, 319 running water and 1336 were ponds (Fig. 5). The catch of signal crayfish is around 250-300 tons.

The number of lakes and rivers is probably an underestimate, since it is difficult to get a total picture of the distribution which involves a large number of illegal stockings. Signal crayfish are now found in most of southern Sweden north to the Dalälven

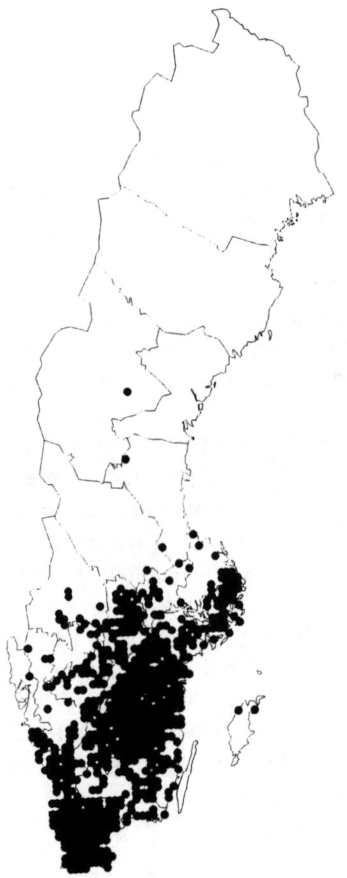

Figure 5. The distribution of signal crayfish *Pacifastacus leniusculus* in Sweden.

river system, and the main area of distribution corresponds to the area struck by the crayfish plague epidemics. Furthermore, signal crayfish have been reported from three localities each at the islands of Gotland and Öland, and at two localities north of the Dalälven river system. All of the 59 main watercourses south of the Dalälven river system except three (Bäveån, Strömsån and Enningdalsälven) close to the Norwegian border harbour signal crayfish.

4.4 *Finland*

Adult signal crayfish (6500 specimens 7-12 cm total length) were imported to Finland from Lake Hennessey and Lake Tahoe in California in 1967-1969. During 1970-1974 stocking continued with stage 2 juveniles (35,000 specimens) imported from Sweden. These crayfish were stocked into 53 separate water bodies. In spite of varying success a stocking program was decided and from 1980 juveniles, stage 2 and summerlings, cultured in Finland have been used for stocking, as well as adult signal crayfish from natural water bodies.

Since 1967, more than 0.6 million signal crayfish have been stocked in Finland into about 300 waterbodies. The interest for signal crayfish has increased and in 1993 about 250,000 signals were stocked into at least 41 new localities. The stocking is confined to lakes previously struck by the plague and mainly in the southern part, i.e. south of 62° N. Only a few stockings have been made in the northern part of Finland

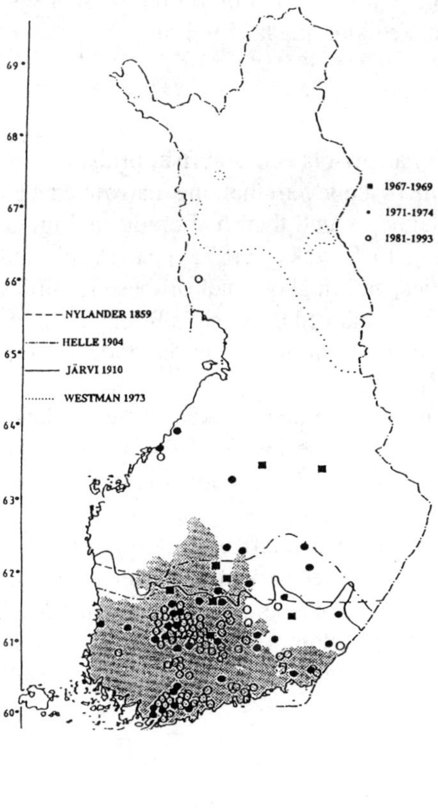

Figure 6. Stocking (Istutusta) of signal crayfish *Pacifastacus leniusculus* in Finland during 1993-1997.

of which Lake Kermajärvi (66° N) represent the world's most northern of the signal crayfish localities. The Finnish Game and Fisheries Research Institute have monitored the development of signal crayfish in a few selected lakes for more than 25 years. Signals reproduce in at least 50 lakes. Annual catch has reached some 100,000 individuals of > 9 cm total length annually. In future, stocking of signal crayfish should be restricted to using only plague-free juveniles in waters were the plague has wiped out the native noble crayfish (Westman 1995).

The narrow-clawed crayfish is presently encountered sporadically in a few south-eastern border waters, where they have spread from introductions in the Karelian part of Russia. The small numbers indicate that this species has not established any permanent populations in Finland yet.

4.5 *Estonia*

There are no alien crayfish species in Estonia at present, but the narrow-clawed crayfish is found close to the border in both Russia and Latvia. An application to import signal crayfish for farming purposes was turned down by the national authorities which stated that introduction of signal crayfish is not desirable.

4.6 *Latvia*

The narrow-clawed crayfish was present in 36 Latvian lakes, especially in the south-eastern part in 1968, and according to Yurane (1973) this species has probably invaded Latvia from stockings in Belarus. We have no information on the present distribution. Signal crayfish have also been introduced into one lake in Latvia in 1983.

4.7 *Lithuania*

Around 1900 several lakes were stocked with narrow-clawed crayfish, probably imported from Latvia and/or Belarus. Landowners believed that the narrow-clawed crayfish were resistant against the crayfish plague, which then had eradicated many of the native noble crayfish stocks. According to Sestokas (1973), narrow-clawed crayfish existed in 8.5% of the crayfish localities, and in about half of these localities it coexisted with noble crayfish. At present narrow-clawed crayfish is found in 22 localities in 11 of the 44 regions (Fig. 7). Most of the localities are in the north-eastern part of Lithuania, in the regions Ignalina and Zarasai.

In 1972, signal crayfish were imported from Sweden and stocked in two isolated lakes. The purpose of the introduction was to replace the losses of the native noble crayfish. Based on official information, signal crayfish is presently found in 10 localities in the mid-eastern part of Lithuania (Fig. 8).

In 1995, the presence of the spiny-cheek crayfish was confirmed. At present, it is found in 10 localities in three regions. It is in the upper part of the river Nemunas draining into Lithuania from Belarus and in the river Sesupe draining from the Russian enclave of Kaliningrad. There is also one observation of spiny-cheek crayfish in the region of Plunge in the northwestern part of Lithuania, which must be the result of stocking (Fig. 9).

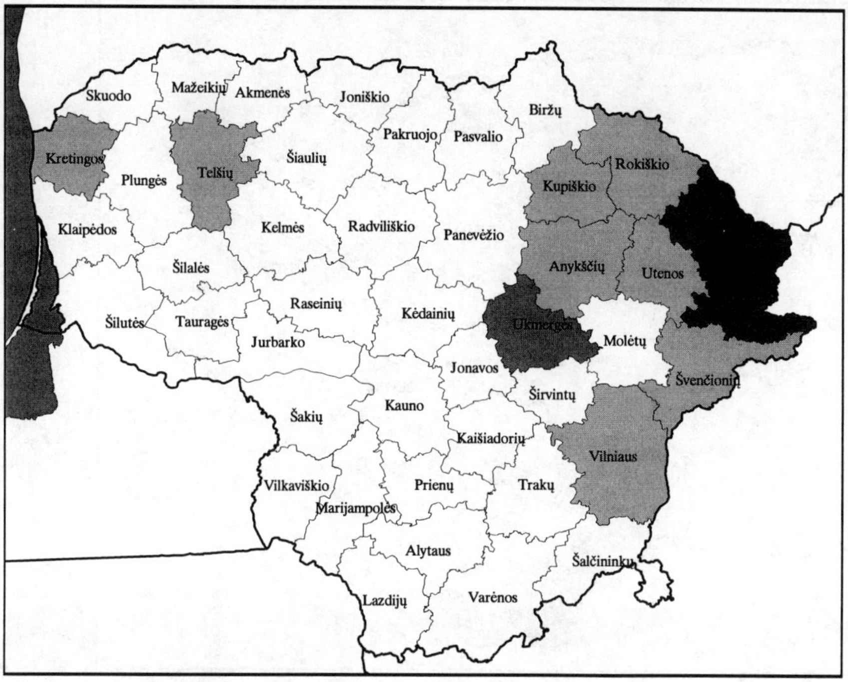

Figure 7. The distribution of narrow-clawed crayfish *Astacus leptodactylus* in Lithuania. Darker shading indicate more crayfish localities.

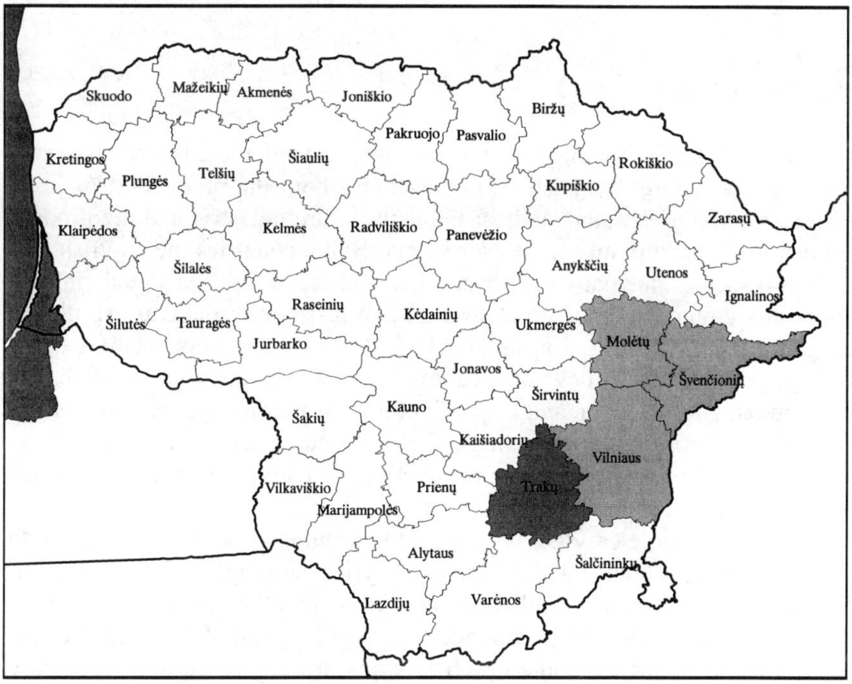

Figure 8. The distribution of signal crayfish *Pacifastacus leniusculus* in Lithuania. Darker shading indicate more crayfish localities.

Figure 9. The distribution of spiny-cheek crayfish *Orconectes limosus* in Lithuania. Darker shading indicate more crayfish localities.

5 CRAYFISH CULTURE

Due to the high market price for crayfish and the high demand, there is a considerable interest for crayfish culture. Crayfish culture includes both the production of juveniles for stocking in natural waters and for the aquaculture industry, and the production of crayfish for consumption. In the Nordic and Baltic countries the crayfish culture involves two species, the native noble crayfish and the introduced signal crayfish (see Ackefors this volume). Signal crayfish are more resistant to the crayfish plague fungus and are therefore preferred in areas where the disease occurs. Many experiments have been undertaken to develop an intensive culture system for crayfish, but due to the cannibalistic behavior and cost of individual rearing systems this is not profitable. Present crayfish culture for the market is based on semi-extensive and extensive production systems with stocking of juveniles or natural reproduction in outdoor ponds of various shape and size.

In Norway only the noble crayfish is farmed, the farming is on a small scale with 11 permissions and the annual yield is small. Crayfish farming is also on a small scale in Denmark and the annual yield is assumed to be less than 5 tons. As a result, most crayfish production, sale and distribution are essentially out of the control of the authorities. Presently, about 15 noble crayfish farms are registered and frequently controlled by the Danish Veterinary and Food Service. These farms only represent a fraction of the total crayfish production. Signal crayfish are also farmed in Denmark.

The crayfish culture industry in Sweden started with the establishment of Simontorp in 1969, producing juvenile signal crayfish for stocking. Signals from Simontorp have been used for stocking in most European countries (Lowery & Holdich 1988). In Sweden the registered harvest of noble crayfish from farms was 12.3 tons whereas the harvest of farmed signal crayfish was 41.8 tons in 1995. However, it is believed that the production is considerably higher (see Ackefors this volume).

Crayfish cultivation in Finland consists mainly of rearing 1-summer old noble and signal crayfish for stocking. Interest in crayfish culture for consumption is however increasing. Rearing trials were initiated by the Finnish Game and Fisheries Institute in 1967, and the first private farm was established in 1969. However, not until the mid 1980s did private farms actually produce juveniles for stocking. At present, there are about 140 crayfish farms in Finland and in addition the Finnish Game and Research Institute produce crayfish at four of their research stations. In 1996, approximately 21,000 noble crayfish juveniles were produced for stocking and 4000 for further rearing. Corresponding figures for signal crayfish are about 227,000 juveniles for stocking and about 6000 for further rearing.

In Estonia, there were some experiments to grow noble crayfish juveniles in indoor conditions during the 1980s. Presently, noble crayfish juveniles is produced in the Härjanurme fish farm, but the production is low, up to 20,000 each year. There is at least one crayfish farm in Latvia (A. Kittelsen, Akvaforsk pers. comm.). In Lithuania, some effort was made to rear crayfish in fish ponds in 1963-1964. Presently, signal crayfish are produced in one private farm and experimental farming of noble crayfish started in 1993 in a governmental fish farm on a small scale, aimed at producing juveniles for stocking. In 1994, 17,000 juveniles were produced and 4000 were stocked into three lakes.

6 LEGISLATION

In the region of Nordic and Baltic countries, the legislation varies between countries. Differences in traditions and the internal organisation of the management in each country strongly influence both the organisation of the legislative system and the aim and regulations themselves. Denmark, Sweden, and Finland are members in the European Union and this of course has a major impact on trade regulations and thus

Table 2. Harvest regulations in the Nordic and Baltic countries.

Country	Season	Size (cm total length)	Comments
Norway	6.8-14.10	9.5	
Denmark	F: 1.8-30.9 M: 1.4-30.9	9	No catch regulations on introduced species
Sweden	No national regulations. A number of local rules with season and minimum size.		
Finland	21.7-31.10	No national regulations.	A number of local rules with season and minimum size.
Estonia	25.7-25.9	10	
Lithuania	15.7-15.10	10	Effort and catch restrictions

crayfish imports. Also various international agreements influence the legislation directly or indirectly. The legislation affecting crayfish consists of various acts which regulate harvesting from the wild, stocking in natural waters, spread of diseases and parasites, import for stocking, aquaculture and the aquarium industry, and establishing and operating crayfish farms. Other acts regulating water management and land use, as well as pollution control, are of importance for crayfish management.

The native crayfish species of Europe are considered vulnerable and special care is recommended in the management (exploitation and conservation) of these species (cf. Bern convention and the EC's Habitats Directive). Three of the species are characterised as 'vulnerable' in the IUCN Red List (IUCN 1996). Due to consideration of the free trade between countries, European Union members have to allow import of live crayfish; however, strict national regulations on the handling of crayfish is thereafter allowed.

6.1 *Norway*

The first crayfish regulations are from 1916 and were designed to prevent the spread of crayfish plague and the over-exploitation of the stocks due to the increased demand and prices. Present regulations include a crayfish harvest season (August 6 from 18:00 h to September 15) and a minimum legal size of 95 mm total length. Stocking crayfish is banned and a special permission is needed. However, permissions will normally be given only for re-establishing lost stocks. The regulations to prevent the spread of diseases and parasites have been revised, as the plague situation has changed. It is not allowed to import unboiled crayfish, crayfish catching gear has to be disinfected and it is only allowed to keep live crayfish in the water of origin. The import ban also includes the aquarium trade. Crayfish culture also needs a special permission, and the conditions for getting permissions and for the operation of the farm are very strict (Taugbøl & Skurdal 1996)

6.2 *Denmark*

Due to its ability to act as a crayfish plague vector the signal crayfish has been the primary target for the Danish legislation on crayfish import, stocking and farming. No plans are currently being undertaken to directly protect noble crayfish and to reduce the spread of the introduced crayfish in the wild. Several steps have been taken, however, to indirectly protect noble crayfish and to minimize the spread of alien crayfish in the wild: stocking with crayfish species other than noble crayfish is now illegal. In 1992, two laws on freshwater fish and nature protection were adopted that prevented the release of alien crayfish including signal crayfish into open water systems. In 1994, a third law was adopted. It disallowed the release of alien crayfish into any freshwater system, including closed lakes and ponds, and at the same time imposed strong restrictions on farming with already introduced alien crayfish. Crayfish can only be imported for consumption, and live crayfish imported from outside the European Union must be killed before the distribution within the country.

6.3 *Sweden*

Before 1994, Sweden had both a restricted harvest season and a minimum legal size of 90 mm total length for both signal and noble crayfish. In 1994, the legislation was changed and there are now no national fisheries restrictions, although local restrictions occur. Even though permits were always needed to stock crayfish and the regulations have aimed at protecting the remaining populations of noble crayfish, the practical policy was to give permits rather liberally during the 1970s and 1980s. The stricter regulations for stocking that have been effective since 1994, means that permits to stock signal crayfish should only be granted in watersheds that already contain signal crayfish and where noble crayfish restoration is not a possible alternative. Thus, permits to stock signal crayfish should not be given on the islands Gotland and Öland, north of the Dalälven river system, for watersheds that extend into Norway, in connection to areas with acute plague and in connection to areas with noble crayfish populations that should be protected, where the signal crayfish is not present today or where noble crayfish restoration is considered a possible alternative.

The import regulations, if applied, should stop foreign crayfish species and diseases from entering Sweden. However, in practice it is almost impossible to check that the regulations are followed, one reason being that the personal at the customs are not trained (the identification of species requires training). When live crayfish are imported for consumption or ornamental use there is no guarantee that they do not escape. Therefore, any import of live crayfish poses a threat to the noble crayfish. In 1997, the situation got worse, since the former veterinary border control was abolished and no permit was needed to import live crayfish for food consumption from the European Community.

6.4 *Finland*

The first national regulation was promulgated in 1875 and included a minimum catch size and a closed season for noble crayfish. The present Fishing Act came in force in 1983. The open season extends from July 21 at 12.00 h to October 31 for both noble and signal crayfish. There is no minimum legal catch size. The holder of the fishing rights can and often do make their own rules regulating the crayfish harvest. Stocking of both noble crayfish and signal crayfish needs a permission from the local fishery authorities (fishery district). A permission is also needed when translocating crayfish from one water body to another. Permission is not needed when cultured juveniles or adults are stocked in waterbodies where a population of the same species already exists or when crayfish are translocated within the same water body.

Introduction of exotic crayfish for aquaculture requires official permission from the national authorities.

6.5 *Estonia*

The harvest regulations in Estonia include an open season from July 25 to September 25 and a legal minimum size of 100 mm total length (Tuusti 1994). Crayfish harvest is based on a small number of permits issued by the county environmental services for recreational fishermen. Import of live crayfish and other aquatic organisms is strictly regulated by the government. Poaching is a problem in many localities.

6.6 *Latvia*

We have no information on the legislation in Latvia.

6.7 *Lithuania*

The harvest regulations were revised in 1994 and now include an open season that lasts from July 15 to October 15, and a legal minimum size of 100 mm total length. The fishery is strictly regulated and a special licence is needed for catching crayfish both for recreational and commercial catching. Both noble crayfish and narrow-clawed crayfish are harvested. Recreational fishing is administered by the Society for Recreational Fishing and Hunting, which have a number of local clubs that rent lakes and rivers. The members can fish with five traps and catch 50 crayfish a day (24 hrs). Altogether 130 lakes and 101 rivers are not for rent due to different causes (National park, Bern Convention species, protected areas, husbandry for fish farm, Red List species or close to the border). Stocking of crayfish is done in a small scale and a permission and a health certificate is needed. Aquaculture regulations are as for fish farming and a permission is needed.

7 CONSERVATION AND RESTORATION

The focus of the management systems of natural waters has turned away from production to conservation. The main objective of the management and the legislative framework is to conserve the natural habitat and the native species. However there are considerable variations between countries in the Nordic and Baltic region.

7.1 *Norway*

A program to restore crayfish in the plague-stricken area has been active since 1989 and includes monitoring of the plague situation with caged live crayfish, restocking with noble crayfish and an annual test fishing programme (Taugbøl 1997). Also efforts have been made to restore previous noble crayfish populations as conditions have improved in polluted rivers and streams. Liming of acidified lakes has improved water quality and improved conditions for crayfish production. Special programs for liming lakes with noble crayfish have been implemented. Habitat improvement is also done in some selected lakes and rivers in order to gain more knowledge about the effects of such measures. The aim of the authorities is to protect the remaining stocks of noble crayfish and efforts will be maintained to prevent further spread of the crayfish plague and to re-establish extinct populations. The ban against importation and stocking with non-native plague-resistant crayfish species will continue.

7.2 *Denmark*

In recent years, efforts have been made to restore some of the lost lakes and rivers. The noble crayfish is assumed to benefit from this restoration. However, even though the noble crayfish is considered a delicacy, it is not well known in Danish cuisine and therefore, has only little commercial value. As a result, a limited amount of data is available on noble crayfish distribution and exploitation and no efforts are made to investigate the importance of nature restoration on noble crayfish populations.

Since 1993, the Danish Fisheries Management Funds have supported stocking of noble crayfish financially by paying 50% of the expenditures (National Forest and Nature Agency 1997). So far, 40 land owners have benefited from this arrangement, stocking a total of 106,700 noble crayfish. The closed season for noble crayfish catching is October 1-March 31 for males and October 1-July 31 for females. Minimum legal size is 9 cm. There is no closed season and no size regulations for the introduced species.

In Denmark, the National Forest and Nature Agency is responsible for the development and updating of the List of Endangered Species, the so called Red List. An appendix to this list is also being published, called the Yellow List. This list covers species that are not presently threatened with extinction, yet to some extent 'demands attention'. A new list will be published soon and the noble crayfish will appear in this appendix (the Yellow List).

7.3 *Sweden*

The interest in the noble crayfish has increased lately and, as a result, a project for the restoration of noble crayfish was started in 1996, and an action plan for the restoration was presented in the summer 1997 (Söderbäck & Edsman1998). The main suggestions in the action plan are: 1. For the protection of the noble crayfish higher priority should be given to the environmental units of the regional administration, 2. Areas with good possibilities for protection of the noble crayfish should be identified, 3. Efforts to protect the noble crayfish should be concentrated to these protected areas, 4. The import of live crayfish should be stopped, 5. Information material, aimed at the general public, about the crayfish situation, the value of the noble crayfish and the threats it faces, the consequences of illegally spreading crayfish and about how to protect the species, should be produced and distributed, 6. High priority should be given in liming projects aiming to restore water quality in those localities affected by acid rain, and 7. Whenever possible, noble crayfish populations hit by the plague should be restored with noble crayfish.

As a practical result of suggestions from the restoration project, several reintroductions of noble crayfish into lakes hit by the plague have been conducted. The development and eventual success of the reintroduction is followed closely.

7.4 *Finland*

Crayfish plague is by far the greatest threat to existing noble crayfish stocks and consequently, the most important management measure is to prevent the disease from spreading. Different regulations and recommendations are given, e.g. concerning disinfecting of traps, transport of crayfish and crayfish trade. Interest in restoration of former noble crayfish waters with noble crayfish is increasing. However, as stocking trials have proved to be ineffective especially in large systems, these chronically plague-infected waters are more and more stocked with the signal crayfish. Studies over the last three decades indicate that signals can become a renewable, permanent natural resource in these waters. In order to stop the spreading of the plague, only plague-free juveniles are recommended for stocking.

Habitat improvement, e.g. liming, supplying shelters, is becoming more important as a management method, however it has been proven difficult to restore the habitat in rivers affected by hydroengineering activities (Westman 1992).

7.5 *Estonia*

During recent years there has been an increasing focus on the restoration of crayfish lakes and rivers. Mature crayfish has been collected from the island of Saaremaa and transplanted to the mainland. Since 1970, more than 200,000 specimens have been transferred from good and strong populations to the old and new crayfish localities on the mainland. A management plan for the crayfish stock is in progress, including an overview of the current crayfish situation (Tuusti et al. in prep.).

7.6 *Latvia*

We have no information.

7.7 *Lithuania*

A management plan for the crayfish stock is included in an overview of the current situation (Taugbøl et al. 1998).

8 DISCUSSION

Four crayfish species have now become established in the Nordic and Baltic countries. The native noble crayfish has been eradicated from many of the former productive areas and present harvest is less than 10% of historic records. Present distribution is confined to the more isolated parts of its former distribution and in Norway, Sweden, and Finland to areas stocked during the last century at higher altitudes and latitudes. Crayfish plague is the primary cause for the change in distribution and abundance, but later habitat alterations, pollution and introduction of alien crayfish and predators have contributed. Currently the management in all countries has increased their focus on conservation and restoration of the native noble crayfish; however the effects of this effort are limited and we expect a further decrease in the distribution and abundance of the native crayfish in the region. Annually a large number of crayfish plague incidents are recorded, many are not reported and many water systems are chronically infected due to the spread of the plague that resistant North American crayfish species are carrying.

The introduced crayfish species are increasing their area of distribution and their abundance. The introduced crayfish have a number of characters which makes them superior to the native noble crayfish, i.e. they are mobile, invasive, aggressive, tolerate a wide range of habitat conditions and are thus capable of spreading rapidly (see Lindqvist & Huner and Holdich this volume). Only Norway and Estonia have at present no introduced crayfish, however the risk of introductions either due to illegal stocking or further natural spread is high. Signal crayfish are used both in aquaculture production and in the cultivation of natural lakes. In spite of an acclimatization of 30 years and intensive stocking and culturing programs, the harvest from many plague-stricken areas is still low, compared to earlier noble crayfish harvest. The optimistic views regarding crayfish production when signal crayfish were introduced do not seem realistic and we cannot expect to reach previous harvest levels in the near future. However, an important reason for the low harvest in many lakes is the poor management of the water systems, e.g. pollution and hydroengineering, which have reduced the potential for crayfish production. In many lakes, signal crayfish produce a considerable harvest and it is expected that harvest of signal crayfish in Sweden and Finland will increase, as more productive lakes are stocked and the populations become more established.

Both signal crayfish and narrow-clawed crayfish can tolerate brackish water (see Holdich et al. 1997) and they have thus a great potential for spreading in the Baltic Sea. Signal crayfish have been registered far from the nearest river outlet in the Baltic Sea. Experiments have demonstrated that signal crayfish cannot molt in brackish water, but adult crayfish obviously can tolerate brackish water for long periods in the intermolt phase. Also in California signal crayfish is found in the estuaries with brackish rivers (Lowery & Holdich 1988).

Noble and signal crayfish coexist in Finland, Sweden, and Lithuania, and also in France (Laurent & Vey 1986, Söderback 1993, Westman 1995). The interpretation is that some of the signal crayfish populations do not carry the crayfish plague fungus and it is also possible to rear plague-free signal juveniles. However, if the crayfish plague fungus strikes, these localities will become chronically infected due to the presence of the signals. There is a substantial overlap in the ecological niche between signal crayfish and noble crayfish, and a number of studies have demonstrated that plague-free signal crayfish outcompetes noble crayfish in co-existing populations and laboratory experiments (Svärdson et al. 1991, Söderbäck 1993, Westman 1995).

Crayfish, generally, have low dispersion abilities and humans are the main cause for increased distribution. Crayfish spread downstream, but upstream the spread is slow and migration barriers such as dams, waterfalls and unfavourable stretches often prevent further upstream spreading. Even though the stocking of crayfish needs an official permit, the public continues to stock crayfish in new localities. The public knowledge on the different species is very limited and also there has been an active promotion of signal crayfish which has resulted in a number of new introductions. Hopefully strict regulations, better enforcement and increased information will reduce the amount of new introductions. Also, the increased focus on the native noble crayfish and the efforts to restore former crayfish localities will have a positive influence.

There is also a risk for the introduction of other crayfish species to the region. The aquarium trade industry uses the North American red swamp crayfish, *Procambarus clarkii*, and this species have been observed in pet stores in many of the countries, i.e. Norway, Denmark, Sweden, and Finland. The species has not been recorded in the wild, and it is presently unknown if this species can survive during our winters. Other crayfish species are used in the aquarium trade. Australian crayfish have been promoted as candidates in the aquaculture industry and some interest for importation have been raised. These are mainly tropical and sub-tropical species and whether they will spread in nature is not known.

The fact that live crayfish for food consumption can be sold between member states of the European Union represents a potential for spreading crayfish species. It is difficult to ensure that the crayfish actually are sold for consumption and not released into ponds, lakes or rivers. Crayfish imported from countries outside the European Union must be killed before distribution within the country. This law can however be legally bypassed if crayfish from outside EU are exported to another member state and then transferred to the next.

We feel that the changing wind in the management of crayfish in the Nordic and Baltic countries is very positive for the native crayfish, and we hope that the efforts to conserve and restore the native stocks will succeed. However, we are greatly concerned for the future. Introduced species continually increase their distribution by spreading from previous areas and by legal and illegal stocking. Crayfish trade still represents a possible source for introductions as do the aquarium trade and the aquaculture industry. It is important to map the distribution of the four crayfish species in the region and monitor changes.

Legislation should be improved and harmonized and should account for the different species. It is also important to improve the knowledge among the public and the management on the crayfish species and the current regulations. Efforts to conserve and restore native crayfish stocks and habitats should be enhanced.

REFERENCES

Bruun, P. 1992. Landbrugsministeriets rapport nr. 69. Alternative husdyrproduktioner - muligheder belyst ved hjorte, slagtekaniner og krebs. Statens Jordbrugsøkonomiske Institut (In Danish).

Burba, A. 1994. Facts about crayfish in Lithuania. *Nordic J. Freshw. Res.* 69:170- 172.

Cukerzis, J. 1970. The biology of crayfish (*A. astacus*). Mintis, Vilnius, 208 pp. (In Russian).

Cukerzis, J. 1989. *Astacus astacus* in Europe. In D.M. Holdich & R.S. Lowery (eds), *Freshwater crayfish: biology, management and exploitation*: 309-340 pp. London: Chapman and Hall.

Fiskeriverket 1993. Möjligheter att öka flodkräftbestånd i svanska vatten. Information från Sötvattenslaboratoriet Drottningholm 2, 66 pp (In Swedish).

National Forest and Nature Agency 1997. Vejledning om regler for udsætning af fisk, krebs og bløddyr i de ferske vande (In Danish).

Holdich, D.M., Harlioglu, M.M. & Firkins, I. 1997. Salinity adaptions of crayfish in British waters with particular reference to *Austropotamobius pallipes*, *Astacus leptodactylus* and *Pacifastacus leniusculus*. *Est. Cstl. Sh. Sci.* 44:147-154.

Köksal, G. 1988. *Astacus leptodactylus* in Europe. In D.M. Holdich & R.S. Lowery (eds), *Freshwater crayfish: biology, management and exploitation*: 365-400 pp. London: Croom Helm (Chapman & Hall).

Larsen, K. 1990. Den ny krebsebog. Skjern, Pinus (In Danish).

Laurent, P.J. & Vey, A. 1986. The acclimation of *Pacifastacus leniusculus* in Lake Divonne. *Freshwater Crayfish* 6:146-155.

Lowery, R.S. 1988. Growth, molting and reproduction. In D.M. Holdich & R.S. Lowery (eds), *Freshwater crayfish: biology, management and exploitation*: 83-113 pp. London: Croom Helm (Chapman & Hall).

Lowery, R.S. & Holdich, D.M. 1988. *Pacifastacus leniusculus* in North America and Europe, with details of the distribution of introduced and native crayfish species in Europe. In D.M. Holdich & R.S. Lowery (eds), *Freshwater crayfish: biology, management and exploitation*: 283-308 pp. London: Croom Helm (Chapman & Hall).

Momot, W.T. 1988. *Orconectes* in North America and elsewhere. In D.M. Holdich & R.S. Lowery (eds), *Freshwater crayfish: biology, management and exploitation*: 262-292 pp. London, Chapman and Hall.

Nylund, V. & Westman, K. 1992. Crayfish diseases and their control in Finland. Report from the EIFAC Workshop on Crayfish Management and Stocking, 22-23 August 1991, Kuopio, Finland. *Finnish Fish. Res.* 14:107-118

Sestokas, J.A. 1973. Distribution of crayfish in Lithuanian SSR - *The works of the Lithuanian Academy of Sciences, Vilnius. Ser. C* 3(63):87-105 (In Russian).

Svärdson, G., Fürst, M. & Fjälling, A. 1991. Population resilience of *Pacifastacus leniusculus* in Sweden. *Finnish Fish. Res.* 12:165-167.

Söderbäck, B. 1993. Population regulation in two co-occurring crayfish species. Acta Universitatis Upsaliensis. Comprehensive Summaries of Uppsala Dissertations from the Faculty of Science 434:1-29.

Söderbäck, B. & Edsman, L. 1998. Åtgärdsprogram för bevarande av flodkräfta. Fiskeriverket och Naturvårdsverket 38 pp. (In Swedish).

Taugbøl, T. 1996. Krepsepestutvalget – Årsrapport 1996. ØF-Notat 11/1996, 28 pp. (In Norwegian).

Taugbøl, T. & Skurdal, J. 1996. Ferskvannskreps i Norge - kunnskapsstatus og forvaltningserfaring. ØF-rapport nr 13/1996, 115 pp. (In Norwegian).

Taugbøl, T., Skurdal, J. & Burba, A. 1998. Freshwater crayfish in Lithuania. I: Action plan for crayfish management. II: Crayfish status report. ØF-rapport nr. 12/1998:83 pp.

Tuusti, J. 1994. On catching of noble crayfish (*Astacus astacus*) in Estonia. *Nordic J. Freshw. Res.* 69:167-169.

Tuusti, J., Paaver, T. & Reier, A. 1993. Status of the noble crayfish (*Astacus astacus*) stocks in Estonia. *Freshwater Crayfish* 9:163-169.

Tuusti, J, Taugbøl, T. Skurdal. J. & Kukk, L. 1998. Freshwater crayfish in Estonia. I. Action plan for crayfish management. II. Crayfish status report. ØF-rapport m22/1998, 92 pp.

Westman, K. 1991. The crayfish fishery in Finland - its past, present and future. *Finnish Fish. Res.* 12:187-216.

Westman, K. 1992. Management of the noble crayfish *Astacus astacus* (L.) and the signal crayfish *Pacifastacus leniusculus* (Dana) in Finland. *Finnish Fish. Res.* 14:19-51.

Westman, K. 1995. Introduction of alien crayfish in the development of crayfish fisheries; experience with signal crayfish *Pacifastacus leniusculus* (Dana) in Finland and the impact on the native noble crayfish (*Astacus astacus* (L)). *Freshwater Crayfish* 10:1-17.

Westman, K., Ackefors, H. & Nylund, V. 1992 Kräftor. Biologi, odling, fiske. 152 pp., Ystad, Kiviksgårdens (In Swedish).

Westman, K., Pursiainen, M & Westman, P. (eds) 1990. Status of crayfish stocks, fisheries, diseases and culture in Europe. Report of the FAO European Inland Fisheries Advisory Commission (EIFAC) Working Party of Crayfish. Helsinki, Riista- ja kalatalouden tutkimuslaitos. Kalatutkimuksia-Fikseundersøkingar 3, 206 pp.

Yurane, A.D. 1973. The distribution of crayfish and the state of crayfish stocks in bodies of water in the Latvian SSR. *Fisheries Research Board of Canada Translation Series No* 2824, 7 pp.

Native and alien crayfish in the British Isles

DAVID M. HOLDICH
School of Biological Sciences, The University of Nottingham, Nottingham, UK

W. DAVID ROGERS
David Rogers Associates, Castle Donington, Derby, UK

JULIAN D. REYNOLDS
Department of Zoology, University of Dublin, Trinity College, Dublin, Ireland

ABSTRACT

The distribution of native and alien crayfish inhabiting the British Isles is described separately for England, Wales, Scotland, Northern Ireland, Irish Republic, and some associated islands. Distribution of the single native species, *Austropotamobius pallipes*, is largely determined by geology, water quality and human intervention. Very large populations exist in England, Wales, and the Irish Republic, but in recent years some of these have been eliminated, particularly by crayfish plague in the 1980s. Active measures are being taken to conserve the remaining stocks of this species in line with the terms of the Rio Convention and European Directives. Since the 1970s, a number of alien crayfish have been introduced into England, Wales, and Scotland. *Pacifastacus leniusculus* is cultivated in a number of extensive and semi-intensive facilities. *Astacus leptodactylus* is not cultivated to any extent but those caught from the wild are used for culinary purposes. Large populations of both species now exist in the wild as a result of escapes and implants, particularly in central and southern England. The wild harvest of these two species is now probably greater than that derived from cultivation. Other alien crayfish such as *Astacus astacus*, *Procambarus clarkii,* and *Orconectes limosus* are also present in the wild in England, and *Cherax quadricarinatus* is kept by aquarists. Due to the environmental impact that wild populations of alien crayfish are having in Great Britain, stringent legislation has recently been introduced to try and stem their spread and that of crayfish plague. Northern Ireland and the Irish Republic ban the import of alien crayfish, although Great Britain does not. Reintroduction programs for *A. pallipes* are being initiated and monitoring of populations of both native and alien crayfish are being carried out at national and local level.

1 INTRODUCTION

Austropotamobius pallipes has a widespread, although somewhat patchy, distribution in Europe, despite the fact that since the last century many populations have been devastated by crayfish plague, alien crayfish and human activity (Holdich 1996). *A. pallipes* is the sole native crayfish species in the British Isles, some of the largest remaining populations in Europe survive there (Reynolds 1988, 1997b, Holdich & Reeve 1991). These too have recently suffered the effects of crayfish plague and, in the case of Great Britain, competitive exclusion by alien crayfish, particularly *Pacifastacus leniusculus* (Reynolds 1988, 1997b, Holdich & Reeve 1991, Holdich et al. 1995a, Holdich & Rogers 1997a). In this paper, the past and present distribution of native and alien crayfish is described and measures currently being undertaken to conserve the native crayfish and manage populations of the alien crayfish are discussed.

2 DISTRIBUTION OF *AUSTROPOTAMOBIUS PALLIPES* IN THE BRITISH ISLES

2.1 *General distribution*

Austropotamobius pallipes mainly inhabits clean, relatively hard, mineral-rich waters with a pH between 7 and 9 and calcium concentrations above 5 mg/l (Gledhill et al. 1993, Foster 1995b). Although there is a strong relationship between clean water and the occurrence of *A. pallipes* (Holdich & Reeve 1991), it has occasionally been found below sewage outfalls and appears to be relatively tolerant of unionized ammonia (Foster & Turner 1993, Foster 1995b). A population was found in the Midlands in 1998 living below a storm drain overflow which was frequently inundated with raw sewage, individuals were covered in a growth of sewage fungus but appeared healthy (W.D. Rogers personal observation). It tends to be absent from areas with igneous and metamorphic rock formations, unless there are pockets of easily-weathered rock such as limestone. It occurs in a wide range of water bodies, as long as they have adequate cover, including lakes, reservoirs and water-filled quarries, as well as in rivers and streams. It can burrow and there were reports of river banks being honeycombed with crayfish burrows in the River Thames catchment last century and early this century (Huxley 1879, Cornish 1902). *A. pallipes* is a long-lived species, at least 11 years in the north of England (Brewis & Bowler 1982).

Very few distribution records exist prior to the 1800s and it was not until the 1970s that detailed records began to be collated (Thomas & Ingle 1971, Reynolds 1982, Lucey & McGarrigle 1987, Holdich & Reeve 1991, Holdich & Rogers 1997b). These revealed an extensive distribution in much of England and the Irish Republic, although less so in Wales and Northern Ireland. Only one record exists for Scotland. Figure 1 shows all the records known for the period 1970-1997, much more extensive survey work having been done in Great Britain than Ireland in recent years (Holdich & Rogers 1997b).

The native crayfish in Great Britain belongs to the same subspecies, *pallipes*, as occurs predominantly in France (Souty-Grosset et al. 1997). The division of *A. pallipes* into three subspecies, *pallipes* (France, part of Switzerland and Austria, and the British Isles), *lusitanicus* (Iberian Peninsula) and *italicus* (Italy, part of Switzerland, and the Dalmatian coast) has recently been confirmed using mtDNA analysis (Souty-Grosset et al. 1997). However, on the basis of their studies using allozyme diversity, Santucci et al. (1997) have suggested that European *A. pallipes* and *A. italicus* (including *lusitanicus*) may well be distinct species, as suggested by some earlier workers.

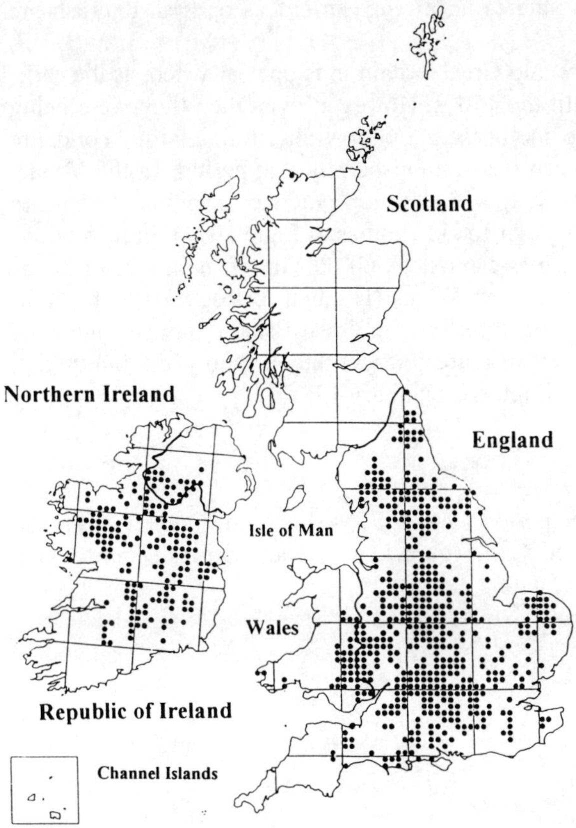

Figure 1. Distribution at the 10-km square level of *Austropotamobius pallipes* in the British Isles for the period 1970-1997 inclusive. In many cases each point on the map represents a number of populations in that 10 km square. Only a few records for Northern Ireland and the Republic of Ireland have been obtained since 1987.

The origins of *A. pallipes* in Great Britain are unclear. Pleistocene fossils of crayfish are reputed to have been found in eastern England (Bell 1920, Karaman 1962). If it was present before the last Ice Age, it is unlikely to have survived it according to Albrecht (1983) and it either recolonized naturally through post-glacial streams connecting Great Britain to rivers such as the Rhine or via human intervention (Holdich & Rogers 1997b). Recent evidence from mitochondrial DNA (mtDNA) analysis has shown that, based on this technique, little genetic diversity appears to exist in English and Welsh populations and that there is a close similarity to some French populations (Grandjean et al. 1997a, b). Santucci et al. (1997) also found a lack of genetic divergence between samples from England and southern France using allozyme analysis. Such studies suggest that *A. pallipes* in Great Britain may have been derived from mainland European stock relatively recently. It may be that when a wider range of populations are studied, *A. pallipes* in Great Britain will be shown to have had diverse origins.

Reynolds (1979) and Laurent (1988) favored a natural post-glacial colonization of Ireland by *A. pallipes* as part of the Lusitanian fauna. However, all Irish crayfish examined belong to the subspecies *pallipes* rather than *lusitanicus*. This does not mean that old native stock has not survived, since Attard & Vianet (1985) found a population in Ireland that was genetically distinct from those in France. The situation should be clearer when ongoing mtDNA studies on a range of stocks of Irish crayfish are completed.

If humans introduced *A. pallipes* into Great Britain, it is unclear when, as the earliest published records were not until the 1500s. However, even then they were being moved from one part of England to another, e.g. by Sir Walter Raleigh into Yorkshire (Fryer 1993), so they must have been well established by that period. In the Middle Ages and into the early 20th century, it was common practice to implant them into suitable waters since they were a popular food (Thomas & Ingle 1971). Implantations this century have shown that *A. pallipes* can exist well outside its main area of distribution, e.g. northern Scotland and western Wales (Holdich & Rogers 1997b), indicating that the potential distribution of *A. pallipes* in Great Britain may be somewhat greater than actually occurs. There are also areas where they should occur but do not, e.g. some chalk streams in southern England (Stanbury 1997).

2.2 *England*

The distribution for England for the period 1990-1997 is shown in Figure 2 and it can clearly be seen that, when compared with Figure 1, many populations seem to have

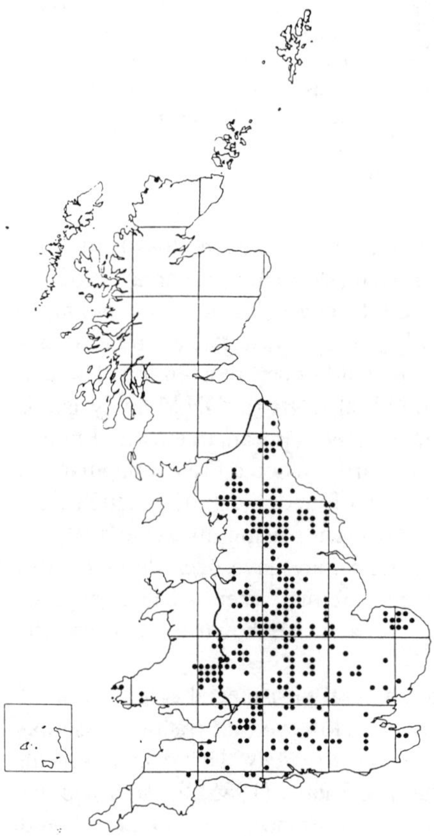

Figure 2. Distribution at the 10-km square level of *Austropotamobius pallipes* in Great Britain for the period 1990-1997 inclusive.

disappeared, particularly in southern England. This is largely due to the impact of crayfish plague (see below). However, recent surveys (Holdich et al. 1995a, Holdich & Rogers 1997b) have shown that there are in fact still many large, healthy populations of *A. pallipes*, particularly in northern and central England and the Welsh borderland, as well as more scattered populations in eastern, southern, southeastern and southwestern England (Fig. 2). Many of these populations are, however, under threat, due to their close proximity to *P. leniusculus* populations, particularly in central and southern areas.

2.3 *Wales*

Austropotamobius pallipes populations are restricted to just a few catchments in Wales, due to unsuitable geology, water quality and human activity (Holdich et al. 1995a). As with England, its range has become contracted in recent years (cf. Figs 1 and 2). Most of the remaining populations are in the catchments of the Upper R. Severn, R. Wye and R. Usk, although two exist in southwestern Wales. At least one of these is the result of a recent introduction.

2.4 *Scotland*

There is only one record of *A. pallipes* for Scotland. This occurs in a coastal, limestone loch in the northwest (Fig. 2), where the climate is ameliorated by the Gulf Stream. The population has survived since the 1940s when it was introduced from England (Thomas 1992).

2.5 *Northern Ireland*

Austropotamobius pallipes is currently known from the lower River Erne catchment and the Ulster Blackwater (Fig. 1). Regular surveys of 300 stream and many lake sites shows no losses apart from isolated pollution incidents. Crayfish have recently been found in two further small streams flowing into Lough Erne (P. Hale personal communication). Reports of crayfish in areas of Antrim and Tyrone have not been substantiated.

2.6 *Irish Republic*

Austropotamobius pallipes appears to be still widespread in much of lowland Ireland (Fig. 1), although there has been no published survey in the past ten years, and there has been no observed recurrence of crayfish plague (see below). There are no reports of range diminution or extension.

2.7 *Isle of Man, Channel Islands*

There are no records of *A. pallipes* being present in the Isle of Man or any of the Channel Islands in recent times.

3 DISTRIBUTION OF ALIEN CRAYFISH

3.1 *Irish Republic and Northern Ireland, Isle of Man and Channel Islands*

Although *A. leptodactylus* and *P. clarkii* have been occasionally seen for sale in Dublin, there are no confirmed records of alien crayfish having been introduced or becoming established in these areas, although *P. leniusculus* was cultured on the Isle of Man in the 1970s.

3.2 *Great Britain (England, Scotland and Wales)*

3.2.1 *Pacifastacus leniusculus*

The successful introduction of the North American, signal crayfish, *P. leniusculus,* from Sweden into southern England in the 1970s was followed by publicity which claimed that crayfish farming was a lucrative business with an expanding future. This led to a demand for it and many implants were made in England and Wales from the mid-1970s onwards (Lowery & Holdich 1988, Holdich & Reeve 1991, Maitland 1996). Most of the early implants were into existing fish farm ponds or lakes, but some were directly into the wild. *P. leniusculus* has similar habitat requirements to *A. pallipes* (Lowery & Holdich 1988). It also has a similar life-history (Lowery 1988), although mating and release of juveniles tend to occur up to four weeks earlier, the actual timing for both species depending on latitude and altitude (D.M. Holdich personal observation). Recent studies using lipofuschin have shown that *P. leniusculus* may be very long-lived, i.e. 16 years in central England (Belchier et al. 1998).

There are very few purpose-built crayfish farms in Great Britain and much of the production comes from extensive systems (Holdich & Rogers 1992, Rogers & Holdich 1995b). Escapes occurred from ponds and lakes and these led to the further development of wild populations. Implants met with variable success, but those that did survive led to a widespread distribution of this species in the 1980s, particularly in southern England (Holdich & Reeve 1991, Holdich et al. 1995a). In the 1990s their rate of colonisation has been reduced somewhat. Both farmed and wild populations of *P. leniusculus* now have a scattered distribution in England and Wales (Fig. 3) that overlaps that of *A. pallipes* (cf. Fig. 2). In addition, *P. leniusculus* occurs in areas not occupied by *A. pallipes*, such as parts of southern, south-western and eastern England (Fig. 3). Recently, wild populations of *P. leniusculus* have been recorded from Scotland (Maitland 1996) (Fig. 3) where there are large areas of suitable aquatic habitat for colonisation.

Over 80 premises were registered as crayfish farms in the early 1990s, reaching a peak of 99 in 1994, but production from such farms has never exceeded 10 tons (Holdich et al. 1995a, Rogers & Holdich 1995a). Few farmers made profits from their ventures and some deregistered voluntarily. Registration was a requirement for anybody producing fish and crayfish for sale or transfer under the Registration of Fish Farming and Shellfish Farming Business Order 1985 and was introduced under the Diseases of Fish Act 1983 as a means of controlling fish and crayfish diseases.

In the mid-1990s, the regulatory authority, the Ministry of Agriculture, Fisheries and Food (MAFF), reconsidered its approach to registration. After 1996, following the introduction of new legislation regarding no-go areas (see below), and in keeping

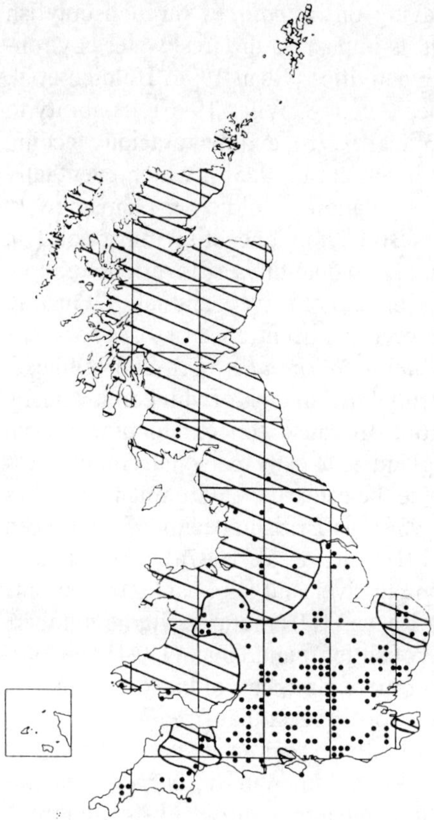

Figure 3. Distribution at the 10-km square level of *Pacifastacus leniusculus* in Great Britain for the period 1990-1997 inclusive. Superimposed on the map is the extent of the no-go areas (shading) for the keeping of *P. leniusculus* imposed by the Government in 1996. In addition, the keeping of all other alien crayfish, with the exception of *Cherax quadricarinatus*, is banned through the whole of Great Britain.

with the registration of (fin) fish farms, many sites were deregistered. These sites were not considered to be farms because they were not producing juveniles or rearing crayfish by feeding, they were simply harvesting crayfish from the wild. What was the 'wild' had never satisfactorily defined until recently and, as far as crayfish are concerned, MAFF have defined it as – any body of water from which crayfish could move to a natural water course. MAFF also considered that there was little difference between a person harvesting from the wild, whether they made the initial introduction to create the wild population or not. The majority of crayfish farms subsequently were considered to be the 'wild' and the number of registered crayfish farms has now reduced to 10 in 1998 (A. Scott personal communication). MAFF has created a distinct division between crayfish farmers who are registered and persons who harvest crayfish from the wild who are not registered.

The introduction of *P. leniusculus* (and *A. leptodactylus* – see below) in the 1980s has led to the establishment of many lacustrine and riverine populations in the wild in the 1990s. Harvesting from these populations was insignificant in 1990, but increased rapidly in the early 1990s and the tonnage harvested exceeded farmed tonnage in 1995 (Rogers & Holdich 1997). The crayfish farms which MAFF has deregistered will now add to this total, so one might expect the farmed production in 1998 to be less than 5 tons (virtually all *P. leniusculus*) and the harvest from the wild (*P. leniusculus* and *A. leptodactylus*) to be in excess of 15 tons.

As well as the effects *P. leniusculus* is having on *A. pallipes* through crayfish plague (see below), there is also concern about its impact on the freshwater environment in general, due to its extensive burrowing activities (Guan 1994, Holdich et al. 1995a, Harris & Young 1996), its high densities (Guan & Wiles 1996), its ability to move over relatively long distances (Guan & Wiles 1997b), and its voracious feeding habits – including predation on *A. pallipes* (Holdich et al. 1995b), which eventually leads to *A. pallipes*' elimination in any mixed populations (Holdich & Domaniewski 1995). The high densities of *P. leniusculus* are also having a negative impact on fish populations by destroying plant cover and eating large quantities of invertebrate food, as well as on anglers by interfering with their sport in both rivers and lakes. Guan & Wiles (1997a) found a negative correlation between the occurrence of *P. leniusculus* and benthic fish such as stone loach, *Noemachelius barbatulus* (L.), and bullhead, *Cottus gobio* L., in a British lowland river. Potential decimation of vulnerable aquatic invertebrate species by *P. leniusculus* is starting to cause concern to professional aquatic biologists and conservationists in England (J. Foster personal communication.). *P. leniusculus* may also be able to colonise the estuarine environment, as it has done in North America (Lowery & Holdich 1988), as British specimens have been shown to be very tolerant of saline conditions (Holdich et al. 1997a). There is also particular concern about the damage being done to river and lake banks by the burrowing activities of *P. leniusculus*. Holdich & Reeve (1991) reported large numbers (20 m^{-2}) of burrows in the banks of a pond on a fish farm and Guan (1994) and Harris & Young (1996) have reported 10 burrows or more per metre of bank in some rivers. Interestingly, there is nothing in the literature to suggest that *P. leniusculus* burrows in other parts of the world, although apparently it does in its native habitat in Oregon, USA (S.D. Lewis personal communication.). Although *A. pallipes* does burrow (see above), the burrows are relatively small compared to those constructed by *P. leniusculus*.

3.2.2 *Astacus leptodactylus*

The narrow-clawed (Turkish) crayfish, *Astacus leptodactylus*, an East European and Near East species (Köksal 1988), was introduced into Great Britain in the 1970s mainly for the restaurant trade. It is not farmed to any extent, but escapes have occurred from holding facilities and implants have also been made into wild waters. Some wild populations have grown to high densities, e.g. the Serpentine lake in Hyde Park, London, and are actively harvested, particularly around London. One population in eastern England has entered a river catchment and has migrated down to its tidal reaches (Holdich et al. 1995a). Like *P. leniusculus*, *A. leptodactylus* has a high tolerance to saline conditions and may also be able colonize the estuarine environment as it has done in its native range (Köksal 1988, Holdich et al. 1995a).

The high densities reached by some populations of *A. leptodactylus* are having similar effects to that of *P. leniusculus*, including a negative impact on angling. Currently, *A. leptodactylus* has replaced *A. pallipes* in a few locations around London, but there is no evidence of direct competition between these two species in mixed populations (Holdich et al. 1995a, b). However, because of the much larger size and higher fecundity of *A. leptodactylus,* it would likely be a superior competitor for resources and have a greater impact on the freshwater environment.

3.2.3 *Other species*

Astacus astacus, a mainland European species (Cukerzis 1988), was introduced from Germany for farming purposes in the early 1980s. However, it has only become established in one river catchment in south-western England, where it occurs on a farm and in the wild. Currently, *A. astacus* is not having a negative impact on *A. pallipes*. However, because of its larger size and higher fecundity, it would probably be a superior competitor in any mixed population with *A. pallipes* and have a greater impact on the freshwater environment.

Procambarus clarkii, a North American species (Huner 1988), has been imported into Great Britain for the restaurant and aquarist trade. It was common at one site, a lake in London, in the 1980s, but its current status is not known. However, individual specimens have been found in the wild in other parts of England, sometimes carrying eggs (Foster 1994, 1995a). As *P. clarkii* appears able to survive British winters, then it may spread if it becomes more established. Currently, *P. clarkii* is not having a negative impact on *A. pallipes*. However, because of its invasive capabilities (see Lindqvist & Huner this volume) and higher fecundity, it would probably be a superior competitor in any mixed population, in addition to the fact that it is a vector of crayfish plague. It is likely to have a greater impact on the freshwater environment than *A. pallipes*, particularly because of its extensive burrowing habits.

Orconectes limosus, a North American species (Momot 1988), although common in mainland Europe, has only recently been confirmed as being present at two lacustrine sites (W.D. Rogers personal observation), but further details have yet to be ascertained. It was known to be on sale at some aquarist centers in the early 1990s (Foster 1995c). It is likely to have a greater impact on the freshwater environment than *A. pallipes,* because of its invasive capabilities and higher fecundity. It would probably be a superior competitor in any mixed population with *A. pallipes*, as well as being a vector of crayfish plague.

Various Australian crayfish belonging to the genus *Cherax* have been introduced into Great Britain in the last 20 years, but no wild populations have yet developed. *C. quadricarinatus* is legally kept by aquarists (see below). *C. destructor* and *C. tenuimanus* (Foster 1994, W.D. Rogers personal observation) were also imported for the aquarium trade in the early 1990s, but this practice is now illegal (see below).

4 CRAYFISH PLAGUE

Between 1980 and 1993, many populations of *A. pallipes* were lost in England and Wales due to the effects of crayfish plague (Holdich & Reeve 1991, Alderman 1993, Holdich & Rogers 1997b). In addition, a severe outbreak in the Republic of Ireland in 1986-1987 eliminated huge numbers of *A. pallipes* in some central lakes, despite the fact that no source of plague, e.g. illegally introduced alien crayfish, were found (Reynolds 1988, Matthews & Reynolds 1992). Its entry into the British Isles completed a clean sweep of the spread of this disease in Europe which started in the second half of the 19th century (Alderman 1996). Outbreaks in Great Britain are closely linked with the importation of *P. leniusculus* from Sweden and North America in the 1970s (Lilley et al. 1997).

Crayfish plague is a very virulent disease and its spores are easily spread by a variety of means. All three North American crayfish introduced into Great Britain can harbor and act as vectors of the disease. *A. astacus*, *A. leptodactylus* and *Cherax* spp. do not naturally harbor the disease and are susceptible to its effects like *A. pallipes* (Alderman & Polglase 1988). *P. leniusculus* is the main vector of crayfish plague in Great Britain, but not all populations harbor the disease as evidenced by the occurrence of mixed populations with *A. pallipes*. However such populations are short-lived, due to the competitive exclusion of *A. pallipes* by *P. leniusculus* (Holdich & Domaniewski 1995).

5 LEGISLATION, CONSERVATION AND MANAGEMENT

5.1 *European legislation*

Austropotamobius pallipes is included in the IUCN Red Data List and in Appendix III of the Bern Convention and Annexes II and V of the European Habitats Directive. The directive requires the designation of protected areas (Special Areas of Conservation – SACs) for species in Annex II (Holdich 1996).

5.2 *Great Britain*

It is illegal to take *A. pallipes* from the wild (without a licence) or sell it as it is listed on Schedule 5 of the Wildlife and Countryside Act, 1981, which gives it protection from such activities.

It is illegal under Section 14 of the Wildlife and Countryside Act 1981 to release alien crayfish into the wild without a licence. It is also illegal to allow alien crayfish species to escape from holding facilities. Reasonable steps and due diligence must be exercised to avoid this. Due to their perceived threat, i.e. as superior competitors and possibly as vectors of crayfish plague, to the native crayfish, *A. pallipes*, the three alien species which are known to be well-established, i.e. *A. astacus*, *A. leptodactylus* and *P. leniusculus*, were classified as pests in 1992 and listed on Schedule 9 of the Wildlife and Countryside Act. If other alien crayfish are found to be breeding in the wild (e.g. *P. clarkii* and *O. limosus*), then they will be added to the list. In addition to the above legislation, persons producing crayfish for sale or transfer are required to register under a 1985 Order of The Diseases of Fish Act (see above).

Recently the MAFF, thc Wclsh Office and the Scottish Office have introduced further measures to protect *A. pallipes* and the freshwater environment by 'The Prohibition of Keeping of Live Fish (Crayfish) Order 1996'. Due to GATT regulations it is not possible for Great Britain to restrict the import of crayfish from mainland Europe, but, by this Order of the Import of Live Fish Act, MAFF can determine what happens to them once they are imported. The new measures (Rogers & Holdich 1997) aim to prevent the spread of *P. leniusculus* and crayfish plague by designating no-go areas for crayfish farming for most of England and Wales and all of Scotland (Fig. 3). In no-go areas the unlicensed keeping of *P. leniusculus* will be banned, although the few crayfish farms in operation will be allowed to continue as long as they adhere to the terms of the Wildlife and Countryside Act. The regulations also ban the unlicensed keeping of all other alien crayfish species throughout Great Britain. There is an exemption from the new regulations for those keeping alien crayfish for direct human consumption. A Code of Practice (Ministry of Agriculture, Fisheries and Food 1996) has been produced for fish markets, hotels and restaurants, advising them how to keep live crayfish securely, and of their legal responsibilities. Outside of no-go areas, *P. leniusculus* can be farmed, sold by aquarist centres and kept in aquaria, but over Great Britain as a whole, only the tropical Australian crayfish, *C. quadricarinatus*, has been given an exemption from the new legislation and can be kept in closed, heated aquaria. Additional legislation, including a ban on the use of crayfish as bait and on harvesting alien crayfish without a licence, are currently being discussed.

The legislation outlined above gives national protection to *A. pallipes* although, unlike the Irish Republic, its habitat is not protected. *A. pallipes* is one of 116 plant and animal species identified as of key importance in the UK Biodiversity Action Plan Steering Group Report (Avery 1995). This is mainly because of the terms of the EC Habitats Directive and the fact that the British Government signed up to the Biodiversity Convention at the 1992 Rio Earth Summit (Holdich & Rogers 1997b). These measures have given added incentive to statutory and voluntary conservation bodies in Great Britain to make greater efforts to protect *A. pallipes* at a local level.

River catchments in England and Wales have recently been categorized on the basis of their crayfish status (Holdich & Rogers 1997a). This information should assist managers in deciding what further measures need to be taken to protect and conserve populations of *A. pallipes*, e.g. restocking native crayfish and eradication of alien crayfish. Restocking of waters previously affected by plague is feasible as long as the source of plague is no longer present. Environment Agency Regions are showing a strong interest in restocking and are writing it into their LEAPS (Local Environment Agency Plans) as are nature conservation bodies in their BAPS (Biodiversity Action Plans). In any restocking program, particular note should be taken of habitat requirements – these have been well documented for *A. pallipes* (Smith et al. 1996, Naura & Robinson 1998). Interest is also being shown in eradicating nuisance populations of alien crayfish, particularly *P. leniusculus* populations which are threatening those of *A. pallipes*. Currently there are no proven methods for eradication (see Holdich, Gydemo & Rogers this volume) and research is needed to ascertain what would be best in terms of cost, effort, and effectiveness.

Unlike most European countries there is no recent tradition of eating crayfish in Great Britain and the public is more concerned about conserving the native species for aesthetic purposes, and anglers because it is part of the diet of many fish species. Educating people about the crayfish problem through the widespread dissemination of information and identification leaflets has been important in ensuring the future survival of *A. pallipes* (National Rivers Authority 1994, Environment Agency in press).

5.3 *Northern Ireland*

Equivalent protection as outlined above operates for *A. pallipes*, and the introduction of alien crayfish into Northern Ireland is an offence under the Wildlife (Northern Ireland) Order 1985. Crayfish populations appear stable, and there are no reports of *P. leniusculus* or of crayfish plague (P. Hale personal communication).

5.4 *Irish Republic*

Austropotamobius pallipes is protected by the Wildlife Act 1976, and it cannot be interfered with, nor can its habitat, without a license. The importation of alien crayfish is prohibited under the Live Fish (restriction of import) Order 1972 of the Fisheries Acts. It is thought that the crayfish plague outbreak which occurred in 1986 was due to infected fishing equipment being brought in by foreign anglers (Reynolds 1988).

Extant crayfish populations are stable and there has been no recent evidence of large-scale mortalities. *A. pallipes* is protected in SACs (e.g. Lough Talt, in Sligo) (C. O'Keeffe pers. comm.) under European legislation (see above). Crayfish have been reintroduced to Lough Lene, a lake depopulated by crayfish plague in 1986 (Reynolds 1997a), and there are plans for further reintroductions.

The main threats to Irish crayfish are indirect, as a result of stream maintenance, pollution and recreational tourism. Water and fisheries management responsibilities and legislation require coordination to protect *A. pallipes*. It is planned to survey the species more regularly in order to monitor its status.

6 CONCLUSIONS

Alien crayfish, despite increasing legislation, continue to spread in Great Britain, e.g. their recent appearance in Scotland (Maitland 1996). Very few catchments in England and Wales are without populations of alien crayfish (Holdich & Rogers 1997a). It is to be hoped that once new legislation and public education takes effect, then the development of new populations of alien crayfish will become less frequent. However, they remain absent from Northern Ireland and the Republic of Ireland due to legislation and cooperative vigilance over many years. Although there have been no reported crayfish plague outbreaks since 1993 there is no room for complacency about this virulent disease which can wipe out a whole river's population in a matter of weeks (Alderman & Polglase 1988).

Three worrying developments in Great Britain are reports of English populations of *O. limosus*, a vector of crayfish plague and probably the most widespread alien crayfish in Europe (Laurent 1988, Momot 1988); the occurrence of *A. leptodactylus* in the lower, tidal reaches of a river (Holdich et al. 1995a); and wild populations of *P. leniusculus* in Scotland (Maitland 1996).

Despite predictions in the 1980s about the possibility of *A. pallipes* becoming extinct in Great Britain (see Holdich & Reeve 1991), it continues to thrive in many areas. However, in England and Wales at least many populations no longer exist, their place having being taken by *P. leniusculus*, or the niche having remained empty.

The conservation and management of crayfish in the fluid situation that exists in Great Britain can only be achieved by a regular program of distribution monitoring, updating of the national distribution database, a better understanding of genetic variation between populations for restocking purposes, an effective eradication program, education of the general public, a ban on the use of crayfish as bait, control of trapping of alien crayfish, and an appreciation of the value of *A. pallipes* by such trappers.

ACKNOWLEDGEMENTS

Thanks are due to the many people who have provided information on crayfish in the British Isles, particularly the more recent information from John Foster, Martin Frayling, Peter Hale, Ciaran O'Keeffe, Milton Matthews, Christopher Moriarty, Robert Rosell and Roy Wiles.

REFERENCES

Albrecht, H. 1983. Besiedlungsgeschichte und ursprhnglich holozäne Verbreitung der europischen Flusskrebse. *Spixiana* 6:61-77.

Attard, J. & Vianet, R. 1985. Variabilité génétique et morphologique de cinq populations de l'écrevisse européenne *Austropotamobius pallipes* (Lereboullet, 1858). *Can. J. Zool.* 63:2933-2939.

Alderman, D.J. 1993. Crayfish plague in Britain, the first twelve years. *Freshwater Crayfish* 9:266-272.

Alderman, D.J. 1996. Geographical spread of bacterial and fungal diseases of crustaceans. *Rev. Sci. Tech. Int. Epiz.* 15(2):603-632.

Alderman, D.J. & Polglase, J.L. 1988. Pathogens, parasites and commensals. In D.M. Holdich & R.S. Lowery (eds), *Freshwater crayfish: biology, management and exploitation*: pp. 167-212. London: Croom Helm (Chapman & Hall).

Avery, M.I. (ed.) 1995. *Biodiversity. The U.K. Steering Group Report, Vol. 2, Action Plans*. London: H.M.S.O.

Belchier, M., Edsman, L., Sheehy, M.R.J. & Shelton, P.M.J. 1998. Estimating age and growth in long-lived temperate freshwater crayfish using lipofuschin. *Freshwater Biology* 39:439-446.

Bell, A. 1920. Notes on the later Tertiary Invertebrata. *Ann. Rep. Yorkshire Phil. Soc.* 1920:1-21.

Brewis, J.M. & Bowler, K. 1982. The growth of the freshwater crayfish *Austropotamobius pallipes* in Northumbria. *Freshwater Biology* 12:187-200.

Cornish, C.J. 1902. The naturalist on the Thames. London: Seeley & Co. Ltd.

Cukerzis, J.M. 1988. *Astacus astacus* in Europe. In D.M. Holdich & R.S. Lowery (eds), *Freshwater crayfish: biology, management and exploitation*: pp. 309-340. London: Croom Helm (Chapman & Hall).

Environment Agency in press. *Freshwater Crayfish in Britain and Ireland.* Bristol: Environment Agency.

Foster, J. 1994. *Procambarus* in England. *Crayfish News* 15(4):4.

Foster, J. 1995a. Crayfish and aquarium pet trade in England. *Crayfish News* 16(4):7.

Foster, J. 1995b. Factors influencing the distribution and abundance of the crayfish *Austropotamobius pallipes* (Lereboullet) in Wales and the Marches, UK. *Freshwater Crayfish* 8:78-98.

Foster, J. 1995c. First record of *O. limosus* in England. *Crayfish News* 16(4):6.

Foster, J. & Turner, C. 1993. Toxicity of field simulated farm waste episodes to the crayfish *Austropotamobius pallipes* (Lereboullet): elevated ammonia and reduced dissolved oxygen concentrations. *Freshwater Crayfish* 9:249-258.

Fryer, G. 1993. *The freshwater Crustacea of Yorkshire*. Yorkshire Naturalists' Union & Leeds Philosophical and Literary Society.

Gledhill, T., Sutcliffe, D.W. & Williams, W.D. 1993. *British Freshwater Crustacea Malacostraca.* 2nd edition. Freshwater Biological Association Scientific Publications 52. Ambleside: Freshwater Biological Association.

Grandjean, F., Souty-Grosset, C. & Holdich, D.M. 1997a. Mitochondrial DNA variation in four British populations of the white-clawed crayfish *Austropotamobius pallipes pallipes*, implications for management. *Aquat. Living Resour.* 10:121-126.

Grandjean, F., Souty-Grosset, C., Raimond, R. & Holdich, D.M. 1997b. Geographical variation of mitochondrial DNA between European populations of the white-clawed crayfish *Austropotamobius pallipes*. *Freshwater Biology* 37:493-501.

Guan, R.Z. 1994. Burrowing behaviour of signal crayfish, *Pacifastacus leniusculus* (Dana), in the River Great Ouse, England. *Freshwater Forum* 4:155-168.

Guan, R.Z. & Wiles, P.R. 1996. Growth, density and biomass of crayfish, *Pacifastacus leniusculus*, in a British lowland river. *Aquat. Living Resour*. 9:265-272.

Guan, R.Z. & Wiles, P.R. 1997a. Ecological impact of introduced crayfish on benthic fishes in a British lowland river. *Conservation Biology* 11(3):641-647.

Guan, R.Z. & Wiles, P.R. 1997b. The home range of signal crayfish in a British lowland river. *Freshwater Forum* 8:45-54.

Harris, R.R. & Young, H.J. 1996. *Distribution, densities and population characteristics of signal crayfish, Pacifastacus leniusculus (Dana), in the Gaddesby Brook, Leicestershire*. Report for the National Rivers Authority.

Holdich, D.M. 1996. *Austropotamobius pallipes*. In P.J. Van Helddingen, I. Willemse & M.C.D. Speight (eds), *Background information on invertebrates of the Habitats Directive and the Bern Convention*. Part 1 – Crustacea, Coleoptera and Lepidoptera. Nature and environment 79. Strasbourg: Council of Europe Publishing.

Holdich, D.M. & Domaniewski, J.C.J. 1995. Studies on a mixed population of the crayfish *Austropotamobius pallipes* and *Pacifastacus leniusculus* in England. *Freshwater Crayfish* 10:37-45.

Holdich, D.M. & Reeve, I.D. 1991. The distribution of freshwater crayfish in the British Isles with particular reference to crayfish plague, alien introductions and water quality. *Aquatic Conservation* 1(2):139-158.

Holdich, D.M. & Rogers, W.D. 1992. Crayfish populations in the British Isles, farming, legislation, conservation and management. *Finn. Fish. Res*. 14:23-32.

Holdich, D.M. & Rogers, W.D.1997a. *Strategy for the management of white-clawed crayfish (Austropotamobius pallipes) populations in England and Wales*. R&D Project 640. Bristol: Environment Agency.

Holdich, D.M. & Rogers, W.D. 1997b. The white-clawed crayfish, *Austropotamobius pallipes*, in Great Britain and Ireland with particular reference to its conservation in Great Britain. *Bull. Fr. Pêche Piscic*. 347(4):597-616.

Holdich, D.M., Harlioglu, M.M. & Firkins, I. 1997. Salinity adaptations of crayfish in British waters with particular reference to *Austropotamobius pallipes*, *Astacus leptodactylus* and *Pacifastacus leniusculus*. *Est. Cstl. Sh. Sci*. 44:147-154.

Holdich, D.M., Rogers, W.D. & Reader, J.P. 1995a. *Crayfish conservation*. Project Record 378/10/N&Y. Bristol: National Rivers Authority.

Holdich, D.M., Rogers, W.D., Reader, J.P. & Harlioglu, M.M. 1995b. Interactions between three species of freshwater crayfish (*Austropotamobius pallipes*, *Astacus leptodactylus* and *Pacifastacus leniusculus*). *Freshwater Crayfish* 10:46-56.

Huner, J.V. 1988. *Procambarus* in North America and elsewhere. In D.M. Holdich & R.S. Lowery (eds), *Freshwater crayfish: biology, management and exploitation*: pp. 239-261. London: Croom Helm (Chapman & Hall).

Huxley, T.H. 1879. *The crayfish. An introduction to the study of zoology*. London: Keegan Paul.

Karaman, M.S. 1962. Ein Beitrag zur Systematic der Astacidae (Decapoda). *Crustaceana* 3:173-191.

Köksal, G. 1988. *Astacus leptodactylus* in Europe. In D.M. Holdich & R.S. Lowery (eds), *Freshwater crayfish: biology, management and exploitation*: pp. 365-400. London: Croom Helm (Chapman & Hall).

Laurent, P.J. 1988. *Austropotamobius pallipes* and *A. torrentium*, with observations on their interactions with other species in Europe. In D.M. Holdich & R.S. Lowery (eds), *Freshwater crayfish: biology, management and exploitation*: pp. 341-364. London: Croom Helm (Chapman & Hall).

Lilley, J.H., Cerenius, L. & Söderhäll, K. 1997. RAPD evidence for the origin of crayfish plague outbreaks in Britain. *Aquaculture* 157:181-185.

Lowery, R.S. 1988. Growth, moulting and reproduction. In D.M. Holdich & R.S. Lowery (eds),

Freshwater crayfish: biology, management and exploitation: pp. 83-113. London: Croom Helm (Chapman & Hall).

Lowery, R.S. & Holdich, D.M. 1988. *Pacifastacus leniusculus* in North America and Europe, with details of the distribution of introduced and native crayfish species in Europe. In D.M. Holdich & R.S. Lowery (eds), *Freshwater crayfish: biology, management and exploitation*: pp. 283-308. London: Croom Helm (Chapman & Hall).

Lucey, J. & McGarrigle, M.L. 1987. The distribution of the crayfish *Austropotamobius pallipes* (Lereboullet) in Ireland. *Ir. Fish. Invest.* A (29):3-13.

Maitland, P.S. 1996. The North American signal crayfish, *Pacifastacus leniusculus* (Dana), established in the wild in Scotland. *Aquatic Conservation* 6:107-110.

Matthews, M.A. & Reynolds, J.D. 1992. Ecological impact of crayfish plague in Ireland. *Hydrobiologia* 234:1-6.

Ministry of Agriculture, Fisheries & Food 1996. Leaflet – Crayfish Alert. Fish Markets, Restaurants, Hotels. London: MAFF Publications.

Momot, W.T. 1988. *Orconectes* in North America and elsewhere. In D.M. Holdich & R.S. Lowery (eds), *Freshwater crayfish: biology, management and exploitation*: pp. 262-282. London: Croom Helm (Chapman & Hall).

National Rivers Authority 1994. *A guide to identifying freshwater crayfish in Britain and Ireland.* Bristol: National Rivers Authority.

Naura, M. & Robinson, M. 1998. Principles of using river habitat survey to predict aquatic species: an example applied to the white-clawed crayfish *Austropotamobius pallipes*. *Aquatic Conserv.* 8:515-527.

Reynolds, J.D. 1979. Crayfish ecology in Ireland. *Freshwater Crayfish* 4:215-220.

Reynolds, J.D. 1982. Notes on the distribution of freshwater crayfish. *Bull. Ir. biogeogr. Soc.* 6:18-24.

Reynolds, J.D. 1988. Crayfish extinctions and crayfish plague in Ireland. *Biological Conservation* 45:279-285.

Reynolds, J.D. 1997a. Successful reintroduction of crayfish to Irish lake. *Crayfish News* 19 (*2*):4-5.

Reynolds, J.D. 1997b. The present status of freshwater crayfish in Ireland. *Bull. Fr. Pêche Piscic.* 347(4):693-700.

Rogers, W.D. & Holdich, D.M. 1995a. Conservation and management of British crayfish populations. *Freshwater Crayfish* 10:92-97.

Rogers, W.D. & Holdich, D.M. 1995b. Crayfish production in Britain. *Freshwater Crayfish* 10:583-596.

Rogers, W.D. & Holdich, D.M. 1997. New legislation to conserve the native crayfish in Britain – will it work? *Freshwater Crayfish* 11:619-626.

Santucci, F., Iaconelli, M., Andreani, P., Cianchi, R., Nascetti, G. & Bullini, L. 1997. Allozyme diversity of European freshwater crayfish of the genus *Austropotamobius*. *Bull. Fr. Pêche Piscic.* 347(4):663-676.

Smith, G.R.T., Learner, M.A., Slater, F.M. & Foster, J. 1996. Habitat features important for the conservation of the native crayfish *Austropotamobius pallipes* in Britain. *Biological Conservation* 75:239-246.

Souty-Grosset, C., Grandjean, F., Raimond, R., Frelon, M., Debenest, C. & Bramard, M. 1997. Conservation genetics of the white-clawed crayfish *Austropotamobius pallipes*: the usefulness of the mitrochondrial DNA marker. *Bull. Fr. Pêche Piscic.* 347(4):677-692.

Stanbury, N.K. 1997. *Feasibility study for the possible reintroduction/introduction of the white-clawed crayfish Austropotamobius pallipes into Sussex river systems*. Report for the Environment Agency. Winchester: Ecological Planning & Research.

Thomas, W.J. 1992. A note on the crayfish of Loch Croispol. *Glasgow Naturalist* 22:107-109.

Thomas, W.J. & Ingle, R. 1971. The nomenclature, bionomics and distribution of the crayfish, *Austropotamobius pallipes* (Lereboullet) (Crustacea, Astacidae) in British waters. *Essex Naturalist* 32:349-360.

The American spiny-cheek crayfish, *Orconectes limosus*, in the fauna of Belarus

ANATOLY V. ALEKHNOVICH, SERGEI E. ABLOV,
VICTOR F. KULESH, & OLEG A. PAREIKO
Institute of Zoology, National Academy of Sciences of Belarus, Minsk, Belarus

ABSTRACT

The crayfish *Orconectes limosus* was first found in the Grodno province of Belarus in June, 1997. It has immigrated from Poland and at present has become established in the rivers Shlyamitsa, Marykha, and Chernaya Gancha, and the Avgustovsky Channel. It is thought to have been in this region for about 6-10 years. Along these rivers, *Orconectes limosus* has penetrated into the Neman River draining to the Baltic Sea. A stable population of the intruder is known in the Shlyamitsa. The single trap catch per day amounted to 22 individuals. Mean sizes were 7.49 ± 0.76 cm for females and 7.42 ± 0.74 cm for males. *Orconectes limosus* has been found in the diet of the perch (*Perca fluviatilis*), burbot (*Lota lota*), and mink (*Mustela vison*).

1 INTRODUCTION

In Belarus, two native crayfish species are found, the noble crayfish *Astacus astacus* and the narrow-clawed crayfish, *A. leptodactylus*. The noble crayfish occurs preferentially in the north of Belarus, in the basin of the Baltic Sea and in rare cases only in waters of the Black Sea basin. The narrow-clawed crayfish is more numerous, being widely distributed in waters of the Black Sea basin and is frequently found in lakes and rivers of the Baltic Sea basin. In 1997, a third species has been revealed in Belarus, the spiny-cheek crayfish, *Orconectes limosus*. The goal of this study was to reveal the spread and the possible migration routes of the latter species in the province of Grodno in Belarus.

2 MATERIALS AND METHODS

The study was performed in October, 1997 in the province of Grodno in the west of Belarus. Crayfish were caught with traps and manually with hoop-nets. Traps were of the eel fyke-net type with 18 mm of mesh size. Traps were not baited. Crayfish were measured from the tip of the rostrum to the end of the telson.

The River Shlyamitsa is a tributary of the R. Marykha, which is in turn a tributary

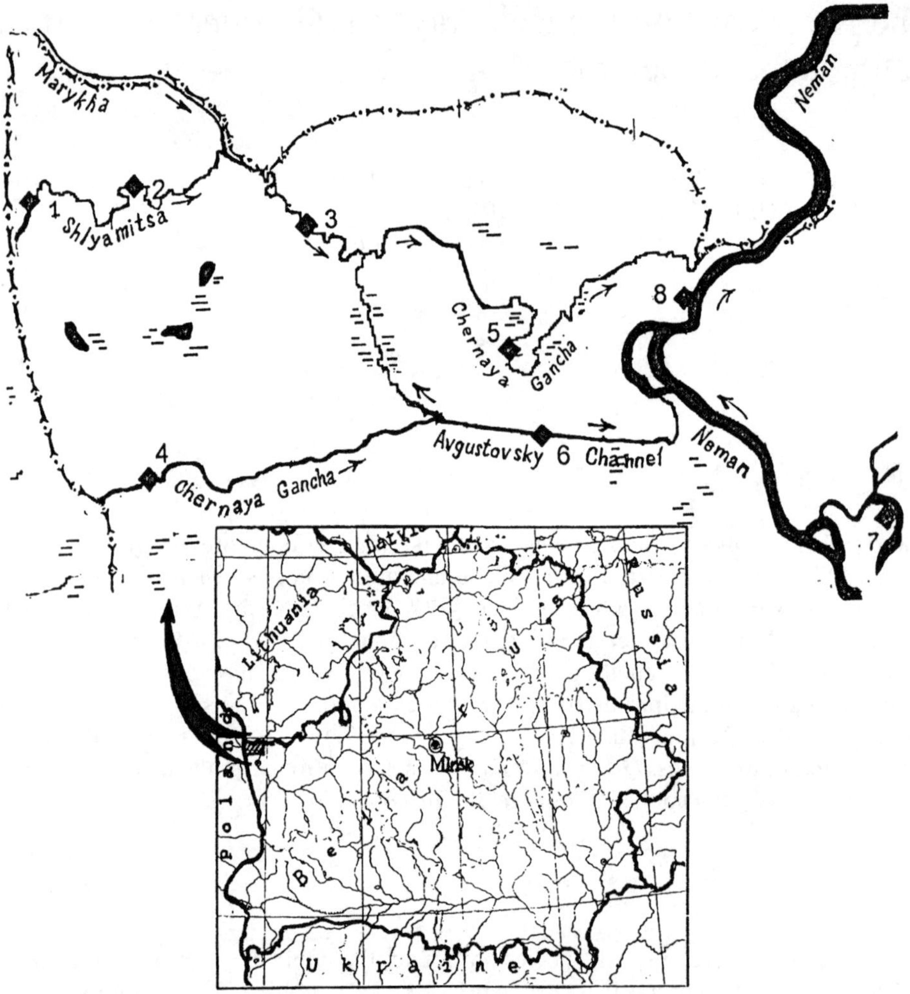

Figure 1. The distribution of the spiny-cheek crayfish, *Orconectes limosus,* in Belarus. 1-8: stations where the samples were taken.

Table 1. Hydrological characteristics of rivers in the area of distribution of *Orconectes limosus* in Belarus.

Rivers	Length* (km)	Width (m)	Depth (m)	Bottom	Flow rate (m/sec)
Shlyamitsa	8	5-10	0.2-1.0	Sand, silt	0.3-0.8
Marykha	5	5-10	0.2-1.0	Sand, silt	0.3-0.8
Chernaya Gancha	35	5-12	0.2-1.0	Sand, silt	0.1-0.5
Avgustovsky Channel	6	5-10	1.2-2.0	Sand, silt	0.1-0.3
Neman	459	110-240	0.7-3.0	Sand, silt, stones	0.7-0.8

*Within the borders of Belarus.

of the R. Chernaya Gancha. The Chernaya Gancha originates in Poland and enters into the Neman River. The Avgustovsky Channel connects the Vistula (Poland) and Neman (Belarus) rivers (Fig. 1). The Neman River is navigable and drains into the Baltic Sea. All other rivers are small and meandering. Banks are preferentially high, in some places low and swampy. The hydrological description of the rivers is given in Table 1.

3 RESULTS

The spiny-cheek crayfish was found in western Belarus not far from the borders with Poland and Lithuania in the Shlyamitsa, Marykha, and Chernaya Gancha rivers, and in the Avgustovsky Channel and the Neman River (Fig. 1).

The rivers are frozen in the middle of December, become clear in the middle of March. However, the ice-cover is unstable, the rivers becoming clear repeatedly in winter and during warm winters the solid ice-cover does not form. The climate is a temperate continental one.

The spiny-cheek crayfish has not been found in isolated lakes of the same region (lakes Endrene, Kavenya, Charne). This situation indicates that the spiny-cheek crayfish has entered the territory of Belarus recently by rivers and has so far become acclimatized in rivers only. According to some personal communications, the spiny-cheek crayfish first appeared in rivers of this region 6-10 years ago. The maximum number of spiny-cheek crayfish has been registered in the Shlyamitsa river, immediately after its outflow from Lake Shlyamy. In some places, crayfish numbered up to 5 individuals per m^2. In contrast to the noble and narrow-clawed crayfish, the spiny-cheek crayfish is very active in full daylight.

In the Shlyamitsa river, a single trap catch per day amounted to 22 individuals. In trap catches, crayfish lengths ranged from 6 to 12 cm. Mean female sizes were 7.49 ± 0.76 cm, male sizes 7.42 ± 0.74 cm (Fig. 2). The trap net mesh (18 mm) allowed crayfish less than 6 cm to escape. At manual harvesting using hoop-nets, individuals from 3 cm were caught. Average individual sizes when harvesting manually have amounted to 7.26 ± 0.8 cm in females and 6.93 ± 1.03 cm in males (Fig. 3). The size structure of spiny-cheek crayfish by different catching methods in different localities is given in the Table 2.

Due to its high numbers in the Shlyamitsa River, the spiny-cheek crayfish has become a food item for many predators. In analyzed stomachs of burbot (*Lota lota*), perch (*Perca fluviatilis*), and mink (*Mustela vison*), only crayfish 3 to 5 cm long were found. In other rivers studied, the number of estimated crayfish as catch per trap per day was approximately 10 times lower than in the Shlyamitsa River. Evidently due to its low density, it was not revealed in stomachs of fish.

October is the mating time for the spiny-cheek crayfish, when we observed mating pairs and females with spermatophores.

4 DISCUSSION

The spiny-cheek crayfish has penetrated into Belarus probably during the last 6-10

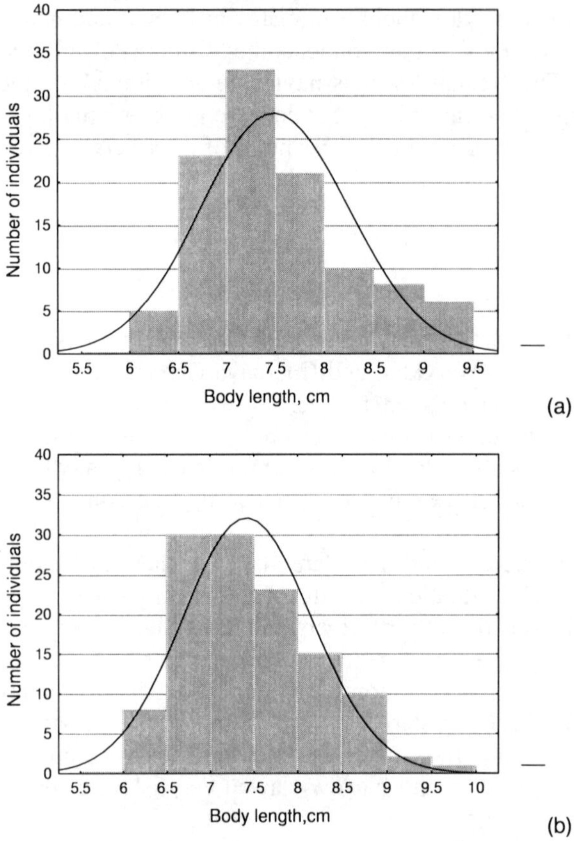

Figure 2. The size structure of females (a), and males (b) of *Orconectes limosus* in trap catches in the Shlyamitsa River.

Table 2. The size structure of spiny-cheek crayfish at different catching methods in different localities of the area of distribution in Belarus.

N of Station*	Locality	Trap catch			Hoop-net catch		
		Ind./day	Mean size		Ind./hour	Mean size	
			Male	Female		Male	Female
1	Shlyamitsa	–	–	–	150	6.6	7.2
2	Shlyamitsa	9.4	7.4	7.5	160	6.9	7.3
3	Marykha	–	–	–	8	7.8	7.6
4	Chernaya Gancha	–	–	–	20	7.3	6.5
5	Chernaya Gancha	0.5	8.6	7.1	–	–	–
6	Avgustovsky Channel	–	–	–	7	6.8	7.5
7	Neman	–	–	–	40	6.6	6.3
8	Neman	0.5	8.6	10.2	–	–	–

*The numbers of the stations follow those seen in Figure 1.

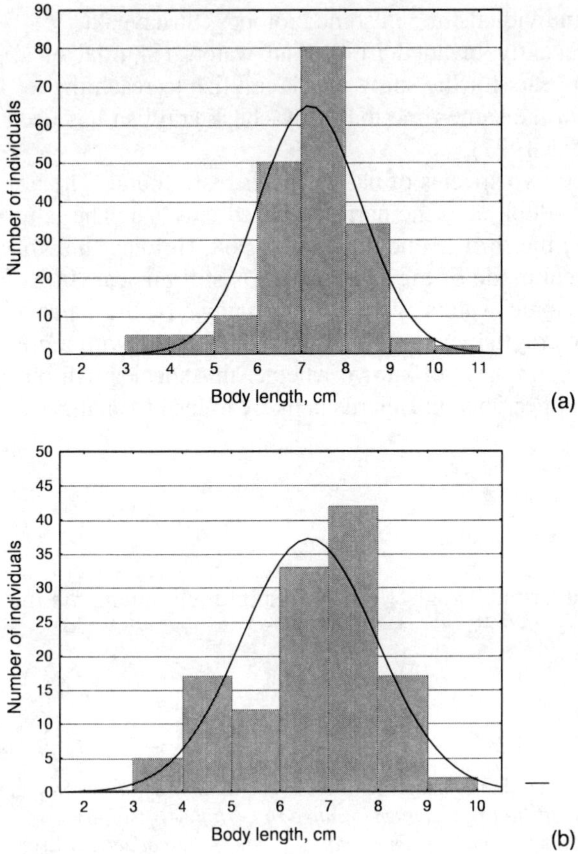

Figure 3. The size structure of females (a), and males (b) of *Orconectes limosus* at manual harvesting with hoop-nets in the Shlyamitsa River.

years from the Polish territory through the Avgustovsky Channel system. The spiny-cheek crayfish migrates along river flows and is still not found in lakes. It seems to be an important source of food for predatory fish and mammal species and according to our data is actively harvested by the local population.

In Europe, this species appeared in 1890 when 100 individuals of spiny cheek crayfish were delivered from the Delaware River (Pennsylvania, USA) to a fishing farm near the Myslya River, the eastern tributary of the Oder (within the territory of modern Poland). In 1911-1913, the spiny-cheek crayfish has acclimatized in the Loire River and was found already in 1925 in both the Seine and the Rhone, and in some lakes in the French Alps. Since then, this species has undergone its extensive expansion throughout European water systems (Boettger 1938, Momot 1988). The area of distribution is still growing. For example, in the Netherlands the spiny-cheek crayfish was registered in 1992, but it penetrated there earlier along the Loire River (Geleen 1978). At present, in that country it occupies practically all the waters. The same situation occurs in Germany, except Bavaria. In Poland, *O. limosus* is a common species now being recorded in more than two-thirds of its territory except the southeastern part and amounting to 70 individuals m^{-2} in some biotopes (Struzynski 1994). The spiny-cheek crayfish have recently invaded Lithuanian waters (Skurdal et al. 1997). Obviously it has taken 100 years for the spiny-cheek crayfish to reach the territory of Belarus. Approximately at the same time the spiny-cheek crayfish has been found in Italy as well (Gherardi et al. 1997).

In the basin of the Neman River, two species of native crayfish are found: The noble crayfish listed in the Red Data Book, and the narrow-clawed crayfish. The latter is thought to gradually outcompete the former one (Cukerzis 1968, Holdich this volume), but as far as we know the real mode of the replacement is still unclear. In Belarus, there are on the one hand some waters with a high number of the narrow-clawed and a low number of noble crayfish, and on the other hand waters with a predominance of noble crayfish. Although it is not known whether this species will outcompete the native one, its recent appearance in Belarus is likely to lead to changes in the aquatic ecosystem.

ACKNOWLEDGEMENTS

We thank Dr Rolf Gydemo and Dr Trond Taugbøl for their helpful comments on the manuscript.

REFERENCES

Boettger, C.R. 1938. Zur Ausbreitung des nordamerikanischen Flusskrebs *Cambarus affinis* Say in Europa. *Sitzungsberichte der Gesellscaft Naturforscender Freunde zu Berlin,* 8-10:307-311.

Cukerzis, J.M. 1968. Interspecific relations between *Astacus astacus* L. and *A. leptodactylus* Esch. *Ecologia Polska Ser. A*., 16(31):629-636.

Geelen, J.F. 1978. The Distribution of the crayfishes *Orconectes limosus* (Rafinesque) and *Astacus astacus* (L.) (Crustacea, Decapoda) in the Netherlands. *Zool. bijdr*. 23:4-19.

Gherardi, F., Barbaresi, S., Raddi, A. & Salvi, G. 1997. Alien crayfish in Europe: the situation of Italy. Abstract Volume. International workshop 'The introduction of alien species of crayfish in Europe. How to make the best of a bad situation?', September 24-27, Florence, Italy, pp. 4-6.

Momot, W.T. 1988. *Orconectes* in North America and elsewhere. In Holdich D.M. & R.S. Lowery (eds). *Freshwater crayfish: biology, management and exploitation*: 262-282. London: Croom Helm.

Skurdal, J., Taugbøl, T., Tuusti, J., Burba, A., Styrrishave, B., Westman, K., Söderbäck B., Edsman, L. 1997. Crayfish introductions in the Nordic and Baltic countries. Abstract Volume. International workshop 'The introduction of alien species of crayfish in Europe. How to make the best of a bad situation?', Florence, Italy, September 24-27, pp. 13-14.

Struzynski, W. 1994. Rak pregowaty rakiem przyszlosci. *Przeglad Rybacki* 5: 9-13 (in Polish).

Conclusions

A review of possible methods for controlling nuisance populations of alien crayfish

DAVID M. HOLDICH
School of Biological Sciences, The University of Nottingham, Nottingham, UK
ROLF GYDEMO
Department of Systems Ecology, Section Gotland, Stockholm University, Fàrösund, Sweden
W. DAVID ROGERS
David Rogers Associates, 9 The Moat, Castle Donington, Derby, UK

ABSTRACT

As alien crayfish continue to spread in many countries throughout the world, there is a need to find a means to control or eradicate those established populations that have become a nuisance, because of their biological and physical effects on the freshwater environment. Such methods should conform to established criteria so as not to make the situation worse. Five broad categories of control are recognized: legislative, mechanical, biological, physical and chemical. Legislative control is in operation in many countries, but is rarely effective. Mechanical methods include the use of traps, fyke and seine nets, and electro-fishing. Each can be used effectively to control nuisance populations if used intensively enough, but eradication by such methods is unlikely. Biological control includes the use of fish predators, disease and microbial insecticides. Of these, only predaceous fish have been used so far. In some cases, there is a clear inverse relation between the presence of fish predators and crayfish and introduction of such fish may lead to a reduction in crayfish numbers, but the situation is not so clear cut in others. Crayfish plague is lethal to non-North American crayfish and even to North American crayfish if they are put under stress. If more specific strains of the disease could be developed then this could be a very effective control agent. Microbial insecticides such as Bti are effective against insect pests and there might be potential in developing a strain against crayfish. Physical methods include draining ponds and lakes, diverting streams and rivers, and creating obstacles such as weirs and electrical barriers. A problem exists with controlling burrowing crayfish during drain-downs as they are known to be able to survive for long periods of time in their burrows. Chemical methods include the use of biocides, surfactants and pheromones. Organophosphate and organochlorine insecticides can be effective against crayfish, but there have been few large scale studies and there are often objections to their use because of problems with bioaccumulation and biomagnification. Baytex and Dipterex have been used effectively in recent experiments in France. Fourth generation pyrethroids are effective in killing crayfish at very low doses, e.g. deltamethrin is effective at levels allowable in European drinking water. The anthelminthic drug, ivermectin, is effective in killing fish lice on farmed salmon but has yet to be tested against crayfish. Rotenone is used as a piscicide in fisheries management, but some crayfish can withstand levels 500 times that required to kill fish, so its use for eradicating crayfish would prove very expensive. Surfactants (wetting agents) have been found to reduce oxygen consumption of crayfish, but are unlikely to prove effective in eradicating crayfish. The use of sexual attractants (pheromones) has not been used as a method to control crayfish, but they might have potential if they could be isolated. No single method is likely to be effective in controlling or eradicating nuisance crayfish populations in all situations. Multiple approaches are needed for each situation and perhaps for each species. Quick kills could be achieved with known organophosphate and pyrethroid insecticides, but in most situations authorities are likely to require more environmentally friendly mechanical, physical and biological methods to be used.

1 INTRODUCTION

The establishment of animals and plants outside their home range is a world-wide problem and one in which humans have played a major part (Holdich & Gherardi this volume). Signatories to the United Nations Convention on Biodiversity at the 1992 Earth Summit in Rio de Janeiro called on governments to prevent the introduction of, control, or eradicate those alien species which threaten ecosystems, habitats, or species (Tibbetts 1997). However, prevention, control or eradication are all extremely difficult to achieve. In fact very few alien animals or plants have ever been controlled let alone eradicated. This is in part due to that fact that many introductions are welcomed by those with commercial and sporting interests, although they may be condemned by environmentalists (Williamson 1996). An excellent example of this dichotomy of interests is the introduction of alien crayfish into various parts of the world, particularly Europe. Holdich (this volume) has outlined the reasons, why crayfish species have been introduced outside their home ranges, and the negative environmental effects that arise if they become established.

The current widespread distribution of alien crayfish in Europe that is documented in this volume is viewed with concern by many ecologists and conservationists. So, how do we make the best of this bad situation? Do we just let it take its course, and, as Taugbøl & Skurdal (this volume) predict, lose our native European crayfish to the invaders, or do we try and do something about it? One course of action is to try and develop effective methods for controlling or eradicating nuisance populations. Control of nuisance populations may be a possibility using available methods, but is it enough? A controlled population is still a threat to the environment and in the long term is likely to be a more expensive option than eradication – if methods could be found.

Any attempt to eradicate nuisance populations must take into account the impact that such actions might have on the freshwater ecosystem as other species may also be eliminated or new ones introduced. This might be in conflict with conservation interests and the pros and cons of doing nothing, trying to control the nuisance population (thus leaving the threat), or trying to eradicate it (and possibly adversely affecting other organisms) must be taken into account.

In order not to make the situation worse, the following criteria, based on the precautionary principle, should be considered in any control or eradication program:

- The method should be safe for the environment and cause as little damage to other biota as possible,
- The method should have a good chance of achieving its goal,
- The method should be inexpensive,
- The method should involve little labor,
- The method should not cause danger to humans, e.g. by contaminating ground water,
- The method should be justifiable to the public.

A survey of relevant literature has revealed very few references directly relating to the control or eradication of nuisance crayfish populations. Those found (see below) have either used predatory fish, intensive trapping or biocides, although there are other options. As far as we can tell, there have been no legal attempts in Europe to control crayfish on a large-scale using predatory fish or biocides, most of the experiments being confined to small ponds or laboratory tanks.

Based on control methods used for insects, there are five broad categories that can be considered (modified from Fernald & Shepard 1955):

1. Legislative – local and national regulations,
2. Mechanical – control by hand or with traps,
3. Biological – parasites, disease and predators,
4. Physical – temperature, environmental manipulation,
5. Chemical – biocides, attractants, repelling agents.

Each of these will be examined separately below. In addition, a number of methods that are theoretically possible, but not, as yet, scientifically tested or evaluated, are discussed.

2 CONTROL CATEGORIES

2.1 *Legislative*

Many European countries have both local and national laws that aim to prevent the spread of alien crayfish (Westman et al. 1990). A complete review of these is beyond the scope of this paper. However, no matter what legislation is in place, illegal introductions are very difficult to police, often being made by anglers or those with an interest in harvesting crayfish for profit. Accidental introductions are also common, as they are with fish (Welcomme 1988).

Some European countries ban the import of live alien crayfish, e.g. Ireland and Norway, but other countries say they cannot do this because of GATT and Single European Market regulations, e.g. England. Germany (European Court of Justice 1994) and Sweden (L. Edsman personal communication), who have tried to stop the importation of live crayfish have also fallen foul of such regulations recently. Countries who had no trade in live crayfish prior to GATT (e.g. Ireland) are able to ban live imports, but those who did not cannot do it retrospectively (A. Scott personal communication). As a consequence of not being able to prevent their import, Great Britain probably has among the most stringent laws regarding what happens to alien crayfish once they are imported (Holdich & Rogers 1997, Holdich, Rogers & Reynolds this volume). For example, the three commonest alien species (*Pacifastacus leniusculus*, *Astacus leptodactylus* and *Astacus astacus*) have been officially declared as pests and much of Britain has been declared a no-go area for the keeping of *P. leniusculus*, and the keeping of all other alien species is banned over the whole of Britain (with the exception of the tropical species, *Cherax quadricarinatus*). Although legislation is in place in Britain, people frequently ignore it (or are unaware of it) and no prosecutions have ever been made against people allowing alien crayfish to escape into the wild. The current policy of free trade in Europe is in conflict with some of the conservation measures that countries have tried to introduce. Some European countries have policies preventing the import of live crayfish from outside Europe, although this is sometimes circumvented by the aquarium trade.

2.2 *Mechanical*

2.2.1 *Trapping*

A considerable body of information exists on the effects of harvesting on the structure of crayfish populations (e.g. Huner et al. 1991, Momot, 1992, 1993, Skurdal & Taugbøl 1994). However, these mostly focus on ways of sustaining production, not reducing it, although there are a number of instances where sustained trapping has been shown to reduce crayfish populations.

In Sweden, Svärdson (1948) found that intensive trapping of an overpopulated crayfish lake reduced the number of *Astacus astacus* by 50% over three years, although the average size of the crayfish increased.

In the USA, Bills & Marking (1988) trapped a nuisance population of *Orconectes rusticus* continuously for six weeks with catches declining from 6500 crayfish to 206 over the trial period. Males dominated the catches. Small-sized crayfish did not enter the traps. They suggested that trapping on its own might suppress the crayfish population but would be unlikely to control it.

In France, Roqueplo et al. (1995) used standard cylindrical traps and also a trap shaped like a tambourine with funnels in the sides, to trap a nuisance population of *Procambarus clarkii* from a pond. They used different mesh sizes and found that they could catch juveniles as well as adults, the color of the trap being important – black catching more crayfish than white. The tambourine-type trap was more effective than the cylindrical one. They found that regular use of traps substantially reduced the population, but did not eliminate it.

In England, a number of attempts have been made to control nuisance crayfish populations. First, a population of *Astacus leptodactylus* that was interfering with angling in a lake was heavily trapped over five months using Swedish trappies and fyke nets (Holdich et al. 1995, Rogers 1996). This reduced the adult population considerably and solved the problem in the short-term. However, at one time 70 fyke nets were being used that involved considerable cost and manpower.

Second, a rapidly expanding, crayfish plague-free, population of *Pacifastacus leniusculus*, which was on the point of eliminating, probably through competitive exclusion a population of *Austropotamobius pallipes* in the same lake, was heavily trapped and seine-netted in February of two succeeding years and all trapped ovigerous females removed (Holdich & Domaniewski 1995, Holdich et al. 1995). This reduced recruitment considerably and, if this had been coupled with the removal of other trappable individuals throughout the year, could have had a significant impact on the population, especially in the face of predation pressure from the fish population. In this population, it proved very difficult to trap *P. leniusculus* below 40 mm carapace length. This has been the experience of other workers in Britain (e.g. Guan & Wiles 1996), even when the mesh size was significantly reduced. This is probably due to the cannibalistic tendencies of this species and small crayfish may try and avoid larger crayfish because of this.

Third, a population of *P. leniusculus* was heavily trapped from carp ponds using Swedish trappies and the population was estimated to have been reduced from 4000 to 1500 over the equivalent of 900 trap nights (Rogers et al.1997). If regular trapping of large individuals had been continued, then fish predation on the smaller individuals may eventually have eliminated the population.

In the cases above, the problems were only solved in the short term and the populations were not eliminated. It is likely that within a couple of breeding seasons they will have returned to their former levels. A considerable amount of time and effort was employed to achieve the results.

Some of the examples above illustrate that improvements in trap design should be considered. Westman et al. (1979) stated that crayfish are very skilful at escaping from standard traps and as a consequence developed the 'Evo-trap,' which has narrow slit-like apertures, that make it much more difficult for crayfish to escape. Westman (1991) reviewed the wide variety of traps used in Finland. Fjälling (1995) assessed the efficiency of seven commercial crayfish traps used in Sweden and found that crayfish catch differed significantly with trap design. He also found that minor modifications of traps could improve their performance significantly. He designed a new collapsible trap, consisting of terylene webbing stretched over a galvanized steel ring and a buoyant plastic ring with three stiff plastic entrance funnels, which he found to be more effective than those already in use, although more expensive to make. Holdich (in Harlioglu 1996) has shown that the Swedish trappy commonly used in Britain is not very effective at retaining crayfish once captured and that traps needed to be emptied frequently.

It is well known that the number of crayfish caught in a trap is dependent on the time of day and season (Laurent 1988). In addition, juveniles and ovigerous females may be trap shy and more effectively caught with seine and fyke nets (Rogers 1996).

There is very little evidence for crayfish stocks being over fished by man (Lodge & Hill 1994). In fact, heavy predation has sometimes been found to increase production and yield in a number of populations (Morgan & Momot 1988, Svärdson et al. 1991, Westman 1991, Momot 1993). For example, populations of *Orconectes virilis* can withstand exploitation rates of 60% of 1+ and older individuals (Momot 1993). This phenomenon acts through reduced competition between YOYs (young of year) and older males. Dominance hierarchies exist in some species, e.g. *P. leniusculus*, with large males dominating females and juveniles. Removing such males by trapping reduces the pressure on the smaller individuals allowing them to grow, often leading to a large population of individuals whose growth is stunted due to competition for resources (Skurdal & Qvenild 1986). Ibbotson et al. (1997) in a study of the *P. leniusculus* in the River Thame in England suggested that intensive trapping of one section of river in the middle of a population acted like a drain on the larger individuals in the population; the areas alongside the section being trapped were thus depleted of larger crayfish and this enhanced population expansion in these adjacent sections by reducing competition for the younger year classes.

As with fish predation (see below), continued pressure through trapping may affect a greater number of individuals in a population and may be more important in controlling population size and production than short-term intensive trapping. However, it is probably impossible to trap out a crayfish population as it is so difficult to trap juveniles. Sustained trapping can be used to reduce the population to a low level, but this must be done regularly over a number of years. Creaming off the large individuals all the time, however, can lead to many small, untrappable individuals. Recruitment can be reduced by removing ovigerous females, but due to feedback mechanisms, crayfish, as with most animals, would probably respond to low numbers in the population by producing more eggs and reaching maturity earlier!

In order to enhance trapping specialized baits can be used. If these are rich in fish oil, then they do often increase numbers in traps (Huner & Paret 1995). However, artificial baits can be expensive to produce. In Louisiana, the cost of bait figures high in the annual operating costs in crayfish culture (Huner et al. 1994). Cange et al. (1986) evaluated 18 artificial crayfish baits against gizzard shad (*Dorosoma cepedianum*), but even the most attractive baits (those containing catfish meal and oil) only caught slightly more than the control bait. In non-commercial operations though the cheapest bait available tends to be used, the type depending on the choice of the trapper, e.g. punctured cans of cat meat, herring, coarse fish, liver, mammal carcasses. Among coarse fish, Westman (1991) listed roach, bream, bleak and white bream as the best crayfish bait, although perch, pike, zander and burbot flesh were not considered good. Spores of the crayfish plague fungus may be transferred by a variety of means, including attachment to fish scales (Alderman et al. 1987), so by using marine fish this possibility would be prevented.

Crayfish are often caught during electrofishing exercises to study fish populations, but the method has also been used specifically for catching crayfish. Huner (1988) reported that special boats with a bow-mounted trawl and a pulsed, direct current electrofishing unit are sometimes used to harvest soft-shelled crayfish in Louisiana, and are particularly effective amongst vegetation. Westman et al. (1978) found that all sizes of crayfish could be trapped by electro-fishing but only in shallow, clear water, when the weather was calm. They recommended using non-pulsating, direct current equipment rather than pulsating, battery-fed equipment as used for fish capture. Laurent (1988) also stated that electrofishing was a good method for catching crayfish but, due to their predominantly nocturnal activity, results varied depending on the time of the day. Westman et al. (1978) suggested that night-time electrofishing can be more productive than during the day-time. Eversole & Foltz (1995) carried out a study of habitat relationships of two species of crayfish in a mountain stream. They used weighted nets to form cells within the river and then successfully electrofished the crayfish out of each cell.

2.3 *Biological*

Although crayfish have many predators (Hogger 1988, Westman 1991, Lodge & Hill 1994), only fish are worth considering as control agents. Likewise, although crayfish may suffer from a number of diseases (Alderman & Polglase 1988), only the crayfish plague fungus, *Aphanomyces astaci*, is of any significance. However, the possible development of microbial insecticides might be worthy of investigation.

2.3.1 *Fish predators*

Many studies suggest that fish predators reduce crayfish populations and that they may consume a large fraction of crayfish production (Westman 1991, Rabeni 1992, Mather & Stein 1993, Roell & Orth 1993, Lodge & Hill 1994, Blake & Hart 1995a). A significant inverse relationship may exist between densities of predaceous fish and crayfish. Svärdson (1972) found that crayfish were less abundant in Swedish lakes with many eels and vice versa. When eels were excluded by dams, then the crayfish populations increased in size. Fürst (1977) suggested that predation by eels and perch were the most important factors in limiting the establishment of newly introduced crayfish in 44 Swedish waters. While European eels have an impact on native crayfish species, it is not known if they would have the same impact on American crayfish, especially ones which burrow. However, there is anecdotal evidence that some signal crayfish farmers in Great Britain have lost much of their stock to eel predation. Westman (1991) mentioned eels, burbot, perch and pike as being partial to crayfish, particularly during moulting, with perch wreaking the greatest havoc amongst the shallows. Eel and pike recently introduced into the Rumensee in Switzerland have had a marked impact on an expanding *Procambarus clarkii* population (A. Frutiger personal communication). Hickley et al. (1994) found that *P. clarkii* introduced into Lake Naivasha, Kenya became the principle food item of another introduced species, largemouth black bass. Similarly, Elvira et al. (1996) found that *P. clarkii* introduced into Spain, became the main prey item of pike, another introduced species. The impact of fish predators on crayfish can be shown if the fish are removed, e.g. an experiment to remove fish using rotenone in Finnish lakes resulted in a dramatic increase of crayfish numbers (Westman, 1991). Under experimental conditions crayfish, densities have been also found to reduce in proportion to the density of the fish (Rickett 1974, Rach & Bills 1989). However, such experiments were relatively short-term.

It is clear from Table 1 that fish predation can have an impact on crayfish populations. However, this is not always the case. For example, very little correlation was found between the presence of largemouth bass and yellow perch and crayfish abundance in 21 lakes in northern Wisconsin (Lodge & Hill 1994). Gowing & Momot (1979) found that brook trout at different densities had little effect on recruitment of *Orconectes virilis* in experimental lakes in Canada. Holdich & Domaniewski (1995) recorded an increasing density of signal crayfish, *P. leniusculus*, with time in an English lake heavily stocked with brown and rainbow trout, perch, and carp.

Cover has been shown to be an important factor in crayfish survival, especially in the young stages when moulting is more frequent (Capelli & Magnuson 1983, Lodge & Hill 1994). Blake & Hart (1993) found under both experimental and field conditions that predation by perch on juvenile *P. leniusculus* was linked to the substrate type, i.e. survival was better the greater the number of hides. Saiki & Tash (1979) found that as weed cover was reduced, then there was increasing predation by largemouth bass on *O. virilis*. Even in the absence of fish predators, cover is important for young crayfish as most species are cannibalistic (Holdich et al. 1995). Reduction in cover, therefore, may increase the chances of crayfish being eaten by predators.

Table 1. Effect of fish predators on adult (A) and juvenile (J) crayfish population size (updated from Lodge & Hill 1994).

Crayfish stage species	Fish	Predator	Type	Impact study	Authors
Astacus astacus	A	Eel	C	+	Svärdson (1972)
Orconectes rusticus *O. sanborni*	A	Smallmouth and rock bass	C	+	Mather & Stein (1993)
Orconectes rusticus *O. propinquus* *O. virilis*	A	Largemouth bass and yellow perch	C	+	Lodge & Hill (1994)
Orconectes nais	A	Largemouth bass	E	+	Rickett (1974)
Orconectes virilis	A	Largemouth bass	E	+	Saiki & Tash (1979)
Orconectes immunis	A	Largemouth bass	E	+	Rach & Bills (1989)
Orconectes virilis	A	Brook trout	E	–	Gowing & Momot (1979)
Pacifastacus leniusculus	J	Perch	C&E	+	Blake & Hart (1995a)

E = experimental and C = field studies. – = no impact; + = impact.

However, despite these observations, Lodge & Hill (1994) have shown that invading *Orconectes rusticus* populations, even when they have eliminated the macrophytes, can still survive at high densities.

Crayfish are known to alter their behaviour in the presence of fish predators by increasing utilisation of shelter and decreasing their post-dusk peak in activity (Hamrin 1987, Blake & Hart 1995a, b). This may have some implication for increasing fish numbers in order to control crayfish populations.

Although the results of the majority of studies seem to indicate that fish predators reduce crayfish populations, the potential exists, as with trapping (see above) for such predation to increase crayfish production (Rabeni 1992). Fish predation may have more subtle sublethal effects which, in the long-term, reduce growth, reproduction and survival. Continued predation pressure may affect a greater number of individuals in a population than direct predation and be more important in controlling population size and production (Lodge & Hill 1994). Very few attempts have been made to control crustaceans with fish, although suggestions have been made that the deep water sculpin, *Myoxocephalus thompsoni*, might be used as a form of biological management of the opossum shrimp, *Mysis relicta*, which has proved an undesirable introduction in some North American lakes and reservoirs (Martinez & Bergersen 1989).

2.3.2 *Diseases*

Crayfish suffer from a number of diseases, mainly caused by bacterial, fungal, protistan and helminth infections (Alderman & Polglase 1988, Edgerton et al. 1995). Thune et al. (1991) also found that some large-scale mortality amongst farmed *P. clarkii* were caused by viral infections. In addition, the enigmatic protistan parasite, *Psorospermium haeckeli* Hilgendorf (see Ragan et al. 1996) may prove to be a threat when more is known about its biology (see Vogt this volume).

Most crayfish diseases are sublethal but one, crayfish plague, caused by the oomycete fungus, *Aphanomyces astaci*, has a dramatic effect and is usually 100% lethal to susceptible species (Smith & Söderhäll 1986). All non-North American crayfish tested so far have proved to be susceptible to crayfish plague (Unestam 1975), and the disease has had a disastrous effect on European crayfish populations since the 1860s (Alderman 1996). Crayfish plague is endemic in North American crayfish, where it occurs as a chronic infection, although they are relatively immune to its effects unless put under stress (Svärdson et al. 1991). As a vector gets older, it presumably carries more of the chronic infection and the immune system is constantly challenged. Additional environmental stressors, such as changes in water temperature and quality, sudden changes in light levels and increased intrapopulational competition (as occurs under rearing conditions) may result in the disease getting the upper hand and death occurring. A chronic infection of crayfish plague may challenge the immune system to such an extent that the crayfish is unable to cope with any additional parasitic load such as *P. haeckeli* which, on its own, may not prove lethal (Thörnqvist & Söderhäll 1993). The same may prove true with another protistan parasite, *Thelohania contejeani* (Henneguy) (see Alderman & Polglase 1988). Indeed, Söderhäll (1989) has suggested that *P. leniusculus* is not suitable for aquacultural purposes because of these possible problems. There are a number of recorded cases of mass mortalities of wild and farmed *P. leniusculus* in Sweden, which have been attributed to stress-induced plus crayfish plague effects. However, not all populations of North American crayfish harbour the crayfish plague fungus. This has led to some mixed species populations becoming established in European waters, at least in the short-term (Westman et al. 1993, Holdich & Domaniewski 1995, Söderbäck 1995).

Crayfish plague could in theory be used to eliminate nuisance populations of susceptible species, e.g. *Astacus leptodactylus*, although this has never been tried. One danger is that the disease could be spread to other populations of susceptible crayfish.

Genetically altering the crayfish plague fungus, so that it can overcome the defense system of North American crayfish, might be a possibility if funds were available (K. Söderhäll personal communication), but at the present time very few laboratories have the knowhow and facilities to do this. Again, this might present an increased danger to native populations as the spores of the crayfish plague fungus are easily transported in water and on damp equipment (Alderman & Polglase 1988). The plague outbreak in central Ireland in the 1980s was thought to have been caused by contaminated fishing equipment belonging to foreign anglers (Reynolds 1988). Also, predators of crayfish such as birds and otters could in theory move infected crayfish from one water course to another. Although this method of transfer is poorly documented, there is anecdotal evidence to indicate that it happens.

2.3.3 *Microbial insecticides*

Varieties of the, bacterium, *Bacillus thuringiensis*, such as *israeliensis* Berliner (H-14) have been developed as natural insecticides to kill insect pests (Pedigo 1989). Crystals of an endotoxin are produced on sporulation and these are dissolved after ingestion and damage the epithelial gut cells. They work by stopping insects feeding so that they starve to death. No crayfish-specific strain has yet been developed, but if it were, it might be a new method for eradicating crayfish. However, it would need to be specific to nuisance crayfish such as *P. leniusculus* or it might also affect the native crayfish species. Jarboe (1988) records a 96-h LC_{50} of 103 mg /l for *P. clarkii* exposed to *B. thuringiensis* var. *israeliensis*, which is used as an insecticide in rice fields. Eversole et al. (1995) gave *B. thuringiensis* a very low toxicity rank for crayfish.

2.4 *Physical*

Physical methods involve draining down a lake/pond, diverting a stream or river, or creating an obstacle such as a weir. However, we can find very little in the literature relating to their actual use in controlling crayfish populations. When a natural event such as a drought occurs, this can be effective in eliminating a crayfish population (Holdich et al. 1995). Crayfish are very adaptable animals (McMahon 1986) and are capable of living out of water for long periods as long as there is some moisture, particularly if they are in burrows. Holdich et al. (1995) reported live *Pacifastacus leniusculus* from under rocks on the bed of a British river three months after it had dried up due to over-abstraction. There is one instance of the same species surviving for over six months, including a British winter, under an outdoor aquarium tank (P.R. Wiles personal communication).

Although burrowing is more associated with cambarid (Huner et al. 1994) and parastacid crayfish (Mills et al. 1994), astacid crayfish can also burrow, although there are only scant references to this activity in native European crayfish in the literature (e.g. Huxley 1879). There is nothing in the literature to suggest that *P. leniusculus* burrows, even in North America (although apparently it does to a minor extent in Oregon – S.D. Lewis personal communication), but when introduced to Britain it did so extensively in suitable substrata. Holdich & Reeve (1991) reported large numbers (26 m^{-2}) of burrows in the banks of a pond on a fish farm, where the owner thought that he had eliminated the crayfish by trapping, seine netting and draining down the pond. The pond was refilled soon after, but when it was drained down again a year later almost as many adult crayfish were found again, having survived the first drain down in their burrows. Guan (1994) and Harris & Young (1996) have also reported extensive damage to river banks caused by the burrowing activities of *P. leniusculus* in England. A lake in Wales was recently (1997) found to be infested with *P. leniusculus* and the banks riddled with burrows (W.D. Rogers personal observation). Healthy, berried females were found in burrows three months after the drain-down. The owner was advised to use a mechanical digger to remove the banks after draining the pond down. This was done and the soil and crayfish deposited some distance away and left to dry out, presumably suffocating the crayfish.

An alternative to using traps is to use barriers such as weirs which might act as a focus for the crayfish to congregate in. Such crayfish can then be manually removed at regular intervals and the weir reconstructed. However, hand trapping is unlikely to be an effective method of controlling crayfish, except after a drain down or isolation of a river bed (see below). Manual removal of crayfish from burrows may prove effective if the labour is available and the population is not too extensive.

Håstein & Gladhaug (1972) described the use of an electric barrier to prevent crayfish plague spreading up rivers. The barrier consisted of two electric fences 100 m apart with one electrical wire on the surface and the other on the bottom. A 220 V alternating current was passed through the fences. With the barrier in place, diseased crayfish migrated away from the area. Such a method could possibly used to concentrate crayfish into areas where they are easily trapped.

Nuisance populations of crayfish in rivers could in theory be isolated if the river was diverted via a channel or pipeline and the remaining water pumped out. The isolated stretch could then be thoroughly searched for crayfish and, if necessary, burrows could be chemically treated (see below) or crayfish could be removed from their burrows by hand.

Various other physical methods have been suggested, but not tested. There is some anecdotal evidence that crayfish in tanks that can feel vibrations from aerators and pumps showed higher mortality than normal. This might have a field application. The use of a curtain of bubbles might provide a means of preventing migration of crayfish in a riverine situation?

2.5 *Biocides*

No biocide has yet been found which is completely specific to crayfish. The effectiveness of those that have been tried appears to be very dependent on water parameters such as pH and temperature and on the age of the crayfish – younger stages usually being more susceptible. Organophosphate, organochlorine, and pyrethroid insecticides, as well as the anthelminthic drug, ivermectin, the piscicide, rotenone, and surfactants are discussed below in connection with their use for crayfish control/eradication.

2.5.1 *Organophosphate and organochlorine insecticides*

As there is a wealth of literature relating to the development and use of organophosphate and organochlorine insecticides (e.g. Matsumura 1985, Pedigo 1989), only information relating to crayfish is reviewed below.

Some of the earlier studies found organophosphate insecticides such as fenthion and methyl parathion to be highly toxic to crayfish (Muncy & Oliver 1963, Grigarick & Lange 1965). One of the first studies on biocides as a means of eliminating crayfish was by Chang & Lange (1967), who tested a variety of insecticides with a view to controlling *Procambarus clarkii* in California rice fields. They too found fenthion and methyl parathion to be the most effective. This study was one of very few to involve large-scale trials, in this case aerial spraying of caged crayfish in rice fields. Fenthion proved effective in that case, but only early on in the rice-growing season when plant cover was low.

Ray & Stevens (1970) tried to control nuisance populations of *Procambarus simulans* and *Orconectes nais* in channel catfish rearing ponds. They found that Baytex (active ingredient fenthion) at 100 μg/ l or greater was very effective at killing the crayfish in a few hours, if sprayed over the surface of a pond. Baytex was found not to be harmful to fish at concentrations as high as 250 μg/l, but at 100 μg/l aquatic insects, cladocerans and copepods were killed, although rotifers survived.

Ludke et al. (1971) assessed the impact of mirex (an organochlorine insecticide) on *Procambarus* spp. They recorded almost 100% mortality in juveniles exposed to 1 μg/l of mirex in six days and 71% mortality with 0.5 μg/l after four days. An important finding was that crayfish bioaccumulated the mirex to levels many thousand times that in the water. In one experiment, the hepatopancreas of four individuals which had survived exposure to mirex had concentrated it to 126,603 times that in the water! As crayfish are eaten by many predators this level of bioaccumulation is obviously important in terms of biomagnification through the food chain.

Hobbs & Hall (1975) provided a review of the effects of various insecticides on crayfish, most relating to cambarid crayfish such as *Procambarus*. Jarboe (1988) listed a wide variety of insecticides, herbicides and fungicides which might have an impact on *P. clarkii* if applied to rice fields. Indiscriminate use of insecticides against invasive *P. clarkii* in Spain led to the loss of other fauna, especially birds (Mackenzie 1986). Huner (1988) cited endrin, malathion (organophosphate) and propanil (amide) as being toxic to crayfish. Other studies in the USA involving *P. clarkii* include those of Brown & Avault (1975) on antimycin, Cheah et al. (1979, 1980) and Ekanem et al. (1983) on the effects of rice pesticides on *P. clarkii*, Gaude (1987) on the thermal effects of pesticide toxicity and Hyde et al. (1975) on the effect of mirex-impregnated bait.

Eversole et al. (1995) and Eversole & Sellers (1997) listed 97 chemical compounds and their associated 96-h LC_{50} values for nine cambarid crayfish taxa. Of these chemicals, hexachloroethane (an organophosphate insecticide) and methoxychlor (an organochlorine insecticide) are listed as being extremely toxic to crayfish with 96-h LC_{50} levels of between 1-10 μg/l. Among the insecticides, the organophosphate and organochlorine compounds were found to be 17 times more toxic than carbamates (broad spectrum insecticides derived from carbamic acid, see Pedigo 1989). The purpose of these experiments was to see what effects pesticides used in agriculture might have on farm-raised crayfish, not the use of such chemicals for eradicating the crayfish. No mention is made of their effect on other organisms, although they are likely to be equally effective on most other arthropods.

Laurent (1995) tested various organophosphate insecticides on *Orconectes limosus* from Lake Geneva in France and found that Baytex PM 40 (active ingredient fenthion) was effective at low doses. Laboratory experiments resulted in a 24-h LC_{50} of 46 μg/l and 48-h LC_{50} of 12 μg/l. Total mortality was achieved after 24 h with concentrations of 90-100 μg/l and after 48 h with 50 μg/l. These levels are much less than those required to kill finish (Anonymous 1979). He also found that the toxicity of the insecticide lasted several weeks. Field trails with Baytex applied from hand or agricultural atomizers over the surface of ponds were equally effective at levels as low as 60 μg/l (total mortality being achieved after 87 hours). Fish, frogs and mammals appeared not to be affected, but insects and other Crustacea were killed with the exception of some species of Copepoda, although many species of Rotifera survived. There appeared to be no ill effects on the molluscs. Laurent (1995) suggested that the relatively long time needed for total mortality of crayfish would limit the use of Baytex as an astacicide in standing waters, as even, at levels harmful to fish, it took several hours to kill crayfish. He also warned that studies on the fenthion residue in the food web (i.e. biomagnification) must be considered before it was widely used for crayfish eradication. Apparently, there are a number of commercial formulations of Baytex, so comparisons with the results of other workers is difficult (cf. Ray & Stevens 1970, Eversole et al. 1995). Indeed, Eversole et al. (1995) stated that the 96-h LC_{50} in their experiments with crayfish was 350 μg/l, although this is still much less than required to kill fish (e.g. carp: 96-h $LC_{50} = 2900$ μg/l, Anonymous 1979).

In France, attempts have been made to eliminate an expanding population of *P. clarkii* from a pond using the organophosphate insecticide, Dipterex (Roqueplo et al. 1995). However, the experiments only involved laboratory tests. Relatively high concentrations of Dipterex were needed to kill crayfish compared to other insecticides, i.e. in the ppt (g /l) range rather than ppm (mg /l) or ppb (μg /l) as reviewed above. However, a short exposure to Dipterex appears to be enough to kill the majority of crayfish some time later, e.g. exposing juveniles to 5 g /l for 30 minutes was sufficient to kill 50% after 20.5 hours. At 25 g /l, exposure of adults for 30 minutes resulted in 100% mortality after 27 hours. Even exposure for 1 minute at 25 g /l resulted in 33% mortality after 60 hours. The effects were found to be temperature dependent, higher temperatures resulting in a faster response time. Despite the apparent effectiveness of the Dipterex, the authors warn against its use as the active molecule persists in the environment and kills other arthropods as well. Presumably like mirex (see above) it can be bioaccumulated and biomagnified through the food chain.

Fornstrom et al. (1997) pointed out that during rain events, pesticides can be transferred into aquatic ecosystems via agricultural runoff. They found that terebufol, an organophosphate insecticide used to control corn rootworm, caused aberrant behaviour in *P. clarkii* at concentrations 1000 x's less than the expected theoretical runoff concentrations. They also found terebufol to be lethal to the crayfish with an 96 h LC_{50} value of 5.9 μg/l in aqueous experiments and LC_{50} value of 4.4 μg/l in 12-h dietary experiments, with aberrant behaviour starting at 50% and 80% of these concentrations, respectively. They concluded that, as crayfish are keystone species, then any decrease in their populations due to pesticides may cause adverse ecosystem changes. This highlights one of the criteria listed in the introduction, i.e. the method (of control/eradication) should be safe for the environment and cause as little damage to other biota as possible.

2.5.2 *Pyrethroids*

As pyrethroids have potential for control of crayfish, some background to their development is given below.

Pyrethroids are synthetic derivatives of pyrethrin I, one of several active components in pyrethrum, which can be extracted from dried flower heads of *Chrysanthemum* spp. grown in countries such as Kenya and Ecuador (Pedigo 1989). Natural pyrethrins in the form of powders are widely used as contact insecticides. The primary action is on the insect's nervous system. Pyrethrum breaks down quickly in the presence of sunlight. The first generation of pyrethrum analogues (e.g. allethrin) was developed in the 1950s and showed higher environmental stability than the natural product and low toxicity to birds and mammals. More effective second generation pyrethroids (e.g. resmethrin) were developed and these were followed by third (e.g. fenvalerate, permethrin) and fourth (e.g. cypermethrin, flucythrinate, fluvalinate, deltamethrin) generation products, with exceptionally low application rates compared to other insecticides (Pedigo 1989).

The toxicity of pyrethroids increases with increasing temperature and they degrades quickly at high environmental temperatures, rapidly becoming transformed to various isomers and binding to dissolved substances in water and soil (Matsumara 1985, Muir et al. 1985, Landrum et al. 1987, National Research Council of Canada 1986, Smith & Stratton 1986, Anderson 1989, Haya 1989, Day & Maguire 1990, Day 1991). Ecosystem recovery is fairly rapid with the toxic effect of pyrethroids lasting from days to months with all major groups recovering within a year (Kaushik et al. 1985, Gydemo 1985, Smith & Stratton 1986).

Pyrethroids are highly toxic to fish, particularly cold-water species (Haya 1989), aquatic insects (Anderson 1989) and crustaceans, although soft-bodies invertebrates such as molluscs are generally unaffected (Anderson 1982, Jolly et al. 1978). During acute and/or chronic exposure, pyrethroids can be bioaccumulated by individual organisms but levels rapidly return to normal after exposure ceases (MacLeese et al. 1980, Ohkawa et al. 1980, Anderson 1982, Spehar et al. 1983, Smith & Stratton 1986). Mammals are more sensitive to pyrethoids than birds, but LD_{50} values are > 1000-fold higher for aquatic insects and crustaceans. Most crustaceans are extremely sensitive to synthetic pyrethroids with toxicity values close to or less than 1 μg/l. For lobsters, deltamethrin was most toxic, than cypermethrin, fenvalerate and permethrin (Zitko et al 1979, McLeese et al. 1980).

Jolly et al. (1977, 1978) found that body size influenced the susceptibility of *Procambarus clarkii* to permethrin, while Thurston et al. (1985) found permethrin to be highly effective against *Orconectes immunis* with a 96-h LC_{50} of < 1.2 μg/l compared to > 89 μg/l for the organochlorine insecticide endrin. Jarboe & Romaire (1991) tested permethrin against *P. clarkii* and found that the 96-h LC_{50} ranged between 0.44 and 0.81 μg/l depending on the size of the crayfish, smaller individuals being less tolerant. Jarboe & Romaire (1995) also tested permethrin against procambarid crayfish in rice ponds and found it to be acutely toxic at concentrations ranging from 1.0 to 3.0 μg/l.

Bills & Marking (1988) tested 19 chemicals on the invasive crayfish, *Orconectes rusticus*, in the USA and found that a synthetic pyrethroid, Baythroid, was by far the most effective, a complete kill being recorded with 0.05 μg/l in laboratory tests. They found it to be fairly selective for crayfish, although they warned that it would be an expensive method of eliminating crayfish as much higher concentrations (25 μg/l) were needed for a complete kill in field trials.

From field analyses of crayfish kills and experiments, Gydemo (1995) reported that deltamethrin has the capacity to kill *Astacus astacus* at concentrations as low as 0.1 μg/l, the level allowable in drinking water by the European Union.

Cypermethrin, used in agricultural spraying and also in sheep dips, has the capacity to eliminate crayfish if it accidentally gets in to a waterbody, as happened in English river recently (J. Foster personal communication).

Eversole et al. (1995) and Eversole & Sellers (1997) listed permethrin as being extremely toxic to cambarid crayfish with 96-h LC_{50} levels of between 1-10 μg/l.

2.5.3 *Ivermectin*

Ivermectin has not been tested on crayfish to our knowledge, but as it has proved effective against other crustaceans, it is reviewed below.

Ivermectin is a synthetic derivative of abamectin, a natural fermentation product of the actinomycete fungus, *Streptomyces avermitilis*. It is used as an anthelminthic drug to kill parasites in cattle, pigs, horses and humans (Campbell 1989). It might have some potential for eradicating crayfish as it is also effective against fish lice on farmed salmon (Palmer et al. 1987). Ivermectin is used by cattle farmers and it is possible that the chemical enters the aquatic environment via their urine (or direct run-off from pour-on applications) when cattle drink from water courses. It may also be present in cattle manure spread and ploughed into land as fertiliser; however, due to its low water solubility and tight soil binding it is unlikely to be translocated very far

(Halley et al. 1989). Halley et al. (1989) have reviewed the environmental effects of ivermectin. They state that ivermectin quickly becomes immobilised by suspended soil particles in water, thus reducing its impact on freshwater organisms. However, they quote laboratory tests, which show that daphniid crustaceans (48-h LC_{50} 0.025 µg/l) and rainbow trout (96-h LC_{50} 3.0 µg) are highly sensitive to low doses of ivermectin.

Ivermectin can now be used under licence on Scottish salmon farms as long as they are not within 3 km of shellfish operations (Scottish Fish Farmer 08/08/96). The Scottish Environmental Protection Agency (SEPA) have stated that ivermectin is not a problem as a food safety or environmental issue. However, others take a different view and workers at the University of East Anglia consider that its use should be banned until its environmental effects are better understood, as even very small amounts can affect crustaceans and worms important in marine food webs (Times 23/02/98). Ivermectin was not licensed or approved for aquaculture in any country in the world until SEPA sanctioned its use in 1996, although it has been widely used in the Republic of Ireland in recent years. Ivermectin is usually applied by injection, by drenching, in feed or as a pour-on. The active ingredient is absorbed into the blood of the animal under treatment and is ingested by the parasite which it then paralyses. It is claimed by some that ivermectin is very persistent in the environment. There are also very narrow safety margins between therapeutic levels and toxic levels in salmon. Irish studies have shown that it takes approximately 120 days at 10°C for all traces of ivermectin to be excreted from salmon flesh. Until more is known about the environmental effects of ivermectin, it is unlikely that permission would be given to add it directly to a water course to kill crayfish.

2.5.4 *Rotenone*

Rotenone is a toxin associated with certain leguminous plants, e.g. *Derris*, and has long been used as a fish poison by tropical communities and as an insecticide (Pedigo 1989). A powdered form was developed in the USA for use as an insecticide, however, it was almost insoluble in water. Later water-soluble emulsions containing 5% rotenone were developed and the modern products are reputed to work well as a piscicide for fishery management at temperatures from 1.5-30°C in both acid and alkaline water (Morrison 1988). Very few studies have mentioned the effect of rotenone on crayfish, although the levels used to kill fish have been reported not to be harmful to them (Berzins 1962, Bills & Marking 1988). Westman (1991) reported an increase in crayfish numbers after fish had been removed with rotenone.

In fish, rotenone acts as a vaso-constrictor narrowing the blood vessels in the gill, thus preventing the normal uptake of oxygen (Hamilton 1941). Biochemically rotenone prevents the transfer of electrons along the respiratory pathway (Oberg 1967). It has been used to reduce recruitment to an overpopulated trout loch (Walker 1975), as well to remove nuisance fish such as perch (Amey 1981) before stocking with game or sport fish. A concentration of 0.5 mg /l for 30 minutes has been successfully used to eradicate trout and 0.05 mg/l has proved effective for pike, perch and eels. Morrison (1988) reviewed the effect of rotenone on vertebrates other than fish and found it had a low toxicity to birds and mammals, although it has been found to be toxic to amphibians at low doses (Chandler 1982), particularly gill-breathing tadpoles.

Studies in Sweden have shown that the efficiency of rotenone is highest in acid waters with a low temperature and sparse aquatic vegetation (Almquist 1959, R. Gydemo personal observation). Marking & Bills (1976), however, found that its toxicity to fish was reduced in cold water. It is thought that rotenone is adsorbed onto silt and organic matter, thus reducing its concentration (Lindgren 1960). Leonard (1939) found that rotenone rapidly lost its toxicity to fish in 41 hours. Rotenone supposedly degrades rapidly (a few days) in strong sunlight, but Bills & Marking (1988) quote unpublished work suggesting that rotenone residues may persists for over 40 days in cold water (5°C), although for only five days in warm water (24°C). They also found that, although Noxfish (5% rotenone) was toxic to fish in laboratory tests at a concentration as low as 30 µg/l, the manufacturer recommended using 5 mg /l for field treatments. They attributed this to the numerous physical and chemical factors that could affect its toxicity in the natural environment.

The effects of rotenone on aquatic invertebrates have been variously reported. They may be affected by rotenone in open water (Almquist 1959, Anderson 1970) but seem to be partly protected if they are benthic (Lindgren 1960). Almquist (1959), in a review of the effect of rotenone on fish food organisms in Sweden, found that most invertebrates, including planktonic crustaceans, were killed by a concentration of 0.5-0.6 mg/l used to kill fish. Lindgren (1960) suggested that, although many benthic invertebrates were likely to be affected by a level of 0.5 mg/l rotenone, they stood a better chance of not being totally eliminated than fish and the survivors would guarantee their continued existence. Even, populations of sensitive species have been found to make a full recovery after a few months (Hockin et al. 1985). Morrison & Struthers (1975) and Morrison (1977) found little effect of rotenone application on the benthos of lakes and streams in Scotland. Amey (1981) found no mortality of invertebrates when exposed to 2-4 mg /l rotenone (0.5 mg /l being lethal to many fish species).

Few studies have been carried out on the effects of rotenone on benthic crustaceans. However, Lindgren (1960) found that *Gammarus pulex* in experimental cages in a lake were not affected by 1-2 mg /l rotenone over 40 hours, and Engstrom-Heg (1987) showed that brown shrimp and blue crab were tolerant of 1.8 and 4.0 mg/l rotenone respectively. As crayfish are also benthic, then higher concentrations may be required to kill them and this may well adversely affect other invertebrates. Leonard (1939) showed that *Cambarus propinquus* and a number of other freshwater invertebrates could survive exposure to 1 mg/l rotenone for 96 hours. Hamilton (1941) found *Cambarus immunis* to be much more tolerant to rotenone than other freshwater crustaceans. Berzins (1962) carried out a series of experiments on *Astacus astacus* and came to the conclusion that a normal dose, i.e. 0.7 mg/l , would not harm crayfish. Bills & Marking (1988) found that 10 mg/l Noxfish (a rotenone-based product) killed 100% of *Orconectes rusticus* by 96 hours in static tests at 12°C, although a dose of only 0.02 mg/l was needed to kill fish under the same conditions.

Recent experiments with *Astacus leptodactylus* and *Pacifastacus leniusculus* have shown that they are relatively tolerant to rotenone, with *A. leptodactylus* even surviving levels of 100 mg /l for 24 hours (D.M. Holdich unpublished).

Morrison (1988) gives a protocol for applying rotenone to still and running waters. However, he warns that it is difficult to forecast precisely what the effect of rotenone treatment will be. As shown above, opinions differ on the effect of temperature. Adequate dispersal is essential and this will be hindered, if there is a lot of plant material in the water body to be treated.

2.5.5 *Surfactants*

Procambarus clarkii has become a serious problem in Portugal, particularly in rice fields and wetlands, since its introduction in the 1980s from Spain. Various pesticides (e.g. parathion) have been used illegally by farmers, but scientists are keen to develop more environmentally friendly control methods (Fonseca et al. 1997). Consequently, they have examined the effect of biodegradable surfactants as an ecotechnological control method. Surfactants are normally used to improve the emulsifying, wetting, and spreading properties of a pesticide formulation (Pedigo 1989).

A number of surfactants were found by Fonseca et al. (1997) to inhibit oxygen consumption through morphological and physiological changes on the surface of the gills, thus leading to decreased activity by the crayfish. Crayfish were found to quickly recover from the treatment. In a related study on surfactants, Cabral et al. (1997) tested the same surfactants on three non-target animals, *Gambusia affinis* (Pisces), *Physa acuta* (Mollusca) and *Daphnia magna* (Cladocera) to assess the risk, if applied to waters containing *P. clarkii*. They found that, at the concentrations required to reduce oxygen consumption in the crayfish, such non-target animals were more adversely affected.

The authors suggested that this method might have some potential in rice fields, which have an impoverished fauna, but would have an adverse effect in the irrigation channels, which have a higher diversity of life. The authors of both papers suggest that the use of biodegradable surfactants might limit the damage being done by the crayfish to rice crops, whilst still enabling an integrated production of crayfish in rice fields.

2.6 *Pheromones*

Crustaceans use a range of water-borne chemicals, pheromones, for many forms of communication, particularly during courtship and mating, agonistic behavior, maternal behavior, aggregation, and stress-related conditions (Ameyaw-Akumfi & Hazlett 1975, Itagaki & Thorp 1981, Hazlett 1985, Rose 1986, Dunham 1988, Bechler 1995).

Male crayfish tend to be very active during the mating season and may locate females by means of pheromones. However, the evidence for the presence of such chemicals in crayfish is inconsistent, having been found in some species but not others (Dunham 1988, Bechler 1995). It has been shown conclusively that newly released juveniles communicate with their mothers by pheromones to prevent them from being cannibalised (Little 1975, 1976). Neither the source nor the chemical nature of such pheromones in crayfish has been characterized (Bechler 1995). However, work with annelids and crabs has shown that this is possible (Zeeck et al. 1988, 1998, J.D. Hardege personal communication).

If a mating pheromone could be isolated and synthesized from crayfish, then it could be used in traps as an attractant. By such a method, many males could be removed from the population at the start of the mating season, so that less mating takes place. Coupled with standard trapping, this could quickly reduce the population, although it should be noted that males are able to mate with many females. An alternative might be to find baits which crayfish dislike, e.g. pike flesh, and to use their odors to drive crayfish into areas where they may be more easily trapped (J.D. Hardege personal communication).

3 CONCLUSIONS

Bills & Marking (1988) stated that no single method is likely to be effective for solving crayfish problems in all situations and that multiple approaches need to be developed for each specific situation. As a result of our literature review, we are of the same opinion. Each nuisance population must be considered separately and the most appropriate techniques applied to it for control or eradication purposes. In insect control, whatever the method and its effectiveness, the cost of control is often considered to be the deciding factor (Fernald & Shepard 1955).

Trapping and electro-fishing have already been proven to be effective methods and would be the most environmentally acceptable. The use of seine and fyke nets is more effective than traps, but involves greater costs and manpower. In addition, their use is not always acceptable as fish and possibly mammals such as otter could be trapped in addition to crayfish. Traps are relatively cheap, although their design could be improved so they are more effective, particularly for catching juveniles. Crayfish traps rarely catch anything other than crayfish, although it is possible that small fish and mammals such as water voles, rats and mink may be caught. The only way to control a nuisance crayfish population through trapping is to have a sustained effort over a number of years. It is unlikely that such methods would eradicate a population, unless coupled with some of the other methods below. Also, not all crayfish species are attracted to traps. Laurent (personal communication), for example, does not consider trapping a suitable method for removing *Orconectes limosus,* at least.

In some cases, there appears to be an inverse correlation between numbers of predaceous fish and crayfish in the natural environment and this is borne out by experiments. However, there have been no large-scale attempts at using predaceous fish to control nuisance crayfish populations. Coupled with trapping, the use of predaceous fish would seem to be the most environmentally acceptable method for attempting to control or eradicate nuisance crayfish populations, although some opposition may be raised about the impact such fish might have on other wildlife, especially when they have consumed the crayfish. One advantage of using eels is that they do not breed in freshwater, another is that they may be able to enter the hiding places of crayfish (Westman 1991). The use of monosex fish would get over the problem of introduced predators breeding (P.J. Laurent personal communication).

In very isolated populations of nuisance crayfish, the use of *Aphanomyces astaci* might be an effective eradication method, particularly if a more virulent strain could be developed. Different strains of the *A. astaci* fungus are known and may differ in their pathogenicity (Lilley et al. 1997); however, it would take time and money to develop one specific and more virulent to *Pacifastacus leniusculus* or other nuisance crayfish. It is unlikely to be a very environmentally acceptable method, due to its possible impact on native crayfish. Although no other organisms are affected by known strains of crayfish plague, this is not to say that a genetically modified strain would be so specific. Varieties of *Bacillus thuringiensis* such as *israeliensis* (Bti) capable of killing insect pests have been developed. Although Bti is relatively ineffective against crayfish, the development of a variety that is effective might be worth exploring.

Draining down and drying out a pond or lake, or isolating a section of river, after intensive netting or electro-fishing, is another possible method for eradicating crayfish. However, in the majority of instances this would be impracticable for burrowing species, unless followed by chemical treatment or a lengthy period (one year plus) before refilling. However, the use of biocides (see below) might be severely hampered by the protection afforded to crayfish in burrows. Control of burrowing crayfish could be effected by dosing individual burrows with chloride of lime, although the density of burrows may make such a method impracticable. Even after intensive trapping a viable population may exist in the burrows that will recolonize the area once the water level is restored. *P. leniusculus*, for example, can live for months in damp conditions and is consequently very difficult to remove by physical and biological methods.

There is clear evidence that a variety of organochlorine, organophosphate, and pyrethroid insecticides are capable of killing crayfish, although few large-scale, controlled experiments have been carried out except in rice fields (e.g. Chang & Lange 1967), and none to our knowledge in rivers. If a quick kill is required, then biocides may be the only answer. However, few field trials have been carried out in Europe, other than in experimental, enclosed ponds. There would likely be opposition from the relevant authorities to the use of biocides to kill crayfish. This was the case in Switzerland recently where a plan was put forward to use an organophosphate insecticide to eliminate a *P. clarkii* population, after trapping was found to be ineffective (A. Frutiger personal communication). However, organophosphate insecticides, some of which have been found to be effective against crayfish, have been successfully used to control other aquatic organisms in the field, e.g. Cladocera and Copepoda (Balvay 1981) and flies (Ray & Stevens 1970, Roqueplo et al. 1995). Pyrethroids would appear to have the greatest potential for eradicating crayfish due to their rapid breakdown and the low doses needed, although their impact on other crustaceans would be equally severe.

The greatest worry about the use of biocides is their potential to be bioaccumulated and biomagnified through the food chain, although this is less of a problem with pyrethroids. The use of such substances, whilst they may be effective, are unlikely to be very acceptable, until more research is done on their environmental impact. However, in the closed pond situation with anti-predator protection, this method would certainly be worthy of consideration. As no biocide known is specific to crayfish, then other invertebrates, particularly arthropods, would also be eliminated from any watercourse to which the poisons were applied. This might not necessarily be a problem as re-seeding could be undertaken, but the residence time of the poison in the environment must be known before this is done.

The use of surfactants to control the activity of crayfish (Cabral et al. 1997, Fonseca et al. 1997) can only be considered as a damage limitation method and is unlikely to be successful in crayfish eradication.

Ivermectin may have some potential for eradicating crayfish, but no work has been carried out and it is unlikely that permission for its general use would be given before much more is known about its environmental impact.

The use of rotenone might be acceptable for crayfish eradication, although it is toxic to fish and amphibians at levels lower than those needed to kill crustaceans, and these would have to be removed before its use was considered. Rotenone is widely used as a piscicide in fisheries management, but has rarely been tested on crayfish (Bills & Marking 1988). As crayfish appear more tolerant to rotenone than fish, then considerable cost would be involved in applying sufficient levels to eradicate them.

The development of more attractive baits has been carried out in Sweden and the USA and these can be obtained at a cost. Comparison would need to be made to see if they are any more effective than using fish or other meat-based baits, as commercially produced baits would greatly add to the cost of any control or eradication program.

The use of sexual attractants is widely used to control insect pests. Crustaceans use similar pheromones, particularly in the mating season. However, this method has not been used as a means of controlling crayfish and research is needed in this area.

There are other methods not considered above which have been used on insect pests, such as introducing sterile males into the population, which have proved very effective. No such work has been done with crustaceans, but it might be worth considering. However, a single male crayfish can mate with many females in a short space of time and, given the propensity of *P. leniusculus*, one of the main problem species, to burrow, it would be difficult to be sure all males had been sterilized. Another method worth considering might be the use of molt-inhibiting or accelerating hormones. However, it is unlikely that hormones specific to crayfish could be developed and consequently other arthropods would also be affected.

Crayfish are very adaptable animals, even being able to move across land if conditions become unfavourable in their environment. Species from North America have quickly adapted to the European environment and have become the dominant freshwater invertebrate in some instances. For example, the adaptability of the supposedly warm-water crayfish, *Procambarus clarkii*, is attested to by its recent colonisation of three small lakes in Switzerland (Stucki 1997, Stucki & Staub this volume). With the increasing spread of alien crayfish in Europe, and elsewhere in the world, there is an urgent need to develop methods for controlling and, if feasible, eradicating nuisance populations. Of course, the best method is not to introduce them in the first place. However, for many countries it is too late and in such cases the cost of eradication (i.e. the ecosystem cost) should be compared both as to the costs and risks of merely attempting to control the problem, as well as not doing anything!

ACKNOWLEDGEMENTS

Thanks are due to the Environment Agency and English Nature for their financial support for the Nottingham part of this review and their permissions to publish it, and to Julie Bywater, Mary Gibson, Jay Huner and Pierre Laurent for their valuable comments.

REFERENCES

Alderman, D.J. 1996. Geographical spread of bacterial and fungal diseases of crustaceans. *Rev. Sci.*

Tech. Int. Epiz. 15(2):603-632.

Alderman, D.J. & Polglase, J.L. 1988. Pathogens, parasites and commensals. In D.M. Holdich & R.S. Lowery (eds), *Freshwater crayfish: biology, management and exploitation*: pp. 167-212. London: Croom Helm (Chapman & Hall).

Alderman, D.J., Polglase, J.L. & Frayling, M. 1987. *Aphanomyce astaci*: pathogenicity under laboratory and field conditions. *J. Fish. Dis.* 10:385-393.

Almquist, E. 1959. Observations on the effect of rotenone emulsives on fish food organisms. *Rep. Inst. Freshwat. Res. Drottningholm* 40:146-160.

Amey, M.J. 1981. The application of liquid derris (5% rotenone) to a spring-fed pond to eradicate perch (*Perca fluviatilis* L.). *Fisheries Management* 12:111-112.

Ameyaw-Akumfi, C. & Hazlett, B.A. 1975. Sex recognition in the crayfish *Procambarus clarkii*. *Science* 190:1225-1226.

Anderson, R.L. 1982. Toxicity of fenvalerate and permethrin to several aquatic animals. *Trans. Am. Fish. Soc.* 107:825-827.

Anderson, R.L. 1989. Toxicity of synthetic pyrethroids to freshwater invertebrates. *Environ. Toxicol. Chem.* 8:403-410.

Anderson, R.S. 1970. Effects of rotenone on zooplankton communities and a study of their recovery patterns in two mountain lakes in Alberta. *J. Fish. Res. Bd Can.* 27:1335-1356.

Anonymous, 1979. BAYTEX (BAY S 17523, OMS 2) Produit pour la lutte contre les parasites de l'hygiène. Information technique. BAYER Pflanzenschutz Leverkusen, Germany.

Balvay, G. 1981. Notes sur une technique d'amélioration de la production dans les étangs d'alevinage de Hongrie. *Bull. Fr. Pisc.* 282:40-42.

Bechler, D.L. 1995. A review and prospectus of sexual and interspecific pheromonal communication in crayfish. *Freshwater Crayfish* 8:657-667.

Berzins, B. 1962. Rotenonförsök med kräftor. *Skr. Södra Sverig. Fisk. Forën Årsskr* 1959-1969:51-52.

Bills, T.D. & Marking, L. 1988. Control of nuisance populations of crayfish with traps and toxicants. *Prog. Fish-Culturist.* 50(2):103-106.

Blake, M.A. & Hart, P.J.B. 1993. The behavioural responses of juvenile signal crayfish *P. leniusculus* (Dana), to stimuli from perch and eels. *Freshwater Biology* 29:89-97.

Blake, M.A. & Hart, P.J.B. 1995a. Habitat preferences and survival of juvenile signal crayfish, *Pacifastacus leniusculus* – the influence of water depth, substratum, predatory fish and gravid female fish. *Freshwater Crayfish* 9:318-332.

Blake, M.A. & Hart, P.J.B. 1995b. The vulnerability of juvenile signal crayfish to perch and eel predation. *Freshwater Biology* 33:233-244.

Brown, R. & Avault Jr, J.W. 1975. Toxicity of antimycin to crayfish *Procambarus* sp. *Freshwater Crayfish* 2:351-369.

Cabral, J.A., Anastácio, P.M., Carvalho, R. & Marques, J.C. 1997. A non harmful chemical method of red swamp crayfish, *Procambarus clarkii*, population control and non target organisms problematics in the lower Modego River Valley, Portugal. *Freshwater Crayfish* 11:286-292.

Campbell, W.C. 1989. *Ivermectin and Abamectin*. London: Springer-Verlag.

Cange, S.W., Pavel, D., Burns, C., Romaire, R.P. & Avault Jr, J.W. 1986. Evaluation of eighteen artificial crayfish baits. *Freshwater Crayfish* 6:270-273.

Capelli, G.M. & Magnusson, J.J. 1983. Morphoedaphic and biogeographic analysis of crayfish distribution in Wisconsin. *J. Crust. Biol.* 3:548-564.

Chandler, J.H. 1982. Toxicity of rotenone to selected aquatic invertebrates and frog larvae. *Prog. Fish-Culturist* 44:78-80.

Chang, V.C.S. & Lange, W.H.1967. Laboratory and field evaluation of selected pesticides for control of the red crayfish in California rice fields. *J. Econ. Entom.* 60:473-477.

Cheah, M-L., Avault, J.W. & Graves, J.B. 1979. Some effects of thirteen rice pesticides to crayfish *Procambarus clarkii*. *Freshwater Crayfish* 4:350-361.

Cheah, M-L., Avault, J.W. & Graves, J.B. 1980. Acute toxicity of selected rice pesticides to crayfish *Procambarus clarkii*. *Prog. Fish-Culturist* 42(3):169-172.

Day, K.E. 1991. Effects of dissolved organic carbon on accumulation and acute toxicity of fenvalerate, deltamethrin and cyhalothrin products to *Daphnia magna* (Straus). *Environ. Toxicol. Chem.* 10:91-101.

Day, K.E. & Maguire, R.J.1990. Acute toxicity of isomers of the pyrethroid insecticide deltamethrin and its major degradation products to *Daphnia magna*. *Environ. Toxicol. Chem.* 9:1297-1300.

Dunham, P.J. 1988. Pheromones and behaviour in Crustacea. In *Endocrinology of Selected Invertebrate Types*. A.R. Liss Inc.

Edgerton, B., Owens, L., Harris, L., Thomas, A. & Wingfield, M. 1995. A health survey of farmed reclaw crayfish, *Cherax quadricarinatus* (Von Martens), in tropical Australia. *Freshwater Crayfish* 10:322-338.

Ekanem, S.B., Avault, J.W., Graves, J.B. & Morris, H. 1983. Effects of rice pesticides on *Procambarus clarkii* in a rice/crayfish model. *Freshwater Crayfish* 5:315-323.

Elvira, B., Nicola, G.C. & Almodovar, A. 1996. Pike and red swamp crayfish: a new case of predator-prey relationship between aliens in central Spain. *J. Fish Biol.* 48:437-446.

Engstrom-Heg, R. 1987. Persistence of rotenone in ponds. *North American J. Fish. Man.* 7:162.

European Court of Justice (1994). Case C-131/93, Free movement of goods. Commission of the European Communities v Federal Republic of Germany.

Eversole, A.G. & Foltz, J.W. 1995. Habitat relationships of two crayfish species in a mountain stream. *Freshwater Crayfish* 9:300-310.

Eversole, A.G. & Sellers, B.C. 1997. Comparison of relative crayfish toxicity values. *Freshwater Crayfish* 11:274-285.

Eversole, A.G., Whetstone, J.M. & Sellers, B.C. 1995. *Handbook of relative acute toxicity values for crayfish*. S.C. Sea Grant Consortium.

Fernald, H.T. & Shepard, H.H. 1955. *Applied Entomology*. London: McGraw-Hill.

Fjälling, A. 1995. Crayfish traps employed in Swedish fisheries. *Freshwater Crayfish* 8:201-214.

Fonseca, J.C., Marques, J.C. & Madeira, V.M.C. 1997. Oxygen uptake inhibition in *Procambarus clarkii*, red swamp crayfish by biogradable surfacants: an ecotechnological approach for population control in rice fields. *Freshwater Crayfish* 11:235-242.

Fornstrom, C.B., Landrum, P.F., Weisskopf, C.P. & La Point, T.W. 1997. Effects of terbufos on juvenile red swamp crayfish (*Procambarus clarkii*): differential routes of exposure. *Environ. Toxicol. Chem.* 16(12):2514-2520.

Fürst, M. 1977. Introduction of *Pacifastacus leniusculus* (Dana) into Sweden: methods, results and management. *Freshwater Crayfish* 3:229-247.

Gaude, A.P. 1987. Thermal effects of pesticide toxicity for Louisiana red swamp crawfish (*Procambarus clarkii*). *Freshwater Crayfish* 6:171-177.

Gowing, H. & Momot, W.T. 1979. Impact of brook trout (*Salvelinus fontinalis*) predation on the crayfish *Orconectes virilis* in three Michigan lakes. *J. Fish. Res. Bd Can.* 36:1191-1196.

Grigarick, A.A. & Lange, W.H. 1965. Rice pest investigations. *Rice J.* 68(7):68-70.

Guan, R.Z. 1994. Burrowing behaviour of signal crayfish, *Pacifastacus leniusculus* (Dana), in the River Great Ouse, England. *Freshwater Forum* 4:155-168.

Guan, R.Z. & Wiles, P.R. 1996. Growth, density and biomass of crayfish, *Pacifastacus leniusculus*, in a British lowland river. *Aquat. Living Resour.* 9:265-272.

Gydemo, R. 1995. *Effect of an insecticide induced crayfish kill*. Report to the Swedish Environmental Protection Agency and the Swedish Board of Fisheries. Dept. of Systems Ecology, Stockholm University, Sweden.

Halley, B.A., Nessel, R.J. & Lu, A.Y.H. 1989. Environmental aspects of Ivermectin usage in livestock: General considerations. In W.C. Campbell (ed.), *Ivermectin and Abamectin*. London: Springer-Verlag.

Hamilton, H. 1941. The biological action of rotenone on freshwater animals. *Proc. Iowa Acad. Sci.* 48:467-479.

Hamrin, S.F. 1987. Seasonal crayfish activity as influenced by fluctuating water levels and presence of a fish predator. *Holarctic Ecology* 10:45-51.

Harlioglu, M.M. 1996. Comparative biology of the signal crayfish, *Pacifastacus leniusculus* (Dana), and the narrow-clawed crayfish, *Astacus leptodactylus* Eschscholtz. Unpublished Ph.D. thesis, University of Nottingham, UK.

Harris, R.R. & Young, H.J. 1996. *Distribution, densities and population characteristics of signal crayfish, Pacifastacus leniusculus (Dana), in the Gaddesby Brook, Leicestershire*. Report for the National Rivers Authority, Lower Trent Division, Nottingham, U.K.

Håstein, T. & Gladhaug, O. 1972. The occurrence of the crayfish plague in Norway and attempts to prevent further spread of the disease. *Freshwater Crayfish* 1:182-184.

Hazlett, B.A. 1985. Chemical detection of sex and condition in the crayfish *Orconectes virilis*. *J. Chem. Ecol.* 11(2):181-189.

Haya, K. 1989. Toxicity of pyrethroid insecticides to fish. *Environ. Toxicol. Chem.* 8:381-391.

Hickley, P., R. North, S.M. Muchiri & Harper, D.M. 1994. The diet of largemouth bass, *Micropterus salmoides*, in Lake Naivasha, Kenya. *J. Fish Biol.* 44:607-619.

Hobbs Jr, H.H. & Hall Jr, E.T. 1975. Crayfishes (Decapoda: Astacidae). In Hart, C.W. & S.L.H. Fuller (eds), *Pollution ecology of freshwater invertebrates*. Academic Press: New York & London.

Hockin, D.C., O'Hara, K., Cragg-Hine, D. & Eaton, J.W. 1985. Fish population estimation: the use of rotenone to evaluate the reliability of a removal exercise. *Aquacult. Fish. Man.* 16:349-357.

Hogger, J.B. 1988. Ecology, population biology and behaviour. In D.M. Holdich & R.S. Lowery (eds), *Freshwater crayfish: biology, management and exploitation*: pp. 114-144. London: Croom Helm (Chapman & Hall).

Holdich D.M. & Domaniewski, J.C.J. 1995. Studies on a mixed population of the crayfish *Austropotamobius pallipes* and *Pacifastacus leniusculus* in England. *Freshwater Crayfish* 10:37-45.

Holdich, D.M. & Reeve, I.D. 1991. *Alien crayfish in the British Isles*. Report for the Natural Environment Research Council, Swindon.

Holdich, D.M. & Rogers, W.D. 1997. The white-clawed crayfish, *Austropotamobius pallipes*, in Great Britain and Ireland with particular reference to its conservation in Great Britain. *Bull. Fr. Pêche Piscic.* 347(4):597-616.

Holdich, D.M., Rogers, W.D. & Reader, J.P. 1995. *Crayfish conservation*. Report for the National Rivers Authority, Northumberland and Yorkshire Region.

Huner, J.V. 1988. *Procambarus* in North America and elsewhere. In D.M. Holdich & R.S. Lowery (eds), *Freshwater crayfish: biology, management and exploitation*. London: Croom Helm (Chapman & Hall).

Huner, J.V. & Paret, J. 1995. Trap harvest of crawfish (*Procambarus* spp.) from a south Louisiana commercial pond: effectiveness of different baits and species composition. *Freshwater Crayfish* 8:376-390.

Huner, J.V., Hentonnen, P. & Lindqvist, O.V. 1991. Observations on noble crayfish, *Astacus astacus* L. (Decapoda, Astacidae), populations in central Finland – management implications. *J. Shellf. Res.* 10:187-193.

Huner, J.V., Moody, M. & Thune, R. 1994. Cultivation of freshwater crayfishes in North America. In J.V. Huner (ed.), *Freshwater crayfish aquaculture in North America, Europe, and Australia*. New York: The Haworth Press Inc.

Huxley, T.H. 1879. *The crayfish. An introduction to the study of zoology*. London: Keegan Paul.

Hyde, K.M., Graves, J.B., Schilling, P.E. & Bonner, F.L. 1975. The influence of Mirex bait on production and survival of the red swamp crawfish *Procambarus clarkii*. *Freshwater Crayfish* 2:349-350.

Ibbotson, A.T., Tapir, G., Furse, M.T., Winder, J.M., Blackburn, J., Scarlett, P. & Smith, J. 1997. *Impact of the signal crayfish Pacifastacus lenuisculus and its associated crayfishery on the River Thame*. Report for the Environment Agency, Thames Division, Reading, UK.

Itagaki, H. & Thorp, J.H. 1981. Laboratory experiments to determine if crayfish can communicate chemically in a flow-through system. *J. Chem. Ecol.* 7:(1):115-126.

Jarboe, J.H. 1988. The toxicity of pesticides to crawfish. *Crawfish Tales* 7(4):22-24.

Jarboe, J.H. & Romaire, R.P. 1991. Acute toxicity of permethrin to four size classes of red swamp crayfish (*Procambarus clarkii*) and observations of post-exposure effects. *Environ. Contam. Toxicol.* 20:337-342.

Jarboe, J.H. & Romaire, R.P. 1995. Responses of procambarid crayfish populations to permethrin applications in earthen ponds. *Bull. Environ. Contam. Toxicol.* 55:58-64.

Jolly, A.L., Avault, J.W., Graves, J.B. & Koonce, K.L. 1977. Effects of Pounce® on newly hatched and juvenile Louisiana red swamp crayfish, *Procambarus clarkii* (Girard). *Freshwater Crayfish* 3:389-395.

Jolly, A.L., Avault, J.W., Koonce, K.L. & Graves, J.B. 1978. Impact of permethrin to several aquatic animals. *Trans. Am. Fish. Soc.* 107(6):825-827.

Kaushik, N.K., Stephensson, G.L., Solomon, K.R. & Day, K.E.1985. Impact of permethrin on zooplankton communities in limnocorals. *Can. J. Fish. Aquat. Sci.* 42:77-85.

Landrum, P.F., Nihart, S.R., Eadie, B.J. & Herche, L.R. 1987. Reduction in bioavailability of organic contaminants to the amphipod *Pontoporeia hoyi* by dissolved organic matter of sediment interstitial waters. *Environ. Toxicol. Chem.* 9:141-150.

Laurent, P.J. 1988. *Austropotamobius pallipes* and *A. torrentium*, with observations on their interaction with other species in Europe. In D.M. Holdich & R.S. Lowery (eds), *Freshwater crayfish: biology, management and exploitation*: pp. 341-364. London: Croom Helm (Chapman & Hall).

Laurent, P.J. 1995. Eradication of unwanted crayfish species for astacological management purposes. *Freshwater Crayfish* 8:121-133.

Leonard, J.M. 1939. Notes on the use of derris as a fish poison. *Trans. Am. Fish. Soc.* 68:269-279.

Lilley, H.J., Cerenius, L. & Söderhäll, K. 1997. RAPD evidence for the origin of crayfish plague outbreaks in Britain. *Aquaculture* 157:181-185.

Lindgren, P.E. 1960. About the effect of rotenone upon benthonic animals in lakes. *Rept. Inst. Freshwat. Res. Drottningholm* 41:172-183.

Little, E.E. 1975. Chemical communication in maternal behaviour in crayfish. *Nature* 255:400-401.

Little, E.E. 1976. Ontogeny of maternal behaviour and brood pheromone in crayfish. *J. Comp. Physiol.* 112:133-142.

Lodge, D.M. & Hill, A.M. 1994. Factors governing species composition, population size, and productivity of cool-water crayfishes. *Nordic J. Freshwat. Res.* 69:111-136.

Ludke, J.L., Finley, M.T. & Lusk, C. 1971. Toxicity of Mirex to crayfish, *Procambarus blandinggi. Bull. Environ. Contam. Toxicol.* 11(1):42-50.

Mackenzie, D. 1986. Crayfish pesticides decimate Spanish birds. *New Scientist* 112(1530):24.

MacLeese, D.W, Metcalfe, C.D. & Ziko, V. 1980. Lethality of permethrin, cypermethrin and fenvalerate to salmon, lobster and shrimp. *Bull. Environ. Contam. Toxicol.* 3:171-182.

Marking, L.L. & Bills, T.D. 1976. Toxicity of rotenone to fish in standardised laboratory test. *US Fish & Wildlife Service Investigations in Fish Control* 72.

Martinez, P.J. & Bergersen, E.P. 1989. Proposed biological management of *Mysis relicta* in Colorado lakes and reservoirs. *North Amer. J. Fish. Manag.* 9(1):1-11.

Mather, M.E. & Stein, R.A. 1983. Direct and indirect effects of fish predation on the replacement of a native crayfish by an invading congener. *Canad. J. Fish. Aquat. Sci.* 50:1279-1288.

Matsumura, F. 1985. *Toxicology of insecticides*. New York: Plenum Press.

McMahon, B.R. 1986. The adaptable crayfish: mechanisms of physiological adaptation. *Freshwater Crayfish* 6:59-74.

Mills, B.J., Morrissy, N.M. & Huner, J.V. 1994. Cultivation of freshwater crayfishes in Australia. J.V. Huner (ed.), *Freshwater crayfish aquaculture in North America, Europe and Australia*. New York: The Haworth Press, Inc.

Momot, W.T. 1992. Stocking and exploitation as management tools for European crayfish. *Finn. Fish. Res.* 14:145-148.

Momot, W.T. 1993. The role of exploitation in altering processes regulating crayfish populations. *Freshwater Crayfish* 9:101-117.

Morgan, G.E. & Momot, W.T. 1988. Exploitation of *Orconectes virilis* in northern climates: complimentary of management options with self-regulatory life history strategies. *Freshwater Crayfish* 7:69-80.

Morrison, B.R.S. 1977. The effects of rotenone on the invertebrate fauna on three hill streams in Scotland. *Fisheries Management* 18:128-138.

Morrison, B.R.S. 1988. *The use of rotenone in fisheries management*. Scottish Fisheries Information Pamphlet No. 15.

Morrison, B.R.S. & Struthers, G. 1975. The effects of rotenone on the invertebrate fauna of three

Scottish freshwater lochs. *J. Inst. Fish. Man.* 6:81-91.

Muir, D.C.G., Rawn, G.P., Townsend, B.E., Lockhart, W.L. & Greenhalgh, R. 1985. Bioconcentration of cypermethrin, deltamethrin, fenevalerate and permethrin by *Chironomus tentans* larvae in sediment and water. *Environ. Toxicol. Chem.* 4:51-61.

Muncy, R.J. & Oliver Jr, A.D. 1963. Toxicity of ten insecticides to the red crayfish, *Procambarus clarkii* (Girard). *Trans. Am. Fish. Soc.* 92(4):428-431.

National Research Council of Canada 1986. *Pyrethroids: Their effects on aquatic and terrestrial ecosystems*. Publ. No. 24376. National Research Council of Canada, Ottawa, Ontario.

Oberg, K.E. 1967. On the principal way of attack of rotenone in fish. *Ark. Zool.* 18:217-220.

Ohkawa, H.R., Kikuchi, R. & Miyamoto, J. 1980. Bioaccumulation and biodegradation of the (S) acid isomer of fenvalerate (Sumiciden) in an aquatic model ecosystem. *J. Pest. Sci.* 5:11-22.

Palmer, R., Rodger, H., Drinan, E., Dwyer, C. & Smith, P.R. 1987. Preliminary trials on the efficacy of ivermectin against parasitic copepods of Atlantic salmon. *Bull. Europ. Assoc. Fish Path.* 7:47-54.

Pedigo, L.P. 1989. *Entomology and pest management*. New York: Macmillan Publishing Co.

Rabeni, C.F. 1992. Trophic linkage between stream centrarchids and their crayfish prey. *Can. J. Fish. Aquat. Sci.* 49:1714-1721.

Rach, J.J. & Bills, T.D. 1989. Crayfish control with traps and largemouth bass. *Prog. Fish-Culturist* 51:157-160.

Ragan, M.A., Goggin, C.L., Cawthorn, R.J., Cerenius, L., Jamieson, A.V.C., Plourdes, S.M., Rand, T.G., Söderhäll, K. & Gutell, R.R. 1996. A novel clade of protistan parasite near the animal-fungal divergence. *Proc. Natl. Acad. Sci.* 93:11910-11912.

Ray, J. & Stevens, V. 1970. Using BAYTEX to control crayfish in ponds. *Prog. Fish-Culturist* 32(1):58-60.

Reynolds, J.D. 1988. Crayfish extinctions and crayfish plague in Ireland. *Biological Conservation* 45:279-285.

Rickett, J.D. 1974. Trophic relationships involving crayfish of the genus *Orconectes* in experimental ponds. *Prog. Fish-Culturist* 36:207-211.

Roell, M.J. & Orth, D.J. 1993. Trophic basis of production of stream-dwelling smallmouth bass, rock bass, and flathead catfish in relation to invertebrate bait harvest. *Trans. Am. Fish. Soc.* 122:42-62.

Rogers, W.D. 1996. The impact of introduced species of crayfish in the British Isles. Unpublished PhD thesis, University of Nottingham, UK.

Rogers, W.D., Holdich, D.M. & Carter, E. 1997. *Crayfish Eradication*. Report for English Nature, Peterborough.

Roqueplo, C., Laurent, P.J. & Neveu, A. 1995. *Procambarus clarkii* Girard (écrevisse rouge des marais de Louisiana). Synthèse sur les problèmes poses par cette espèce et sur les essais pour contrôler ses populations. *L'Astaciculteur de France* 45:2-17.

Rose, R.D. 1986. Chemical detection of 'self' and conspecifics by crayfish. *J. Chem. Ecol.* 12:271-276.

Saiki, M.K. & Tash, J.C. 1979. Use of cover and dispersal to reduce predation by largemouth bass. *Amer. Fish. Soc. Spec. Publ.* 6:44-48.

Skurdal, J. & Qvenild, T.1986. Growth, maturity and fecundity of *Astacus astacus* in Lake Steinsfjorden, S.E. Norway. *Freshwater Crayfish* 6:182-186.

Skurdal, J. & Taugbøl, T. 1994. Do we need harvest regulations for European crayfish? *Rev. Fish Biol. Fish.* 4:461-485.

Smith, T.M. & Stratton, G.W. 1986. Effects of synthetic pyrethroid insecticides on non-target organisms. *Residue Rev.* 97:93-120.

Smith, V.J. & Söderhäll, K. 1986. Crayfish pathology: an overview. *Freshwater Crayfish* 6:199-211.

Söderbäck, B. 1995. Replacement of the native crayfish *Astacus astacus* by the introduced species *Pacifastacus leniusculus* in a Swedish lake: possible causes and mechanisms. *Freshwater Biology* 33:291-304.

Söderhäll, K. 1989. The crayfish plague fungus *Aphanomyces astaci* and other diseases during aquaculture of crayfish. In J. Skurdal, K. Westman & P.I. Bergan (eds), *Crayfish culture in Europe*. Report from the workshop on crayfish culture, 16-19 Nov. 1987, Trondheim, Norway.

Spehar, R.L., Tanner, D.K. & Nordling, B.R. 1983. Toxicity of the synthetic pyrethroids, permethrin and AC 222,705 and their accumulation in early life stages of fathead minnows and snails. *Aquat. Toxicol.* 3:171-182.

Stucki, T.P. 1997. Three American crayfish species in Switzerland. *Freshwater Crayfish* 11:130-133.

Svärdson, G. 1948. Stunted crayfish populations in Sweden. *Rep. Inst. Freshwat. Res. Drottningholm* 29:135-144.

Svärdson, G. 1972. The predatory impact of eel (*Anguilla anguilla* L.) on populations of crayfish (*Astacus astacus* L.). *Rep. Inst. Freshwat. Res. Drottningholm* 52:149-191.

Svärdson, G., Fürst, M. & Fjälling, A. 1991. Population resilience of *Pacifastacus leniusculus* in Sweden. *Finn. Fish. Res.* 12:165-177.

Thörnqvist, P.O. & Söderhäll, K. 1993. *Psorospermium haeckeli* and its interactions with the crayfish defence system. *Aquaculture* 117:205-213.

Thune, R.L., Hawke, J.P. & Siebeling, R.J. 1991. Vibriosis in the red swamp crayfish. *J. Aquat. Anim. Hlth.* 3:188-191.

Thurston, R.V., Gilfoil, T.A., Meyn, E.L., Zajdei, R.K., Aoki, T.I. & Veith, G.D. 1985. Comparative toxicity of ten organic chemicals to ten common aquatic species. *Water Resources* 9:1145-1155.

Tibbetts, J. 1997. Exotic invasion. *Envir. Hlth. Persp.* 105(6):590-593.

Unestam, T. 1975. Defense reactions in and susceptibility of Australian and New Guinean freshwater crayfish to European crayfish plague fungus. *Aust. J. Exp. Med. Biol. Sci.* 53:349-359.

Walker, A.F. 1975. The use of rotenone to control recruitment of juvenile brown trout (*Salmo trutta* L.) into an 'overpopulated' loch. *J. Inst. Fish. Manag.* 6:64-72.

Welcomme, R.L. 1988. *International introductions of inland aquatic species*. FAO Fisheries Technical Paper 294. Rome: FAO.

Westman, K. 1991. The crayfish fishery in Finland – its past, present and future. *Finn. Fish. Res.* 12:187-216.

Westman, K., Pursiainen, M. & Vilkman, R. 1979. A new folding trap model which prevents crayfish from escaping. *Freshwater Crayfish* 4:235-242.

Westman, K., Pursiainen, M. & Westman, P. (eds) 1990. *Status of crayfish stocks, fisheries, diseases and culture in Europe.* Report of the FAO European Inland Fisheries Advisory Commission (EIFAC) Working Party on Crayfish.

Westman, K., Savolainen, R. & Pursiainen, M. 1993. A comparative study of the growth and moulting of the noble crayfish, *Astacus astacus* (L.), and the signal crayfish, *Pacifastacus leniusculus* (Dana), in a small forest lake in southern Finland. *Freshwater Crayfish* 9:451-465.

Westman, K., Sumari, O. & Pursiainen, M. 1978. Electric fishing in sampling crayfish. *Freshwater Crayfish* 4:251-255.

Williamson, M. 1996. *Biological invasions*. London: Chapman & Hall.

Zeek, E., Hardege, J., Bartels-Hardege, H. & Wesselmann, G. 1988. Sex pheromone in a marine polychaete: determination of the chemical structure. *J. Exp. Zool.* 246:285-292.

Zeek, E., Müller, C.T., Beckmann, M., Herdege, J.D., Papke, U., Sinnwell, V., Schroeder, F.C. & Francke, W. 1998. Cysteine-glutathione disulphide, the sperm-release pheromone of the marine polychaete *Nereis succinea* (Annelida: Polychaeta). *Chemoecology* 8:33-38.

Zitko, V., McLeese, D.W., Metcalfe, C.D. & Carson, W.G. 1979. Toxicity of permethrin, decamethrin and related pyrethroids to salmon and lobster. *Bull. Environ. Contam. Toxicol.* 21:338-343.

The future of native crayfish in Europe: How to make the best of a bad situation?

TROND TAUGBØL & JOSTEIN SKURDAL
Eastern Norway Research Institute, Lillehammer, Norway

ABSTRACT

Wild populations of crayfish in Europe include five native species and four species introduced from North America. Three of the native species are considered as 'vulnerable' in the IUCN Red List of Threatened Animals. Prospects for the future of crayfish in Europe appear different, depending on which crayfish group one considers. For the alien species the situation is good as they steadily increase their distribution, whilst for the native species the situation in general is very bad. The alien species are competitive superior and also carry the crayfish plague which kills the European species. This paper looks at the future from the native species point of view and focuses on necessary management actions and measures. Actions include: restore aquatic habitats, determine and monitor crayfish distribution and abundance, establish 'Native Crayfish Areas', secure fulfilment of national crayfish protection aims in different management authorities and levels, and prevent spread of alien crayfish. The spread of alien crayfish is the main threat to the native, European crayfish. The different actions are discussed and elucidated with a number of examples.

1 INTRODUCTION

Wild populations of crayfish in Europe include five native species, *Astacus astacus*, *A. leptodactylus*, *A. pachypus*, *Austropotamobius pallipes*, and *A. torrentium,* and four species introduced from North America, which invaded the wild with reproducing populations, *Pacifastacus leniusculus*, *Orconectes limosus*, *Procambarus clarkii*, and *P. zonangulus* (Henttonen & Huner this volume). Of the introduced species, *P. leniusculus*, *P. clarkii*, and *O. limosus* are most widespread and were reported to occur in at least 7-12 countries in 1992 (Westman & Westman 1992). The introduced species continue to increase their distribution at the sacrifice of the native ones, as the foreigners all carry the crayfish plague (Diéguez-Uribeondo & Söderhäll 1993) and also otherwise outcompete the European species (e.g. Söderbäck 1995, Holdich this volume). Three of the native species are considered as 'vulnerable' in the Red List of Threatened Animals (IUCN 1996). Thus, prospects for the future of crayfish in Europe appear different, depending on which crayfish group one considers. For the alien species, the situation is good as they steadily increase their distribution, while for the native species the situation, in general, is very bad (Lindqvist 1988). In this paper, we will look at the future from the native species point of view and focus on necessary management measures and actions.

Looking ahead one hundred years, two possible scenarios may exist for the worst and best situations. Scenario 1 describes the worst situation: almost all watersheds in Europe suitable for crayfish are inhabited by alien species. All native species are considered critically endangered and survive in a few protected localities within national parks/restricted areas. Scenario 2 describes the best situation: there are still European countries (that is Ireland, Estonia, Norway) free from alien crayfish species (Holdich, Rogers & Reynolds this volume, Skurdal et al. this volume), and in most European countries there are specific areas with numerous, viable native crayfish populations considered as valuable and protected by the authorities as well as local people. Local people, as argued later, may well be a key factor for the conservation success.

All of us who have a devotion for nature, biological diversity and crayfish in particular would, of course, like scenario 2 to be fulfilled rather than scenario 1. The question is: How to make the best of a bad situation? Which actions and measures should be undertaken by authorities at local, regional, national and international levels?

2 ACTIONS

A conceptual model of important actions related to the protection of native crayfish is outlined in Figure 1. A brief discussion of the different actions follows below.

2.1 *Restoring aquatic habitats*

The first action is restoration of aquatic habitats. This action will of course also benefit alien crayfish, but nevertheless, it is an important action in order to strengthen and re-establish native crayfish populations. Habitat alterations and pollution are considered to be a major reason for the decline in crayfish populations in many areas (Westman 1985). In Norway, the 75% reduction in crayfish yield during the last 25-30 years is mostly due to these factors (Taugbøl & Skurdal 1993). The habitat alterations in streams are mainly caused by encroachments like canalization, dredging and embanking. In many lakes and slow flowing rivers, mud accumulation due to erosion from agriculture areas have turned previously good crayfish bottoms into unfavorable habitats (Waters 1995). Important hiding places, especially for juveniles, disappeared. Pollution from agriculture and other domestic and industrial sources may act directly on crayfish survival (toxic effects) or more indirectly through mud accumulation and a change in flora and fauna composition (e.g. the effects of eutrophication).

In degraded areas, it is possible to make habitat improvements. During the last decade the pollution situation, in general, has improved. When habitat and water quality are adequate, it may be possible to re-establish a crayfish population by re-stocking. There are, however, few documented examples of crayfish re-introductions into previously polluted and/or physically degraded areas. A review of current knowledge, as well as more controlled experiments, especially regarding effects of different habitat improvement techniques, are greatly needed.

2.2 *Protection of native species – a national aim*

Now, let us concentrate on actions more specifically directed to the protection of the native species. First of all, the protection of native species must be a national aim in all countries. This seems to be an aim in all countries, as three of the native crayfish species (*A. astacus, A. pallipes*, and *A. torrentium*) are included in the Bern Convention, the EC Habitat Directive and in the Red List of the World Conservation Union (Rogers & Holdich 1995, IUCN 1996). Most European countries signed up the Biodiversity Convention at the 1992 Rio Earth Summit and this has resulted in local and national efforts to conserve the native fauna and control aliens. In Britain, *A. pallipes* is one of 116 organisms identified as of key importance in the UK Biodiversity Action Plan (Department of Environment 1995). However, in most countries the international commitments are not properly incorporated into national legislation. And even if legislation is in place, the enforcement is often neglected or the crayfish loose the battle with other sector interests (aquaculture, infrastructure, industrial development, etc.). In most countries, strong interests like the aquaculture business are in favor of alien species. Also many landowners or fishermen are convinced that alien crayfish are the only alternative for natural crayfish production. This might of course be true in areas where alien crayfish are well established. Different authorities (e.g. agriculture/food production vs. environmental protection authorities), authority levels (national vs regional/local levels) and management officers have often different opinions on the value and possibility of protecting native species. Furthermore, changes in national policy need time to be adopted by local authorities and local people. For instance, in Sweden, landowners were a few years ago liberally given permits to stock signal crayfish into waters that were void of crayfish or had very low populations of native crayfish. Today, stocking of signal crayfish is allowed only in waters where it already exists (Fiskeriverket 1998). This is hard to accept by many people.

So, even though national goal to protect native crayfish exist, decisions are often taken that counteract this aim. To be effective, it is a prerequisite that the national aim is fulfilled and understood by relevant authorities at all levels. Good communication and agreement on a common strategy are needed between the different authorities. The importance of co-operation and co-ordination between national authorities are discussed further in a later section.

2.3 *Status of crayfish distribution and abundance*

In order to manage and protect crayfish, it is imperative to have a good knowledge of the status of crayfish distribution and abundance. Many countries or regions have quite a good overview of the situation (see description by country, this volume), but there are also many areas with very little information. The situation is dynamic and needs continuous monitoring. It is especially difficult to get reliable information from areas with expanding alien species, because local people, who often are important information sources, are not capable of identifying the different species.

2.4 *Identify and establish native crayfish areas (NCA)*

'Native crayfish areas' (NCA) should be established in all countries based on the status of crayfish distributions. This is one of the most important actions to be taken. As we define it, NCA is a very flexible concept. Basically, NCAs are established and recognized by the authorities as areas where the native crayfish should have a reasonable chance to survive and to which the main protective efforts are directed. An NCA can include the whole country, a region, or even a watershed or a single lake. The actual situation with the distribution and the spreading possibilities of the different crayfish species is the basic information when establishing such areas. The NCAs can differ in regulations; e.g. from strict restrictions within a National Park or Special Areas of Conservation (see below) to areas where exploitation of the crayfish populations is desirable.

In Norway we have only one crayfish species, the noble crayfish, *Astacus astacus*. No alien species exist. Therefore the whole country may be considered a 'native crayfish area'. This is strongly focused in a proposed national management plan for crayfish (Taugbøl & Skurdal 1998). Also, Estonia and Ireland are in the same situation with only one native crayfish species, thus, the whole country could be considered an NCA. In the Swedish 'Action Plan for the Noble Crayfish' (Fiskeriverket 1998), identifying areas with good possibilities for protecting the noble crayfish is a major goal. In Britain, a concept of 'no-go areas', i.e. specified areas where keeping and releasing of non-native species are forbidden, has been implemented by the government (Rogers & Holdich 1995, Holdich, Rogers and Reynolds this volume). In a restoration plan for native crayfish in Navarra, Spain, (Diéguez-Uribeondo et al. 1997), specific 'native crayfish areas' are designated and comprise watercourses where the native crayfish, *Austropotamobius pallipes,* occurs. A major goal is to prevent invasion of alien crayfish into these watercourses. As mentioned above, the European Habitat Directive (Annexes II and V) includes three of the European crayfish species. The directive requires the designation of protected areas (SACs; Special Areas of Conservation) for species in Annex II that includes *A. pallipes* and *A. torrentium*. For instance, in England, Wales and Ireland, a number of SACs are being set up (Holdich, Rogers and Reynolds this volume).

2.5 *Prevent further spread of alien species*

Undoubtedly, the main threat to native crayfish is the spread of alien species. These North American species are all resistant carriers of the crayfish plague which kills the European species (Diéguez-Uribeondo & Söderhäll 1993). Disease-free American species may also otherwise outcompete the European species (Holdich & Domaniewski 1995, Holdich et al. 1995, Söderbäck 1995). The spread of the alien species is by both natural migration and, more importantly, by man. So, in a protection effort it is imperative to prevent further spread of alien crayfish, especially to the designated NCAs. This struggle must include many different actions (Fig. 1)

2.5.1 *Legislation*

Good legislation is an important measure in order to prevent further spread of alien species. The best would be if one could ban the live crayfish trade between nations, but this is almost impossible within the European Community framework. So far, EC

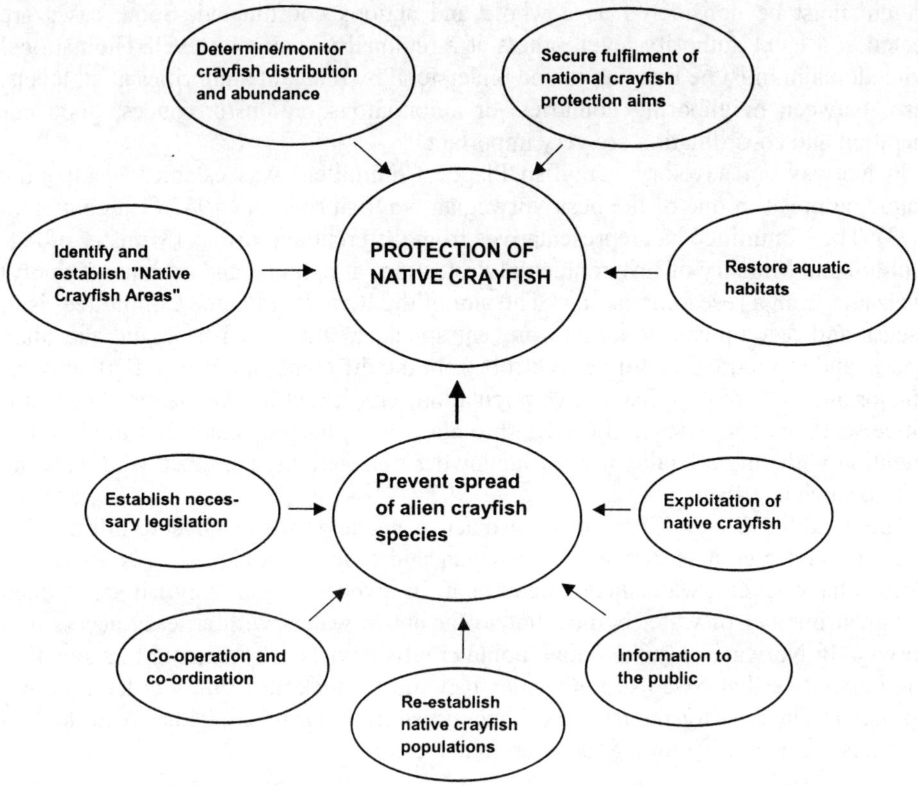

Figure 1. Conceptual model of different actions necessary for the protection of native crayfish.

member Ireland has managed to stop live crayfish imports, but in general, free trade regulations make it impossible to restrict this import (European Court of Justice 1994, Holdich, Rogers & Reynolds this volume). Within a nation or a province, however, possibilities exist to establish legislation to control possession and releasing of live crayfish, assuming that politics will exist. As mentioned above, the international commitments regarding protection of vulnerable species are not properly incorporated into the national legislation in many countries. Politicians and other decision-makers should be approached to incorporate these commitments into legislation.

Good laws alone do not solve any problems. Legislation is one thing – getting people to obey is another matter. However, good legislation is a necessary basis for further actions. If the legislation is to be effective, it requires public education as well as control and enforcement (Rogers & Holdich 1997).

2.5.2 *Co-operation and co-ordination*

Good co-operation and co-ordination are other important actions, necessary for an effective and unified protection effort. Different authorities may deal with different matters within the same complex problem. For instance, in Norway regulations to prevent spreading of diseases are under the jurisdiction of the Ministry of Agriculture and associated authorities while regulations for catching, exploitation, and stocking belong to the Ministry of Environmental Protection. In a protection context, all these matters must be considered as a whole and actions coordinated. Some cases are treated at a local authority level, others at a regional or national level. The national protection aim must be lived up to and understood by relevant authorities at all levels. Also, between neighboring countries, or autonomous regions/provinces, good co-operation and co-ordination are very important.

In Norway, an advisory 'Crayfish Plague Committee' was established after the plague outbreak in one of the best Norwegian crayfish rivers in 1987 (Taugbøl et al. 1993). The Committee has representatives from different authorities (Ministry of Agriculture and Ministry of Environmental Protection) at national and regional (County) levels and from a research institute. The aim of the 'Crayfish Plague Committee' is to discuss and recommend actions to prevent spread of the crayfish plague and alien species and to secure a coordinated effort from the different authorities. Furthermore, a major aim is to re-establish native populations eradicated by the plague. The Committee secretariat initiates and carries through restocking projects, often in close co-operation with single landowners or landowner associations, i.e. those who have the right to catch crayfish.

The 'Crayfish Plague Committee' also act as a catalyst with respect to information exchange and meetings between Norwegian and Swedish authorities. Norway and Sweden have several watersheds in common. In Sweden, signal crayfish are stocked in a great number of water bodies, but so far not in waters with an easy access into Norway. In Norway, only the native noble crayfish exists. To maintain this situation it is imperative that Swedish authorities ban signal stocking in the border area and promote the interest for the native noble crayfish. Regular meetings between the two countries are necessary for a good coordinated effort.

2.5.3 *Re-establish native crayfish*

If a crayfish population is eradicated by the plague, it is possible to re-introduce the native crayfish. The crayfish plague fungus does not appear to have any secondary hosts, no resting structures and spores with very limited viability outside the crayfish host (Unestam 1973, Smith & Söderhäll 1986). The plague often causes 100% mortality in European crayfish populations, thus there is a great chance that the plague will 'burn' itself out.

Most stockings of plague-resistant signal crayfish have occurred by landowners/fishermen after a plague attack in order to replace the lost crayfish population. To forestall a possible alien introduction (legal or illegal), the authorities should encourage and support a re-stocking with native crayfish. Information to the public (especially landowners/fishermen) about the re-introduction possibilities and the future prospects of exploitation is of great importance in this context.

2.5.4 *Exploitation*

Exploitation of the native crayfish as a protective effort may sound like a contradiction. However, in our opinion, exploitation and protection are closely linked. Those who exploit are, in most cases, also concerned about the exploited resource. If local people are allowed to catch the crayfish, if children grow up with the magic of catching crayfish, they will also be interested in protecting the crayfish population against extermination.

Two examples to show the close link between exploitation and protection: Lake

Steinsfjorden is the best crayfish locality in Norway. Unlike most water bodies in Norway, in which the right to catch fish and crayfish belongs to the landowners, Lake Steinsfjorden is public water, i.e. everybody is allowed to catch crayfish during the catching season. The local fishermen are very concerned about the crayfish; they especially fear an introduction of crayfish plague by more distant fishermen. The local fishermen's association received the authority from the County Environmental Office to establish a disinfection station and a system for marking the crayfish traps. Those who want to catch crayfish must disinfect and mark their traps. The fishermen's association also make sure that the fishery regulations are obeyed. A ban on crayfish catching in Lake Steinsfjorden in order to protect the noble crayfish would lead to a loss of interest from the fishermen in protecting the native crayfish. Undoubtedly, many fishermen would prefer a catchable signal crayfish population rather than a totally protected noble crayfish population. With signal crayfish accessible within a few hours of driving distance (in Sweden) and little control at the border, the chances of illegal stockings by fishermen interested in catching crayfish would be great. If catching of a native species is banned but not the catching of an alien species, it is no wonder that local people interested in catching crayfish could be tempted to illegally stock alien crayfish, just to get a legal opportunity to catch crayfish later.

Another example, clearly showing that the possibility for exploitation is of major importance for the protection effort is from the Halden watercourse in Norway. Crayfish plague eradicated the noble crayfish population in the watershed in 1989-1991 (Taugbøl et al. 1993). The authorities started immediately to discuss re-establishing the native crayfish with the landowners' associations. Since 1995, more than 20,000 noble crayfish adults and juveniles have been stocked, with costs equally shared between the authorities and landowners. In order to prevent illegal stocking of signal crayfish, it has been imperative to keep the focus and interest high on the native noble crayfish. What makes the landowners interested in re-establishing and protecting the noble crayfish is undoubtedly the prospect of exploitation in near future.

It should, however, also be mentioned that crayfish exploitation traditions differ between countries. For example, in Great Britain there is no recent tradition for eating or catching crayfish. The public is more concerned about conserving the native species for aesthetic purposes, and anglers because it indicates good, clean water and provides a source of food for their fish (Holdich, Rogers and Reynolds this volume).

2.5.5 *Information to the public*

Information to the public, in general and especially directed to landowners, fishermen and other groups interested in recreational use of rivers and lakes, is also a very important action to implement and maintain. The information should increase public knowledge and awareness, increase the status or reputation of the native crayfish and decrease the status/reputation of alien crayfish. Most people probably know about crayfish, but only specialists are able to distinguish between species. Thus, in areas where it is important to be aware of which species one is dealing with, simple identification brochures or leaflets are necessary. Excellent examples of such leaflets have been made by the National Rivers Authority in Britain ('A Guide to Identifying Freshwater Crayfish in Britain and Ireland') and Fischereiforschungsstelle des Landes Baden-Württemberg in Germany ('Flusskrebse in Baden-Württemberg. Gefährdung und Schutz').

Among many landowners and fishermen in Europe there has been, and still is, a belief that stocking of signal crayfish is the only way to get back the crayfish production. In Norway, where the signal crayfish has not yet been introduced, information that provides good arguments against the introductions of the signal crayfish has been directed to fishermen and landowners. Arguments that not only relate to biological diversity, protection of a vulnerable species and illegality of introductions (such arguments have often minor effect), but that also directly relate to the recreational and economical value of the waters. From Sweden, it is known that many signal crayfish populations have developed into stunted populations. Large numbers of small-sized crayfish cause problems for the gill-net fishery, because the crayfish eat the fish. Also, the crayfish are too small to have a commercial or recreational interest (M. Persson personal communication). The 'scientifically correct nature' or generality of these observations are unknown, but the objections are used as good arguments against the stocking of signal crayfish. In the same breath, the great demand and higher price achieved for the native noble crayfish are stressed.

3 FINAL REMARKS

Ultimately, we are convinced that the knowledge and attitude of local people is a key factor in the protection of the native crayfish species. With an easy access to alien crayfish, it is impossible to prevent spreading of such species if local people along the watersheds want otherwise (Rogers & Holdich 1997). An active and co-operative attitude and effort from the managing authorities, i.e. good information, support for restocking/restoration, possibilities for exploitation, etc., are major factors shaping the attitude of local people.

However, when speaking about the protection of native species it should not be forgotten that the alien crayfish species also represent a valuable resource. No methods have been developed for eradication of unwanted crayfish populations without causing harmful effects on other biota (Holdich, Gydemo & Rogers this volume), and there may never be. That means that the alien species have to be accepted as part of the European fauna. Also the best should be made out of this situation, i.e. alien crayfish should also be exploited as best as possible. Accepting this does not mean giving up the native species. Discussions on the future of crayfish in Europe often become very polarized, with very little mutual understanding between diverse groups of interest, i.e. those who are very concerned about the native crayfish and 'hate' the alien crayfish per se, and those who see a future only in the signal or red swamp crayfish. A more balanced view with minds open for different solutions in different areas is a more fruitful approach. In fact, such policies have or will be adopted in several European countries or regions (Diéguez-Uribeondo et al. 1997, Holdich, Rogers and Reynolds this volume, Fiskeriverket 1998). If those who are advocates for the native crayfish also accept the existence and exploitation of alien species outside the 'Native Crayfish Areas', this may perhaps, in return, lead to more understanding for the necessary native crayfish conservation actions.

REFERENCES

Department of the Environment 1995. Biodiversity. The UK Steering Group Report, Vol. 2, Action plans. London: H.M.S.O.

Diéguez-Uribeondo, J. & Söderhäll, K. 1993. *Procambarus clarkii* Girard as a vector for the crayfish plague fungus, *Aphanomyces astaci* Schikora. *Aquacult. Fish. Mgmt.* 24:761-765.

Diéguez-Uribeondo, J., Rueda, A., Castien, E. & Bascones, J.C. 1997. A plan of restoration in Navarra for the native freshwater crayfish species of Spain, *Austropotamobius pallipes*. *Bull. Fr. Pêche Piscic.* 347:625-638.

European Court of Justice 1994. Case C-131/93. Free movement of goods. Commission of the European Communities vs Federal Republic of Germany.

Fiskeriverket 1998. Åtgärdsprogram för bevarande av flodkräftan (*Astacus astacus* L.) (Action plan for the noble crayfish). Report from Fiskeriverket. (In Swedish).

Holdich, D.M. & Domaniewski, J.C.J. 1995. Studies on a mixed population of the crayfish *Austropotamobius pallipes* and *Pacifastacus leniusculus* in England. *Freshwater Crayfish* 10:37-45.

Holdich, D.M., Rogers, W.D., Reader, J.P. & Harlioglu, M.M. 1995. Interactions between three species of freshwater crayfish (*Austropotamobius pallipes*, *Astacus leptodactylus* and *Pacifastacus leniusculus*). *Freshwater Crayfish* 10:46-56.

IUCN 1996. *1996 IUCN Red List of Threatened Animals*. Gland, Switzerland: IUCN.

Lindqvist, O.V. 1988. Restoration of native European crayfish stocks. *Freshwater Crayfish* 7:6-12.

Rogers, W.D. & Holdich D.M. 1995. Conservation and management of crayfish in Britain. *Freshwater Crayfish* 10:92-97.

Rogers, W.D. & Holdich, D.M. 1997. New legislation to conserve the native crayfish in Britain – will it work? *Freshwater Crayfish* 11:619-626.

Smith, V. & Söderhäll, K. 1986. Crayfish pathology: an overview. *Freshwater Crayfish* 6:199-211.

Söderbäck, B. 1995. Replacement of the native crayfish *Astacus astacus* by the introduced species *Pacifastacus leniusculus* in a Swedish lake: possible causes and mechanisms. *Freshw. Biol.* 33:291-304.

Taugbøl, T., Skurdal, J. & Håstein, T. 1993. Crayfish plague and management strategies in Norway. *Biol. Cons.* 63:75-82.

Taugbøl, T. & Skurdal, J. 1993. Noble crayfish catching in Norway: legislation and yield. *Freshwater Crayfish* 9:134-143.

Taugbøl, T. & Skurdal, J. 1998. Forslag til forvaltningsplan for kreps. (Proposal to a national management plan for crayfish). Utredning for DN 1998-1. (In Norwegian).

Unestam, T. 1973. Significance of diseases in freshwater crayfish. *Freshwater Crayfish* 2: 136-150.

Waters, T. F. 1995. *Sediment in streams. Sources, biological effects, and control.* American Fisheries Society Monograph 7, 251 pp.

Westman, K. 1985. Effects of habitat modification on freshwater crayfish. In J.S. Alabster (ed.), *Habitat modification and freshwater fisheries*. Butterworth, London: EIFAC Symposium,.

Westman, K. & Westman, P. 1992. Present status of crayfish management in Europe. *Finnish Fish. Res.* 14: 1-22.

Native and alien crayfish in Europe: Some conclusions

DAVID M. HOLDICH
School of Biological Sciences, The University of Nottingham, Nottingham, UK
HANS ACKEFORS
Department of Zoology, University of Stockholm, Stockholm, Sweden
FRANCESCA GHERARDI
Department of Animal Biology and Genetics 'Leo Pardi', University of Florence, Florence, Italy
W. DAVID ROGERS
David Rogers Associates, Castle Donington, Derby, UK
JOSTEIN SKURDAL
Eastern Norway Research Institute, Skurva, Lillehammer, Norway

ABSTRACT

The purpose of this volume is to bring together what is known about the impact of established populations of alien crayfish in Europe, both from the positive and negative points of view. This has been done to highlight various aspects of their biology relevant to the topic and to discuss what might be done about the problem. The situation is summarized below and some possible solutions and conclusions are reached.

1 INTRODUCTION

In Europe there have been five main types of introductions involving crayfish. First, there have been deliberate (solicited and unsolicited) stockings made with native species but outside their natural range. This practice has been going on for hundreds of years up until the present day and has mainly involved *Astacus astacus* in Scandinavia, *Astacus leptodactylus* in many parts of Europe, and *Austropotamobius pallipes* in the British Isles. In the first and third cases, the stockings have usually been made into water bodies with no crayfish, and there have been no environmental problems reported. However, in the second case, introductions have often been made into areas with a resident crayfish species, such as *A. astacus,* and this has usually resulted in the displacement of the native species (see Holdich this volume). Consequently, *A. leptodactylus*, is considered to be an alien crayfish in many European countries (see Holdich & Gherardi and Case Studies in this volume).

Second, there have been reintroductions into areas where the native crayfish species has been eliminated, often by crayfish plague, but usually with individuals from a different gene pool, e.g. in Norway (Taugbøl & Skurdal 1993) and England (Holdich & Rogers 1997). Where possible, restockings should be made with crayfish from the same area in the hope of maintaining the gene pool. However, our knowledge of the genetics of European crayfish populations is still in its infancy, and it is not known how different they are with any certainty (Grandjean et al. 1997a, b). In Norway, attempts have been made to maintain live gene pools from important populations threatened by crayfish plague outbreaks for restocking purposes.

Third, there have been introductions (solicited and unsolicited) made from other continents, particularly North America (see below), and it is these that are the main focus of this volume. Alien crayfish have been deliberately introduced from North America into Europe for two main reasons: first, to supplement stocks of native crayfish decimated by crayfish plague; and second, for aquaculture, usually by those with entrepreneurial interests. In some cases, the North American species have been introduced and then cultured in one European country, and then supplied to other European countries, as has happened with *Pacifastacus leniusculus* from Sweden (Ackefors & Lindqvist 1994). They have then spread within those countries, sometimes naturally but often aided by humans.

Fourth, there have been many accidental releases of crayfish into the wild, often escapes from holding facilities, and usually involving alien species. Accidental release due to a natural event such as flooding comes under this category, and has probably aided the dissemination of alien crayfish in some countries, as has the building of new waterways such as canals.

Fifth, there is the question of disposal of unwanted specimens. Although usually unrecorded, these last two methods of introduction are probably more widespread than appreciated. Of particular concern in this respect is the aquarium trade, which is very difficult to police. As long as crayfish are in demand in Europe then there will always be people who will cut corners and ignore legislation to try and make money out of them. Other geographical areas are facing the same problem, e.g. according to Mikkola (1996) those crayfish that now occur in African freshwaters (where there are no native crayfish species) have been introduced without the knowledge and the permission of the relevant authorities.

Moyle et al. (1986) have stated that introductions are often made to solve some local or regional problem, but, if the broad-scale consequences of each introduction are not considered, it may ultimately cause more problems than it solves. They described this as the Frankenstein Effect and illustrated how it could be applied to fish introductions in North America. Apparently, more than 25-50% of freshwater fish caught by anglers in the continental US are from introduced stocks. The term 'Frankenstein Effect' relates to the fact that attempts to improve on nature – in Mary Shelley's story (1818) Count Frankenstein tried to create an improved human – may not always turn out as expected – in Frankenstein's case, a monster. There are many examples of this phenomenon in biology, and for the invertebrates, Spencer et al. (1991) have described the wide-ranging, negative, cascading effects within food webs caused by the introduction of the opossum shrimp, *Mysis relicta*, into some North American lakes and rivers as another example. Could alien crayfish in Europe be yet another?

It is clear from some of the papers in this volume that, from the point of view of those with an interest in harvesting, farming and selling crayfish, the introduction of alien crayfish in Europe has been relatively successful. Such people would not consider the introduction of alien crayfish in Europe as an example of the Frankenstein Effect. However, from a conservationist's point of view it is and the majority of papers reflect the view that established populations of introduced crayfish, particularly those from North America, are having an adverse environmental impact, not only on native European crayfish but also on the freshwater environment. This is largely due to their invasive tendencies, aided by their higher fecundity, faster growth rates, and better physiological tolerance to changing environmental conditions than European crayfish (see Holdich and Lindqvist & Huner this volume). This problem is not restricted to Europe alone as it seems that wherever crayfish with r-selected characteristics (see Lindqvist & Huner this volume) become established outside their natural range, then similar effects are seen (Holdich this volume).

Introducing a large omnivorous macroinvertebrate such as a crayfish into any body of water is initially going to alter the ecological balance if it becomes established (Holdich this volume). Ideally, over time the introduction becomes part of the ecosystem and the balance is restored. This is presumably what happened in Western and Northern Europe when crayfish from more southerly climes colonized these regions in post-glacial times, probably aided by humans, and also with more recent stockings outside their natural range (see above). However, the situation with *A. leptodactylus* and North American crayfish introductions in Europe has not in many cases followed this pattern. When these species have been moved from their natural environment and from their normal suite of predators, they have undergone huge increases in numbers, and the ecosystem is still trying to come to terms with the invasion. Perhaps it will in time, but will the native crayfish in Europe survive the onslaught (see Taugbøl & Skurdal this volume)? As pointed out by Hobbs III et al. (1989), introduced crayfish may impose considerable environmental stress on the freshwater ecosystem such that irreparable shifts in species diversity may occur (this is mainly due to the fact that they become the dominant species). They went on to say that 'every effort should be taken to prevent unsolicited transplantations of any crayfish outside of its native range' and 'that resource managers should realize that successful crayfish transfers for aquacultural purposes will almost certainly lead to the establishment of breeding populations in the natural environment.'

Native crayfish in Europe are not only under threat from crayfish plague and alien crayfish but from activities of humans that cause pollution and destruction of habitats. Alien crayfish might be better at coping with such changes, particularly acid and turbid waters, as crayfish chosen for commercial purposes tend to be good colonizers due to the fact that they are r-strategists and exhibit a wide range of physiological tolerance (see Lindqvist & Huner this volume). For example, *Procambarus clarkii*, is a good colonizer of disturbed aquatic habitats, e.g. rice fields, and can survive in almost anoxic conditions in burrows, so it is no small wonder that it has become the most successful and widely-spread introduced crayfish species. In Europe, at least, other crayfish species do not live in rice fields and these have been an unexploited habitat as far as crayfish are concerned. The colonization of rice fields in Europe by *P. clarkii* has provided an additional food source for many predators, including humans. *Orconectes limosus* can occupy waters of lower quality than native European species, and this may be one of the reasons for its success. Similarly, *A. leptodactylus* is more hardy than other European species and can also live in brackish water.

2 ALIEN CRAYFISH IN EUROPE – THE CURRENT SITUATION

Three North American crayfish species have become successfully established in European freshwaters: *Orconectes limosus*, first introduced into Germany in the 1890s and now the most widespread alien crayfish Europe, *Pacifastacus leniusculus*, first introduced into Sweden in the 1960s and now also widely spread in Europe, and *Procambarus clarkii* introduced into Spain in the 1970s, where it soon became the dominant crayfish species, and which has more recently been found in a number of other European countries. The Australian yabby, *Cherax destructor*, is currently proving popular amongst aquaculturists and has started to crop up in a number of European countries, e.g. Italy and Spain, although only in the latter is there a wild population (Gutiérrez-Yurrita et al., this volume). In addition, *Astacus leptodactylus*, has gradually spread westward, usually aided by humans, into Eastern, Northern, and more recently into Western Europe. Many western and northern European countries have only *Astacus astacus* or *Austropotamobius pallipes* as their native crayfish species and even the introduction of other native European crayfish should be considered a threat to them. A case in point is the recent establishment of *A. astacus* in England, where *A. pallipes* is the sole native species (see Holdich, Rogers & Reynolds this volume).

The introduction of crayfish plague into Europe, originally last century but probably a number of times since then with the introduction of various North American crayfish species, has had a devastating impact on native crayfish populations and still continues to do so. The introduction of North American crayfish and, in some cases, *A. leptodactylus*, as alternatives has met with mixed success. In some water bodies production may be back to former levels but in Northern Europe the crayfish harvest is still much less than it was at the turn of the century despite numerous stockings with alien crayfish. For example, pre-plague catches of *A. astacus* in Sweden were 600-1000 tons per annum, but the present wild harvest averages only 52 tons plus 12 tons from farming, and the wild harvest of *P. leniusculus* is around 264.5 tons plus 42 tons (in 1995) from registered farms (Ackefors 1997). However, in Spain, production is greater due to the introduction of *P. clarkii* (see Ackefors this volume). In the region of 90% of the European crayfish harvest is based on alien crayfish! Demand in Western Europe outstrips supply, however, and imports from Spain, Eastern Europe, and North America are still needed. The lack of crayfish for human consumption in Europe has therefore been partly solved by establishing alien crayfish populations that can be harvested from the wild or farmed – but at what price? Ackefors (this volume) has shown that there have been positive commercial, socioeconomic, recreational, and biological effects to crayfish introductions, but on the other hand, others in this volume have highlighted the negative aspects of such introductions (see Holdich & Nyström this volume). These include displacement of native crayfish species, transfer of disease, consumption of fish eggs, reduction of fish stocks, consumption of large amounts of macrophytes, indirect and direct effects on other invertebrates, upsetting production in rice fields, and displacement of amphibians. In addition, the burrowing activities of some species can cause physical damage to irrigation structures and banks of rivers and lakes. All these factors are causing environmental degradation of European freshwater habitats (see Holdich this volume).

The known occurrences of alien crayfish in European countries is given in Table 1, but it is probable that unpublished records exist for a number of the blanks. What is clear is that the majority of countries in Europe have at least one alien species of crayfish. New occurrences are also being reported. *Orconectes limosus* has been reported from Belarus (Alekhnovich et al. this volume) and England (Holdich, Rogers & Reynolds this volume), and *Orconectes immunis* from Germany (Dehus et al. this volume); the latter species making its first appearance in Europe. *Procambarus clarkii* is now abundant in some lakes in Switzerland (Stuki et al. this volume) and Italy (Gherardi et al. this volume). *Astacus leptodactylus* continues to be stocked by entrepreneurs in France (Arrignon et al. this volume) and other countries, as does *Pacifastacus leniusculus*. *Cherax destructor* occurs in the wild in Spain (Gutiérrez-Yurrita et al. this volume) and is being cultured in Italy (Gherardi et al. this volume) with no apparent protection against its escape into the wild.

3 THE FUTURE FOR THE NATIVE EUROPEAN CRAYFISH

Although they have a fairly wide-spread distribution three native European crayfish, *Astacus astacus*, *Austropotamobius pallipes*, and *Austropotamobius torrentium* are all considered to be threatened species by the IUCN (Groombridge 1993). They are protected to some extent by the Bern Convention on the Conservation of European Wildlife and Natural Habitats and the European Habitats Directive. *A. astacus* is considered vulnerable and *A. pallipes* rare by the IUCN, but not enough is known about *A. torrentium* to be sure of its status.

Much mention has been made of the European Habitats Directive (Directive 92/43/EEC on the conservation of natural habitats and of wild fauna and flora) in this volume. One outcome of the Directive is the Natura 2000 network of protected sites, Special Areas of Conservation (SACs) (see Holdich, Rogers & Reynolds this volume), which are supposed to be identified by 1998 and put in place by 2004. The Directive lists 168 natural habitat types in Europe whose conservation requires designation as SACs because they are rare and becoming rarer. The Directive also requires all Member States to set up an effective system to prevent the capture, killing, injuring or damaging disturbance of certain endangered species. The Directive lists 632 animal and plant species whose conservation requires the designation of SACs. Among European crayfish only *A. pallipes* is listed as warranting SAC status by being listed on Annex II of the Directive. For such species, data banks of information have been compiled (see Holdich 1996). Appendix III of the Bern Convention and Annex V of the Directive cover species where exploitation should be subject to appropriate control in member states and this may give further protection to *A. pallipes*, as well as to *A. astacus* and *A. torrentium*. Although populations of *A. leptodactylus* have been affected by crayfish plague, the species does not appear to be threatened. Virtually nothing is known about the status of *Astacus pachypus*, although it apparently common in the Caspian Sea as well as some northern rivers entering the Black Sea (Starobogatov 1995). There may in fact be many more species of astacid crayfish in the Ponto-Caspian region as Starobogatov (1995) supports the views of Bott (1950), which were abandoned by Albrecht (1982), as well as supporting those of Brodksi (1983), who placed crayfish species from the Ponto-Caspian region in the genus *Pontastacus* rather than *Astacus* and upgraded a number of sub-species to specific rank. In addition, Starobogatov (1995) has placed *A. pachypus* in a new genus *Caspiastacus*.

Taugbøl & Skurdal (this volume) have suggested that within 100 years almost all watersheds suitable for crayfish in Europe might be inhabited by alien species and that all native species may be considered critically endangered and survive in a few protected localities within national parks and restricted areas. However, in 2100 there may still be European countries free from alien crayfish species, as there are now, i.e. Estonia, Ireland, and Norway, as long as current legislation is continued. Some countries in Eastern Europe are also currently free from alien crayfish, e.g. Croatia, but pressure from entrepreneurs could soon change that! It is likely that with the continuing interest there is in conserving the native biota in most countries that there will still be specific areas with numerous, viable native crayfish populations that are considered as a valuable asset to be protected by the authorities as well as local people well into the next century.

The known occurrences of native crayfish in European countries is given in Table 1, but, as with the alien crayfish species, it is likely that unpublished records exist for a number of the blanks. It may even be that as some of the records have not been confirmed in recent years that some of the species have become extinct in certain East European countries (Y. Machino, pers. comm.).

Table 1. Known occurrences (x) of established, wild populations of native and alien crayfish in European waters.

Country	Aa	Al	Ap	Aup	Aut	Oi	Ol	Pl	Pc	Cd	Cq	Ct
Albania	X											
Andorra												
Austria	X	x		x	x		x	x				
Belarus	X	x					x					
Balearics									x			
Belgium	X	x*					x	x				
Bosn.-Herz.												
Bulgaria	X	x	x									
Corsica			x*									
Croatia	X	x		x	x							
Cyprus	x*							x	x			
Czech Rep.	X	x*			x		x	x				
Denmark	X	x*						x				
England	x*	x*		x			x	x	x			
Estonia	X											
Finland	X	x*						x				
France	X	x*		x	x		x	x	x			
Georgia		x										
Germany	X	x*		x	x	x	x	x	x			
Greece	X	x						x				
Hungary	X	x			x		x					
Italy	X	x*		x	x		x	x	x			
Kazakstan		x	x									
Latvia	X	x*						x				
Liechtenstein	x*			x*								
Lithuania	X	x*					x	x				
Luxembourg	X						x	x				
Macedonia												
Malta												
Moldova												
N. Ireland				x								
Netherlands	X	x*					x		x			
Norway	X											
Poland	X	x*					x	x				

Table 1. Continued.

Country	Aa	Al	Ap	Aup	Aut	Oi	Ol	Pl	Pc	Cd	Cq	Ct
Portugal				x					x			
R. Ireland				x								
Romania	X	x			x							
Russia	X	x	x				x	x				
Scotland				x*				x				
Serb.-Mont.												
Slovakia	X	x			x		x					
Slovenia												
Spain		x*		x				x	x	x		
Sweden	X							x				
Switz.	X	x*		x	x		x	x	x			
Turkey		x										
Ukraine		x										
Wales				x				x				

*European species but introduced in recent times. Key to abbreviations: Aa = *Astacus astacus*, Al = *Astacus leptodactylus*, Ap = *Astacus pachypus*, Aup = *Austropotamobius pallipes*, Aut = *Austropotamobius torrentium*, Oi = *Orconectes immunis*, Ol = *Orconectes limosus*, Pl = *Pacifastacus leniusculus*, Pc = *Procambarus clarkii*, Cd = *Cherax destructor*, Cq = *Cherax quadricarinatus*, Ct = *Cherax tenuimanus*.

4 POSSIBLE SOLUTIONS

Before conservation and restoration or control programs are initiated, it is essential to know the current distribution of the native and alien species – regionally, nationally, and Europe-wide. Table 1 is only an overview, and what is needed is more detailed and accurate knowledge of their distribution at least on the 100 km^2 level, but preferably at the 10 km^2 level or even better. Various papers in this volume give examples of good practice in this regard. There is need for a European databank into which people can feed records and access information. The distribution of crayfish, especially of alien species, is not static, and regular monitoring must be carried out so that threats can be highlighted.

Some countries are doing a good job at educating government authorities and the public on the dangers of disease and alien crayfish transfer but much more needs to be done. Every European country should have education leaflets for general distribution as has been done in the countries forming the British Isles and in Germany.

Legislation aimed at conservation of native crayfish and control of alien crayfish is in place in many countries, but does it really work? Where more than one crayfish species is present in a country it is important to have species-specific regulations. Under Article 22 of the Habitats Directive (see above) Member States are obliged to ensure that deliberate introductions to the wild of non-native species is regulated, so as not to prejudice habitats or native fauna or flora. Unfortunately for native crayfish, the political situation in Europe has led to trade barriers being lowered and countries who thought that they had laws to stop the entry of live crayfish such as Germany and Sweden now find that when they are challenged they do not stand up. More needs to be done to persuade the EU and national governments that allowing live, alien crayfish to spread will only cause problems in the future. Even when countries have laws to protect their crayfish people often do not know about them or they take no notice of them – but how often does anybody ever get prosecuted? They should be – more often.

Protection of the remaining native crayfish fauna in Europe should be a national aim of all countries as the ratification of the Rio Convention implies. The relevant national and regional authorities should be made aware of the dangers of allowing alien crayfish to become established. They should be encouraged to set up 'native crayfish reserves', not just as part of SACs, but perhaps whole countries, e.g. Croatia, Estonia, Ireland, and Norway, where the species is protected and alien crayfish are forbidden. This should not only be for *Austropotamobius pallipes* but also for other native crayfish species, particularly *Astacus astacus* and *Austropotamobius torrentium*. Islands, such as those in the Baltic Sea, and isolated water bodies with native crayfish populations, should be highlighted in this respect as they may have to form 'Noah's Ark' populations for restocking programs in the future.

It is considered by some that those who are allowed to exploit crayfish themselves have a strong interest in their protection and that exploitation and protection are closely linked. This is true in those countries allowing their native species to be harvested. The wise utilization of such populations is to be encouraged as long as there are harvest regulations to prevent them being over fished. However, Momot (1992) considers that some of the size regulations in operation in some European countries are counterproductive to getting the best out of a population. Size regulations vary considerably between countries and Skurdal & Taugbøl (1994) have recommended a minimum harvest size of 90-95 mm total length for *A. astacus*, although others disagree with this and consider the limit should be lower. The situation is different in those countries where the native species is protected and not harvested, e.g. England. However, the public should still be encouraged to view the crayfish as a valuable freshwater resource, but from the aesthetic and biological diversity points of view.

There was a strong feeling among European astacologists that crayfish plague should be made a notifiable disease and movements of alien crayfish should be restricted – not made easier! Although the crayfish plague fungus entered Europe in the 1860s, clones of *Aphanomyces astaci* isolated from *Pacifastacus leniusculus* in Sweden, Finland, and England are different, indicating that the introduction of this North American species from the 1960s onwards is at least partly responsible for the continuing spread of the disease (Huang et al. 1994, Lilley et al. 1997, Vennerström et al. 1998). A third clone has recently been isolated from *Procambarus clarkii* (Dieguez-Uribeondo et al. 1995). Because it is difficult to identify in situ, *Aphanomyces astaci* is on List 3 of Directive 91/67. Upgrading it to List 2 would mean that outbreaks would have to be notified. Consequently, if an area had an outbreak then crayfish could not be moved out of the area. As crayfish plague is still rampant in Europe then the whole of the geographical area could in theory come under this heading! Work is in progress at the University of Uppsala and the University of Munich to develop a PCR probe that could be used to identify *A. astaci* (K. Söderhäll and B. Oidtmann, personal communications), which would go a long way toward getting the fungus regraded. It would also enable a rapid means of finding out if there is such a thing as a plague-free population of North American crayfish in Europe? It could also be used to find out more about how crayfish plague is spread.

A number of papers in this volume mention restocking programs for native crayfish eliminated by crayfish plague and other factors. In some cases, e.g. where alien crayfish are prevalent, trying to restore native populations would be a waste of time and money (Taugbøl & Skurdal this volume). Often crayfish populations eliminated by catastrophes, e.g. drought, and even pesticide spills, will recover of their own accord, if individuals remain in the headwaters and tributaries, but there are few records of natural recovery by plague-affected populations. Although restoration programs are to be encouraged, very little is known about the population characteristics of European crayfish, particularly their genetic structure, and care must be taken not to upset what biodiversity there is by mixing stocks up. Lessons can be learned from studies on Atlantic salmon and brown trout, e.g. Hindar et al. (1991) and Hindar (1992), who found large genetic differences between populations.

The review of control and eradication methods has revealed a number of ways by which the alien crayfish problem might be alleviated (see Holdich, Gydemo & Rogers this volume). However, whether or not the relevant authorities would allow anything other than the environmentally friendly methods, e.g. intensive trapping and use of fish predators, to be used remains to be seen. Using environmentally friendly methods is very unlikely to eradicate nuisance crayfish populations, only control them, and only then if continued effort and money is put in. Although the use of fourth generation pyrethroids is not to be encouraged, if a quick kill is required then these may be the only answer, and are probably more environmentally acceptable than using organophosphate or organochlorine insecticides.

5 CONCLUSIONS

Despite increasing interest in conserving native crayfish stocks in Europe the threat of crayfish plague, pollution, habitat destruction, the continued spread of wild populations of alien crayfish, and even 'a more open Europe', continue to hamper efforts to ensure their future. It seems inevitable that alien crayfish are here to stay in Europe and are likely to continue to be spread by those with entrepreneurial and sporting rather than conservation interests.

In order to conserve the remaining stocks of European crayfish we need to translate our knowledge to the governments and the public as well as those involved in trapping crayfish. As shown in this volume and in Westman & Westman (1992), there is a plethora of regulations concerning crayfish in Europe. What is needed is a more unified approach within Europe and between countries, particularly with regard to the alien crayfish species. We need more detailed knowledge on the distribution of both native and alien crayfish and we need to monitor the situation on a regular basis. We need to devise ways of managing nuisance populations of alien crayfish, including developing ways of eradicating them. We need safe areas for the remaining stocks of native crayfish. More research should be carried out on the characteristics of populations and the way by which crayfish plague is spread before restoration programs are initiated. There needs to be a method of identifying whether or not alien crayfish stocks harbor the crayfish plague fungus. Crayfish plague should be made a notifiable disease so that movement of crayfish out of affected areas is controlled.

Europe has less than 1% of the world's crayfish species and those we have occur nowhere else. So far we have done very little to ensure their future survival. The message is that as Europeans we need a much more unified approach to protect native crayfish and manage alien crayfish populations. We should be doing more to highlight the International Association of Astacology Resolution drawn up at the Lausanne Symposium in 1987 (see Freshwater Crayfish 7, p. xi, 1988), particularly in the light of the European Habitats Directive and the Rio Convention on Biological Diversity. This resolution highlighted the dangers to the indigenous European crayfish species of allowing live imports of crayfish from one country to another, especially as there are no guarantees that such crayfish do not carry communicable parasites and diseases, particularly crayfish plague. It was recommended to Governments that means be found to stop the importation of living crayfish for any purpose, except those approved by them, and that they should be responsible for assuring that any introductions were disease and parasite free. Finally, Governments were encouraged to restore native crayfish stocks wherever possible. The situation has hardly improved in the last ten years and the aims of this Resolution were reconfirmed at the 12th Symposium (see below).

It was encouraging at the Florence Workshop (September 1997) and more recent meetings (4th International Crustacean Congress in Amsterdam, July 1998; 12th Symposium of the International Association of Astacology in Germany, August 1998), to find that so many people have an interest in conserving their native crayfish fauna not only in Western, but also in Eastern Europe. It is to be hoped that their information will gradually lead to a much better understanding of crayfish distribution in Europe. Arrignon et al. (this volume) have suggested setting up a European Astacological Committee. However, it is not only native crayfish that need protecting in Europe, but many other crustaceans as well. To this end it was agreed at the Amster-

Table 2. Crayfish management scheme.

Priority	Aim	Species	Actions	Research
1	Conservation of native species	*Astacus astacus Austropotamobius pallipes, torrentium*	Prevent spread of alien fish	Map and monitor distribution
			Eradication of nuisance populations of alien crayfish	Develop environmentally friendly methods
			Disease control	Epidemiological studies
			Protection of habitats	
			Establish native crayfish reserves	
			Re-introduction of native crayfish	Genetic studies
			Education	
2	Wise utlization of native and alien crayfish	*Astacus astacus, leptodactylus, Pacifastacus leniusculus, Procambarus clarkii*	Species-specific legislation	Life-history and ecological studies
			Harvest regulations	Studies of population dynamics and effects of harvest strategies

dam Congress that 'The European Council for the Conservation of Crustacea' should be set up to encourage such endeavors. Its first job will be to try and set up a database of indigenous and alien crustaceans in Europe. For crayfish this should be a relatively straightforward task. It is to be hoped that the information provided in Table 1, which is based on Westman et al. (1990) and other sources, will stimulate astacologists to verify and add to what is currently known. The senior author of this chapter (david.holdich@nottingham.ac.uk) undertakes to keep it up to date if provided with information.

The contents of this volume should provide sufficient information on the dangers of allowing alien crayfish introductions so that people will think twice before doing it! However, as it is too late for many European countries then the volume should contain enough examples of 'good practice' to allow the best to be made of a bad situation. Table 2 shows a crayfish management scheme that could be used for this purpose.

REFERENCES

Ackefors, H. 1997. The development of crayfish culture in Sweden during the last decade. *Freshwater Crayfish* 11:627-654.

Ackefors, H. & Lindqvist, O.V. 1994. Cultivation of freshwater crayfish in Europe. In J.V. Huner (ed.), *Freshwater crayfish aquaculture in North America, Europe and Australia.* New York: The Haworth Press, Inc.

Albrecht, H. 1982. Das System der europäischen Flusskrebse (Decapoda, Astacidae): Vorschlag und Begründung. *Mitt. hamb. zool. Mus. Inst.* 79:187-210.

Bott, R. 1950. Die Flusskrebse Europas. *Abh. senckenberg. naturf. Ges.* 483:1-36.

Brodski, S. Ya. 1983. On the systematics of Palaearctic crayfishes (Crustacea, Astacidae). *Freshwater Crayfish* 5:464-470.

Dieguez-Uribeondo, J., Cerenius, L. & Söderhäll, K. 1995. Physiological adaptations in an *Aphanomyces astaci* strain from the warm-water crayfish *Procambarus clarkii. Mycol. Res.* 99:574-578.

Grandjean, F., Souty-Grosset, C. & Holdich, D.M. 1997a. Mitochondrial DNA variation in four British populations of the white-clawed crayfish *Austropotamobius pallipes pallipes*, implications for management. *Aquat. Living Res*our. 10:121-126.

Grandjean, F., Souty-Grosset, C., Raimond, R. & Holdich, D.M. 1997b. Geographical variation of mitochondrial DNA between European populations of the white-clawed crayfish *Austropotamobius pallipes. Freshwater Biology* 37:493-501.

Groombridge, B. 1993. *1994 IUCN Red List of threatened animals.* World Conservation Monitoring Centre, Cambridge, UK.

Hindar, K. 1992. Conservation and sustainable use of Atlatic salmon. In O.T. Sandlund, K. Hindar & A.H.D. Brown (eds), *Conservation of biodiversity for sustainable development.* Oslo: Scandinavian University Press.

Hindar, K., Ryman, N. & Utter, F. 1991. Genetic effects of cultured fish on natural fish populations. *Can. J. Fish. Aquat. Sci.* 48:945-957.

Hobbs, H.H. III, Jass, J.P. & Huner, J.V. 1989. A review of global crayfish introductions with particular emphasis on two North American species (Decapoda, Cambaridae). *Crustaceana* 56:300-316.

Holdich, D.M. 1996. *Austropotamobius pallipes.* In van Helddingen P.J., I. Willemse & M.C.D. Speight (eds), *Background information on invertebrates of the Habitats Directive and the Bern Convention.* Part 1 – Crustacea, Coleoptera and Lepidoptera. Nature and environment 79. Strasbourg: Council of Europe Publishing.

Holdich, D.M. & Rogers. W.D. 1997. The white-clawed crayfish, *Austropotamobius pallipes*, in

Great Britain and Ireland with particular reference to its conservation in Great Britain. *Bull. Fr. Pêche Piscic.* 347:597-616.

Huang, T-s., Cerenius, L. & Söderhäll, K. 1994. Analysis of genetic diversity in the crayfish plague fungus, *Aphanomyces astaci*, by random amplification of polymorphic DNA. *Aquaculture* 126:1-10.

Lilley, J.H., Cerenius, L. & Söderhäll, K. 1997. RAPD evidence for the origin of crayfish plague outbreaks in Britain. *Aquaculture* 157:181-185.

Mikkola, H. 1996. Alien freshwater crustacean and indigenous mollusc species with aquaculture potential in eastern and southern Africa. *Sth. Afr. J. aquat. Sci.* 22:90-99.

Momot, W.T. 1992. Stocking and exploitation as management methods for European crayfish. *Finn. Fish. Res.* 14:145-148.

Moyle, P.B., Li, H.W. & Barton, B.A. 1986. The Frankenstein Effect: impact of introduced fishes on native fishes in North America. In R.H. Stroud (ed.), *Fish culture in fisheries management.* Bethesda, MD: American Fisheries Society.

Shelley, M.W. 1818. Frankenstein, or the Modern Prometheus. Edited with introduction and notes by M. Butler, 1994. Oxford: Oxford University Press.

Skurdal, J. & Taugbøl, T. 1994. Do we need harvest regulations for European crayfish? *Rev. Fish Biol. Fish.* 4:461-485.

Spencer, C.N., McClelland, B.R. & Stanford, J.A. 1991. Shrimp stocking, salmon collapse, and eagle displacement. *BioScience* 41(1):14-21.

Starobogatov, Ya.I. 1995. Taxonomy and geographical distribution of crayfishes of Asia and East Europe (Crustacea Decapoda Astacoidei). *Arthropoda Selecta* 4(3/4):3-25.

Taugbøl, T. & Skurdal, J. 1993. Crayfish plague and management strategies in Norway. *Biol. Conserv.* 63:75-82.

Vennerström, P., Söderhäll, K. & Cerenius, L. 1998. The origin of two crayfish plague (*Aphanomyces astaci*) epizootics in Finland on noble crayfish, *Astacus astacus*. *Ann. Zool. Fennici* 35:43-46.

Westman, K., Pursiainen, M. & Westman, P. (eds) 1990. *Status of crayfish stocks, fisheries, diseases and culture in Europe.* Report of the FAO European Inland Fisheries Advisory Commission (EIFAC) Working Party of Crayfish.

Westman, K. & Westman, P. 1992. Present status of crayfish management in Europe. *Finn. Fish. Res.* 14:1-22.

List of contributors

Hans Ackefors, Department of Zoology, Stockholm University, S-10691 Stockholm, Sweden.

Sergei E. Ablov, Institute of Zoology, National Academy of Sciences of Belarus, Academicheskaya str. 27, 220072 Minsk, Belarus.

Anatoly V. Alekhnovich, Institute of Zoology, National Academy of Sciences of Belarus, Academicheskaya str. 27, 220072 Minsk, Belarus.

Jacques, C.V. Arrignon, Association Français de Limnologie, 24 rue de la 8ème Divisione, F 60200 Compiègne, France.

Gilberto N. Baldaccini, A.R.P.A.T., Servizio Locale della Versilia, Via Pontenuovo 22, Località Pollino, I-55045 Pietrasanta (LU), Italy.

Silvia Barbaresi, Dipartimento di Biologia Animale e Genetica 'Leo Pardi', Università di Firenze, Via Romana 17, I-50125 Florence, Italy.

João M. Bernardo, Departamento de Ecologia, Universidade de Évora, Portugal.

Erik Bohl, Bavarian State Office for Water Resources Management, Institute for Water Research, D-82407 Wielenbach, Germany.

Miguel Á. Bravo-Utrera, Departamento de Ecología, Facultad de Ciencias Biológicas, Universidad Autónoma de Madrid, Cantoblanco, E-28049 Madrid, Spain.

Aloyzas Burba, Institute of Ecology, Akademijos 2, LT-2600 Vilnius, Lithuania.

Peter Dehus, Fisheries Research Station, D-88085 Langenargen (Lake Constance), Germany.

Giorgio De Luise, Via XXIV Maggio, I-33010 Reana del Rojale (UD), Italy.

Lennart Edsman, Institute of Freshwater Research, Fisheries Board of Sweden, S-178 93 Drottningholm, Sweden.

Paolo Ercolini, A.R.P.A.T., Servizio Locale della Versilia, Via Pontenuovo 22, Località Pollino, I-55045 Pietrasanta (LU), Italy.

Pierre Gérard, Station Recherche Forest, Avenue Maréchal Juin 23, B 5030 Gembloux, Belgium.

Francesca Gherardi, Department of Animal Biology and Genetics 'Leo Pardi', University of Florence, Via Romana 17, I-50125 Florence, Italy.

Pedro J. Gutiérrez-Yurrita, Departamento de Ecología, Facultad de Ciencias Biológicas, Universidad Autónoma de Madrid, Cantoblanco, E-28049 Madrid, Spain.

Rolf Gydemo, Department of Systems Ecology, Stockholm University, Section Gotland, Ar Flering, S-620 35 Fàrösund, Sweden.

Paula Henttonen, Department of Applied Zoology and Veterinary Medicine, University of Kuopio, FIN-70211 Kuopio, Finland.

David M. Holdich, School of Biological Sciences, University of Nottingham, Nottingham NG7 2RD UK.

Jay V. Huner, Crawfish Research Center, University of Southwestern Louisiana, Lafayette, Louisiana 70504-4650, USA.

Maria Ilhéu, Departamento de Ecologia, Universidade de Évora, Portugal.

Max Keller, First Bavarian Crayfish Hatchery, D-86199 Augsburg, Germany.

Ady Krier, Service de la Chasse et de la Pêche, 67 rue Michel Welter, L 2730 Luxembourg.

Victor F. Kulesh, Institute of Zoology, National Academy of Sciences of Belarus, Academicheskaya str. 27, 220072 Minsk, Belarus.

Pierre J. Laurent, Association des Astaciculteurs de France, Avonnex, Port à Marin, F 74200 Thonon les Bains, France.

Sandra Lechleiter, Fish Health Service, D-70174 Stuttgart, Germany.

Ossi V. Lindqvist, Department of Applied Zoology and Veterinary Medicine, University of Kuopio, FIN-70211 Kuopio, Finland.

José. M. Martínez, Departamento de Ecología, Facultad de Ciencias Biológicas, Universidad Autónoma de Madrid, Cantoblanco, E-28049 Madrid, Spain.

David Mazzoni, Dipartimento di Protezione e Valorizzazione Agroalimentare, Università di Bologna, Via del Guasto 5/b, I-40126 Bologna, Italy.

Carlos Montes, Departamento de Ecología, Facultad de Ciencias Biológicas, Universidad Autónoma de Madrid, Cantoblanco, E-28049 Madrid, Spain.

Mario Mori, Istituto di Anatomia Comparata, Università di Genova, Viale Benedetto XV 5, I-16132 Genova, Italy.

Per Nyström, Limnology, Department of Ecology, University of Lund, Ecology Building, S-223 62 Lund, Sweden.

Birgit Oidtmann, Institute of Zoology, Fish Biology and Fish Diseases, University Munich, D-80539 Munich, Germany.

Oleg A. Pareiko, Institute of Zoology, National Academy of Sciences of Belarus, Academicheskaya str., 27, 220072 Minsk, Belarus.

Simon Phillipson, Fisheries Research Station, D-88085 Langenargen (Lake Constance), Germany.

W. David Rogers, David Rogers Associates, 9 The Moat, Castle Donington, DE74 2PD Derby, UK.

Julian D. Reynolds, Department of Zoology, University of Dublin, Trinity College, Dublin 2, Ireland

Jostein Skurdal, Eastern Norway Research Institute, P.O. Box 1066 Skurva, N-2601 Lillehammer, Norway.

Björn Söderbäck, Institute of Freshwater Research, Fisheries Board of Sweden, S-178 93 Drottningholm, Sweden.

Erich Staub, Swiss Environmental Protection Agency (BUWAL), Fisheries Section, CH-3003 Bern, Switzerland.

Thomas P. Stucki, Zoological Museum, University of Zürich, Winterthurerstr. 190, CH-5057 Zürich, Switzerland.

Bjarne Styrishave, Department of Life Science & Chemistry, University of Roskilde, Box 260, DK-4000 Roskilde, Denmark.

Trond Taugbøl, Eastern Norway Research Institute, P.O. Box 1066 Skurva, N-2601 Lillehammer, Norway.

Jaanus Tuusti, Sagadi Training Centre, EE-2123 Sagadi, Estonia.

Günter Vogt, Department of Zoology 1 (Morphology/Ecology), University of Heidelberg, Im Neuenheimer Feld 230, D-69120 Heidelberg, Germany.

Kai Westman, Finnish Game and Fisheries Research Institute, P.O. Box 6, SF-00721 Helsinki, Finland.

Index